500 DO-IT-YOURSELF WOOD MOULDING PROJECTS

Everything from bird houses to bookshelves,
from door trim to drawer dividers, and from planters to picture frames
can be made from wood mouldings. And anyone can do it.
The over 500 ideas in this book are just a beginning,
so put your imagination and creative talents to work
and have fun on these easy-to-build projects.

WOOD MOULDING
AND MILLWORK PRODUCERS

PO BOX 25278, PORTLAND, OR. 97225

Library of Congress Cataloging in Publication Data

Main entry under title:

500 do-it-yourself wood moulding projects.

1. Woodwork. 2. Moldings. 3. Do-it-yourself work.
TT185.F58 684'.08 78-3733
ISBN 0-668-04619-8 (Library Edition)
ISBN 0-668-04625-2 (Paper Edition)

Printed in the United States of America

ARCO PUBLISHING, INC.
NEW YORK

CONTENTS

BASIC WOOD MOULDING PROFILES

More than 350 standard wood moulding profiles, plus many special shapes and sizes, are produced throughout the United States. Some are sold only in certain areas due to the architectual history of that area. Because of this, you may not be able to find every type of moulding shown in this book in your vicinity. Many mouldings can be combined to simulate larger, more detailed patterns. With a little imagination and some suggestions from your wood moulding salesperson, you'll be able to substitute available moulding patterns for those that cannot be found in your area. The common names of the standard mouldings are given below as a reference for purchasing the types you need.

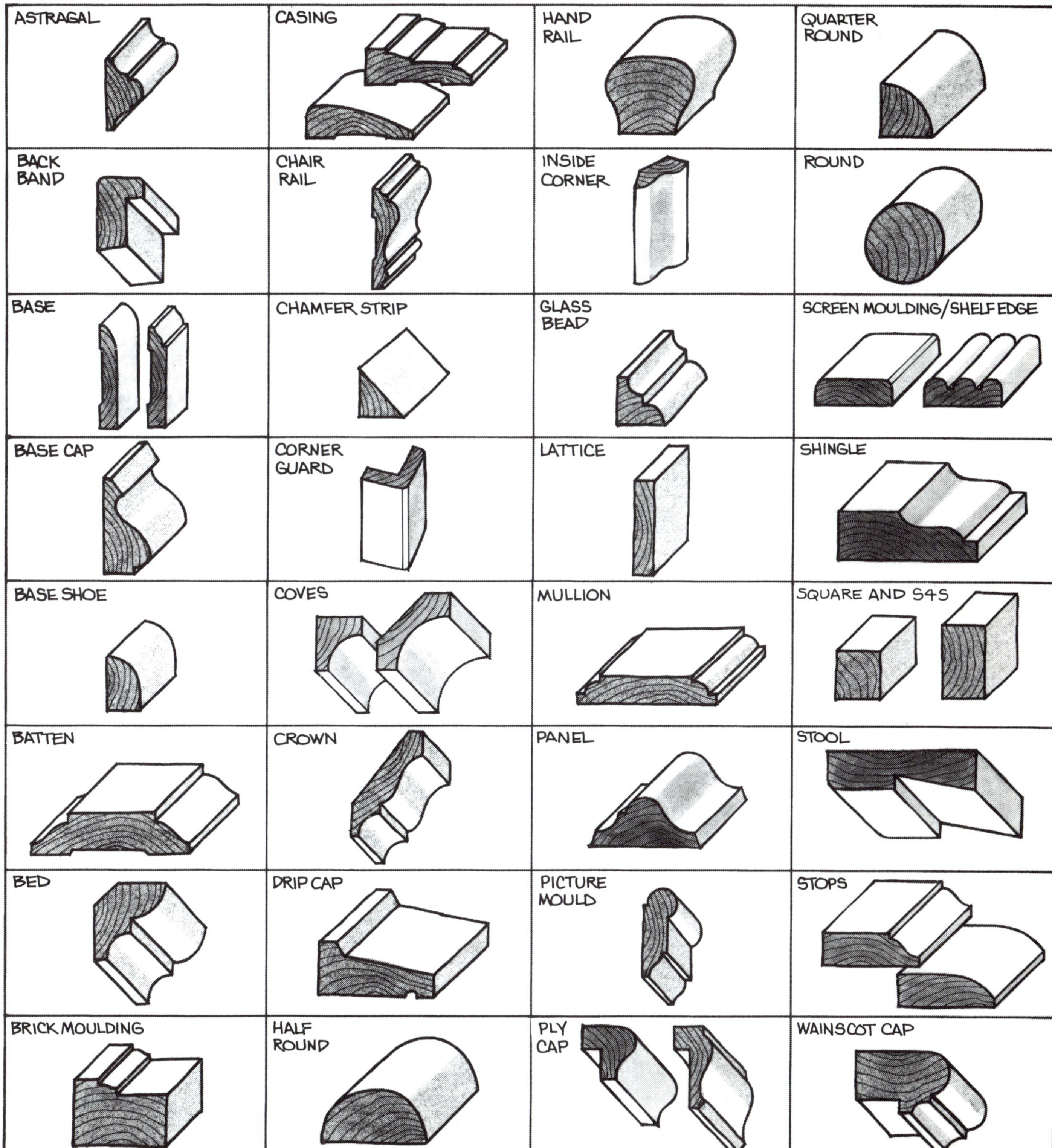

HOW TO WORK WITH WOOD MOULDINGS

Wood mouldings can be easily worked. The ease depends on planning, the right tools, and practice.

Planning is, of course, the first step in any project. The overall project needs to be evaluated and broken down into its individual parts. A list of materials should be made that will contain all of the parts needed for the project. It is also a good idea to list the tools and supplies that you will need. If you are in doubt as to what tools or supplies are needed, ask the salesman where you buy mouldings.

MOULDING TOOLS

The correct tools are important for any job to be done easily and successfully. For most moulding work you will need a simple miter box so that accurate 90° and 45° angles can be cut. A small, fine-toothed saw to use in the miter box will give clean, accurate cuts. A small hammer, nail set, and nails are also almost always necessary. A coping saw is sometimes needed to cope a joint so it will fit together tight. Other important tools include a tape measure, glue, sandpaper, and wood filler. Power tools are handier, of course, but not essential.

Practice is the last tip to successful projects. Try cutting 45° or coping a joint on some scrap pieces before cutting the actual piece you need. Try different techniques to see which suits you and will get the job done in the best and easiest way.

A corner clamp is a very helpful tool when clamping mitered corners together. The clamp allows each side of a 90° corner to be clamped individually so that a nice tight corner can be constructed. Once the corner is glued and clamped, it should be nailed in from each side with small finishing nails or brads to provide a strong corner once the clamp is removed.

When gluing pieces together, be careful not to get glue on the surface to be finished. It is best to not wipe glue away if it oozes out from a joint. Wait until it dries and then cut or chip it away. If glue is smeared on the surface of the wood it will impair the finish.

HOW TO MEASURE MOULDINGS

Some helpful hints on measuring mouldings can save expense and trouble. When mitering mouldings to make a square or frame, use the outside dimension as the length you need. This is done by adding the width of the mouldings to the inside dimension as shown below, or measuring the overall outside dimension of the frame and cutting to that size.

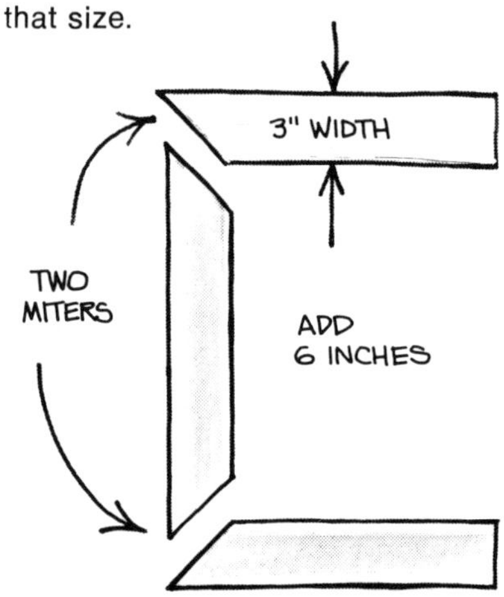

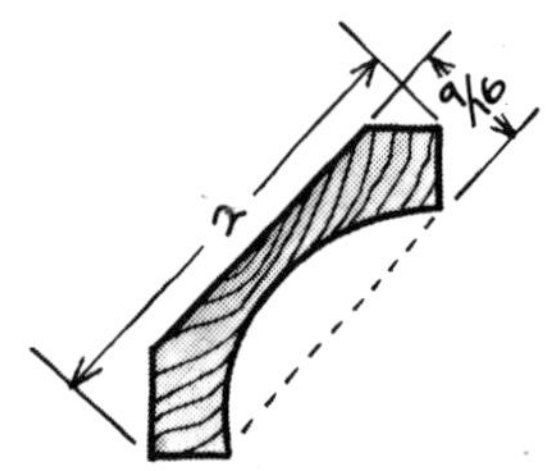

Individual moulding size is determined by the thickest and widest part of each moulding shape (profile). An example would be a 9/16 x 2″ cove.

TIPS ON BUYING MOULDINGS

Mouldings are available in lengths from 3 to 16 feet. Try to use short pieces whenever possible. It is often unnecessary to buy one long piece when several small pieces would be easier to handle and would do the job just as well.

When figuring the amount of moulding you need, round up to the next largest foot to allow for trimming and cutting.

HOW TO FINISH

HOW TO MITER A MOULDING

"Mitering" a moulding usually means to cut the ends at 45° angles so the two pieces form a 90° corner. It is important to do this carefully so a good joint results. A miter box is necessary since it has guides for the saw at a 45° angle. The moulding is either placed flat on the bottom or against the back of the miter box depending on how the moulding is to be used. It helps to clamp the moulding in place when cutting, so it will not slip in the hand, resulting in an uneven cut.

To form a 90° corner, one 45° angle is cut from the left hand side of the miter box and one from the right hand side of the miter box.

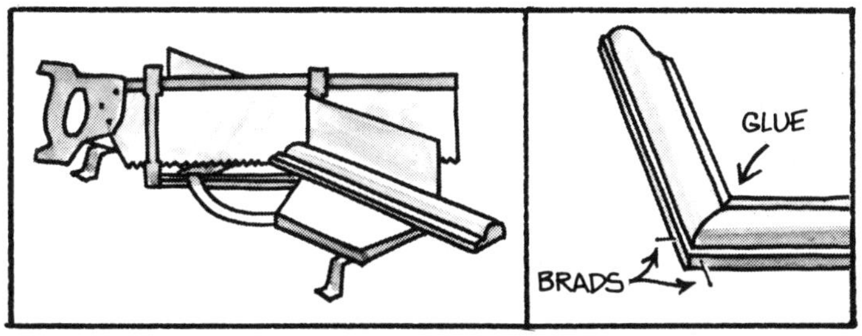

HOW TO COPE A MOULDING

It is sometimes necessary to cope mouldings in order to get a tight fit at an inside corner. Coping means to cut the profile of the moulding being butted up against the first piece.

To cope a joint, first cut a 45° miter on the piece. It should be positioned in the miter box just as if the back of the miter box were the wall or whatever surface this piece is being put against. The resulting cut exposes the profile of the moulding, serving as a template. Use the coping saw and cut along the profile made by the miter holding the saw perpendicular to the piece. The result will be a profile that will fit tightly against the first piece.

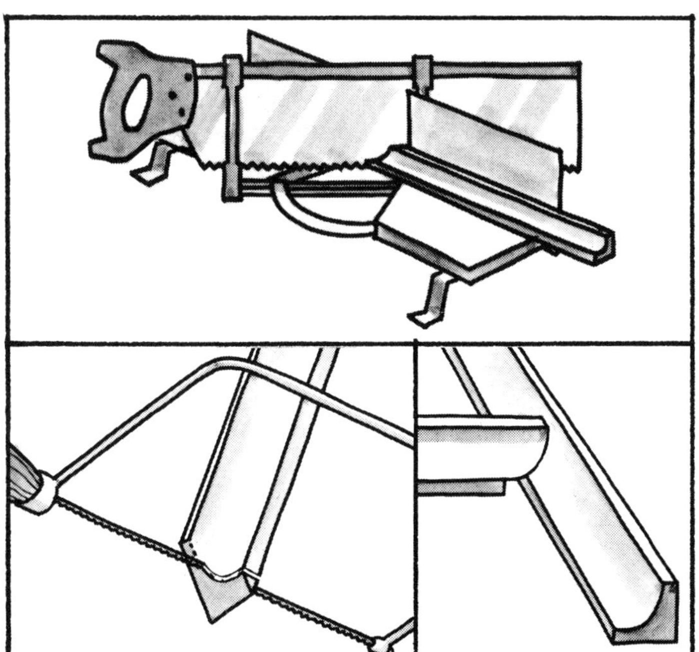

The finish is one of the most important parts of any project. A finishing job well done can really enhance the project you have just completed.

There are many finishes on the market and the most important thing to do is to read and follow the directions given for each individual type. Some examples of the types of finishes that can be applied are stain, varnishes, lacquers, antiquing kits, water base paint, oil base paint, wax, and oils. With such a variety it is obvious that there is not one single finish that is best for all occasions. Your local retailer should be able to give you some tips once he knows the type of project you have and where and how it will be used.

There are some points to remember that can be used in almost all circumstances. First, you must prepare the surface of your wood properly before any finish is applied. This means that cracks, nail holes, etc. must be filled with wood filler. This may be done after a stain is applied so it will match the stain properly. The entire surface should be sanded with a fine grit sandpaper to remove any rough spots. Smooth the wood filler and remove any grease or dirt that has accumulated by handling the pieces.

Textures can be applied to the wood for special effects by any number of methods. Some common ones are brushing with a wire brush, or even scorching the wood with a torch or flame and then brushing out some of the softwood to give a contrast. Scrapers, dog's wire brushes, combs, can openers, etc. can be used to rake the finish and give it a rough surface. A very light sanding may be done after these techniques to eliminate any splinters, but be careful not to over-do and ruin your texturing job.

Water base paints do not work very well on raw wood because the moisture tends to raise the grain of the wood resulting in a rough finish. The surface should be smoothed between coats of finish with steel wool or a very fine sandpaper. Take extra precaution when using sandpaper so that you don't sand through the finish on the edges.

You can use a tack rag to remove dust from your project after sanding, but try to do your finishing in a dust free area.

The beauty of wood can only be enhanced by you and a finishing job well done. There are many other special techniques that can be used to create unusual finishes, so consult a finish or paint specialist in your local area.

ACCENT WALL DISPLAYS

A simple wood moulding backdrop can help accent flowers, plants, or those special articles you treasure. If the unit is to be used as a display area for pictures, it can be made with or without the shelf. The sample designs shown here are built with lattice, but if more variety is desired, battens, squares, S4S stock, mullions, flat astragals, half rounds or screen moulding can be used. If a shelf is added, it can be a simple piece of ¾″ lumber cut to the desired size and can be held in place with decorative angle brackets. When finished, just paint or stain it to accent whatever you wish to display.

TOP VIEW

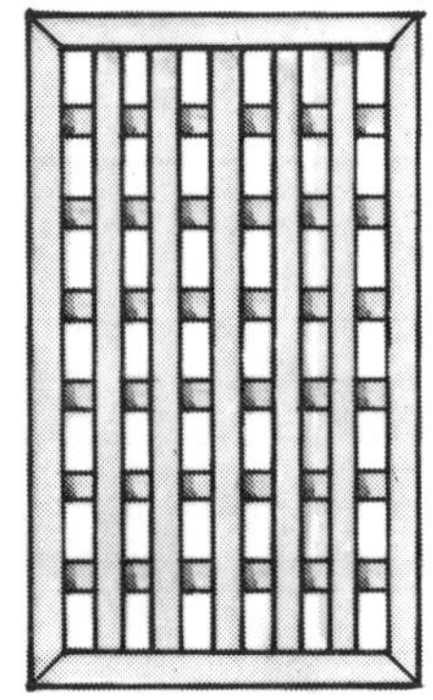

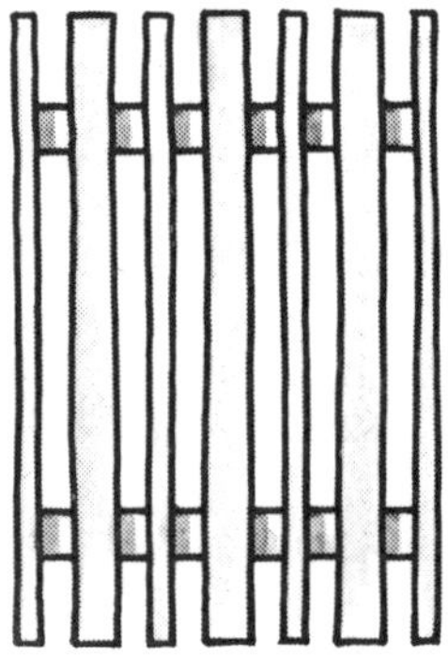

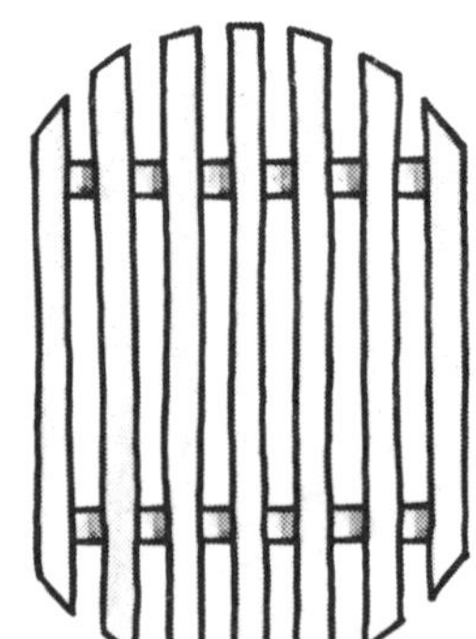

BATHROOM TISSUE HOLDERS

Who says your bathroom tissue holder has to look like everybody else's bathroom tissue holder! Be unique by making yours out of wood mouldings. The styles shown here are each so simple they can be put together in minutes.

Remember to drill the holes slightly larger than the round so it can be slipped out easily. And here's another idea. By just lengthening the design you can build matching towel racks. How's that for individualizing your bathroom!

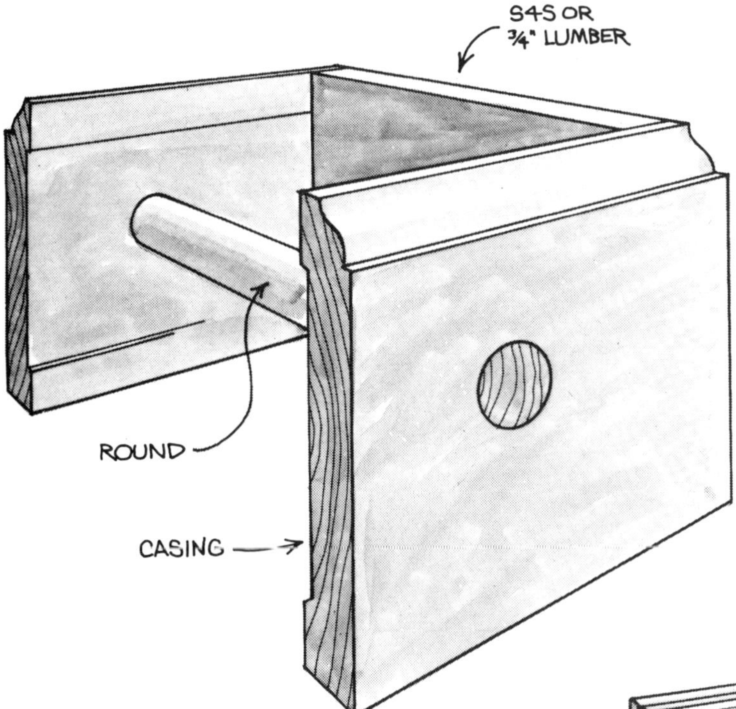

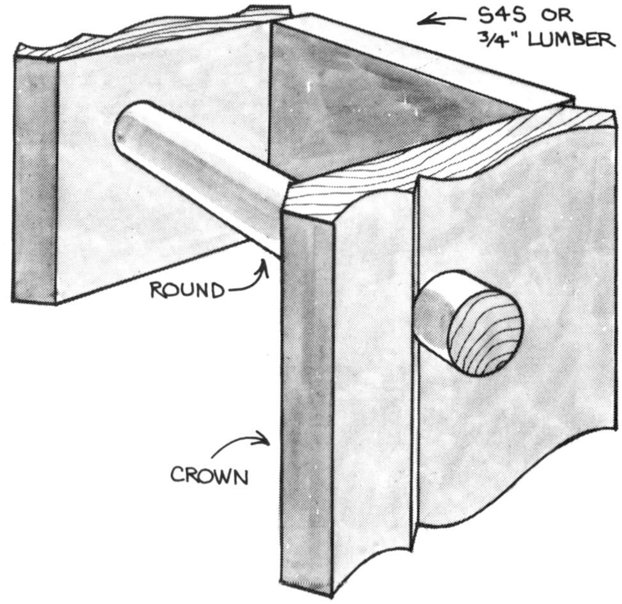

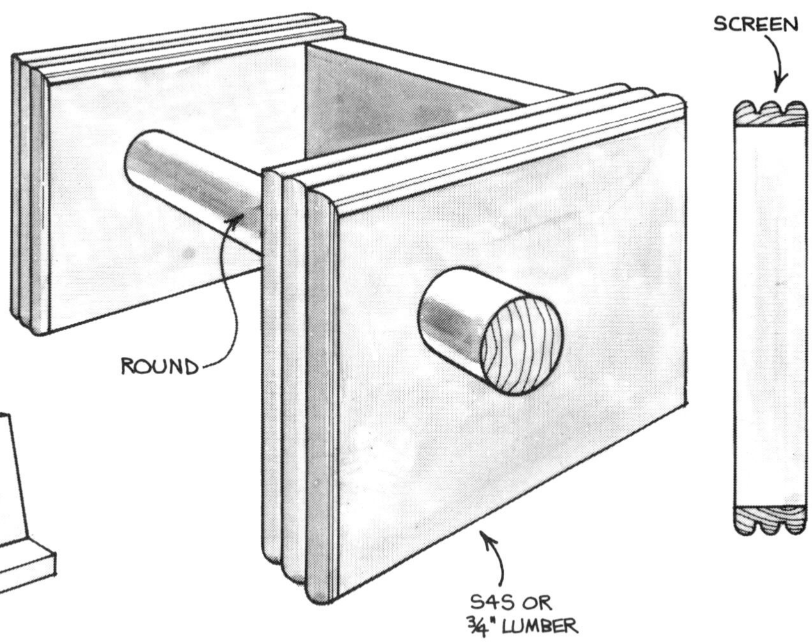

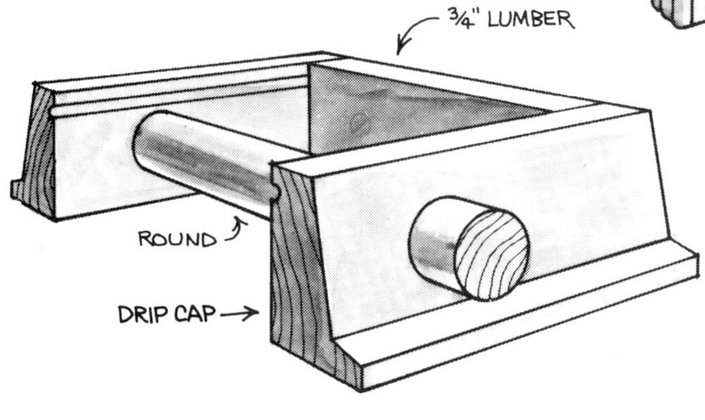

BIRD FEEDER

Want to make your friendly neighborhood birds happy? Build'em a bird feeder. And to make them *really* happy, put feed in it. All it takes to construct this roosting restaurant is a framework of S4S and squares attached to a plywood floor with an apron of stop moulding around the outside. The plywood roof has a covering of cedar shingles which is topped off with a piece of corner guard to keep out the rain.

The whole thing can then be put on a post and set in the ground or it can be hung from a tree or your porch. If you choose to hang it, use chains attached to hooks screwed in the framework. Drill holes in the corner guard to prevent cracking by the hooks. Since this structure will be exposed to the elements, it's a good idea to be extra generous with the sealer, stain, or paint.

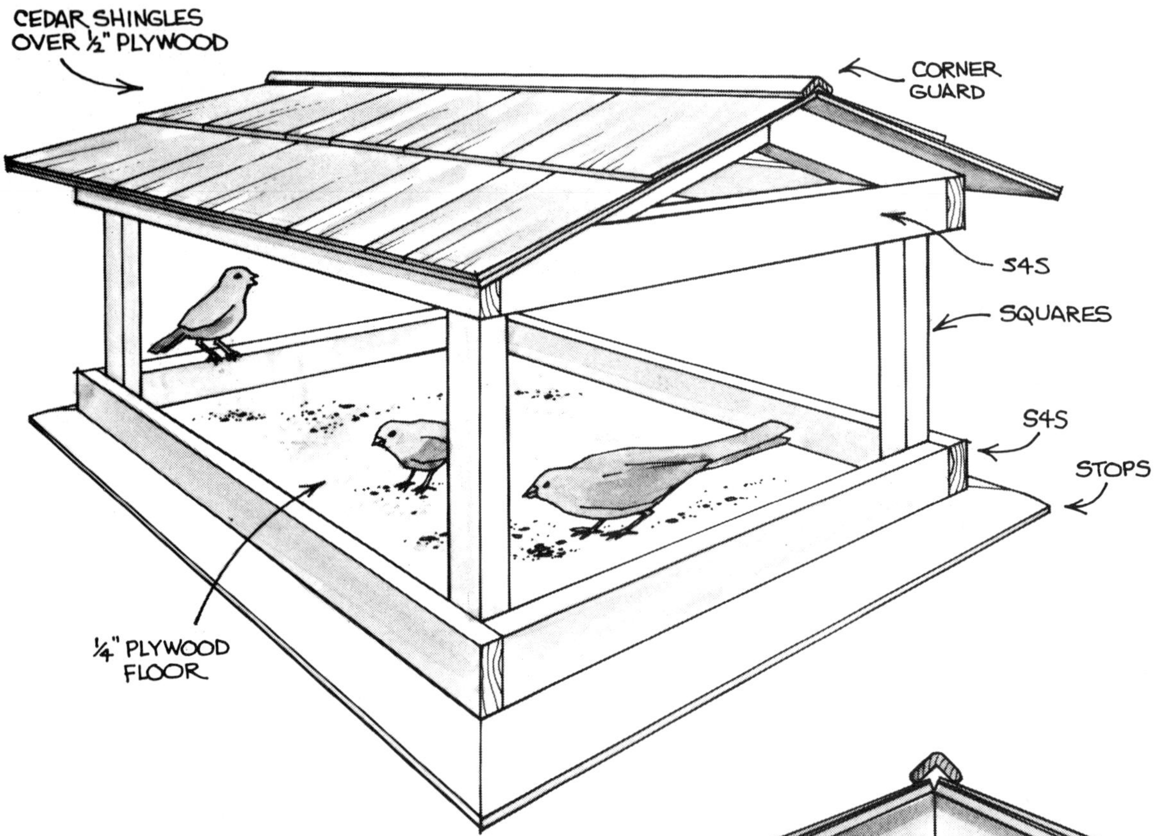

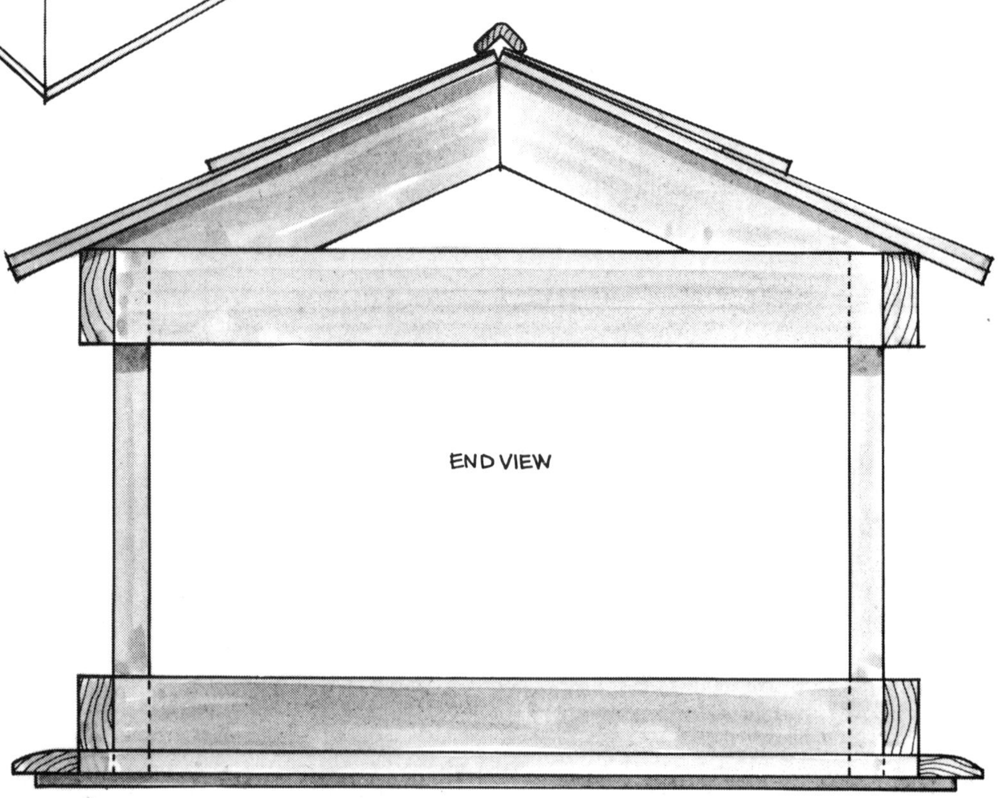

BIRD HOUSES

Housing may be going skyhigh, but in this case it's the placement, not the price. Building one of these bright little bird bungalows will cost you but a few dollars. (Downright cheep, you might say.) The size of the abode is left up to your bird watching judgement. Whichever house you choose to build, most birds will be delighted to take up housekeeping in it as long as it's not placed too close to the local admiring feline population. One added note, wherever plywood is called for in construction, be sure to use an exterior grade.

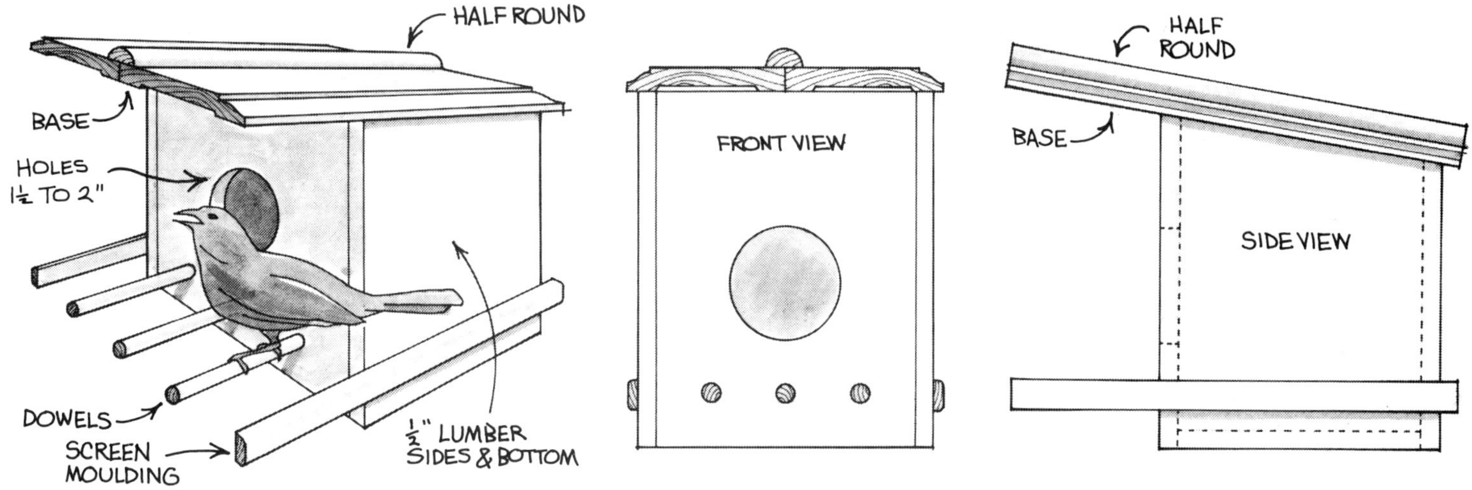

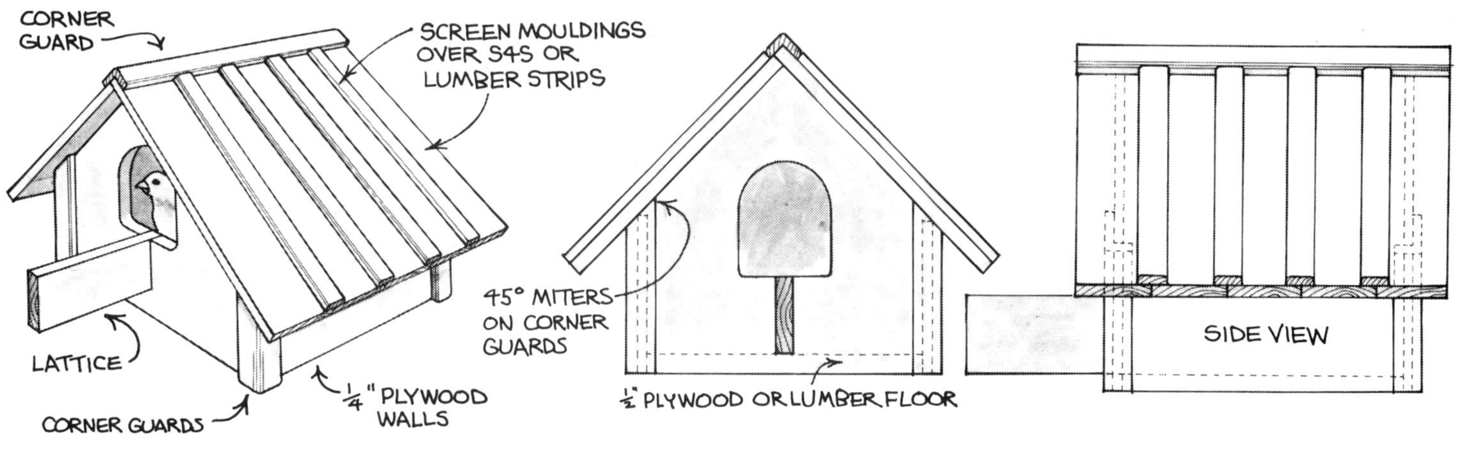

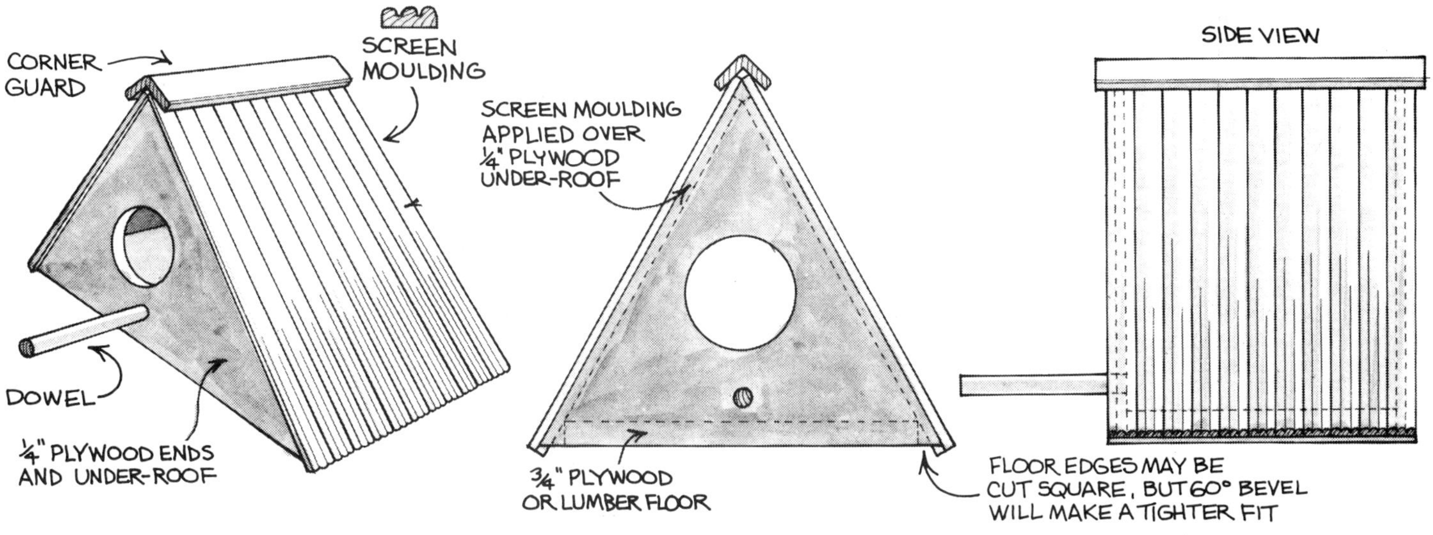

BLACKBOARD/BULLETIN BOARD

Here's a handy communication center designed for your kitchen or game room. Jot down a quick note on the blackboard. Tack up that weird cartoon on the bulletin board. Say whatever needs to be said. Tell it like it is! Its size depends on how communicative your family is. To build it, all you need is a piece of hardboard painted with chalkboard paint or a piece of *real* blackboard if you can find such an item these days and a piece of ½″ corkboard. Since the blackboard is thinner than the corkboard, it'll need some plywood backing to make them equal in thickness. The blackboard and corkboard are then butted together and both pieces are glued onto another piece of plywood backing. The whole thing is then framed with ply cap moulding. Next, apply a strip of half round or screen moulding over the raw edge where the blackboard and corkboard come together. Then paint it or stain it as desired and attach picture frame hooks to the back for hanging.

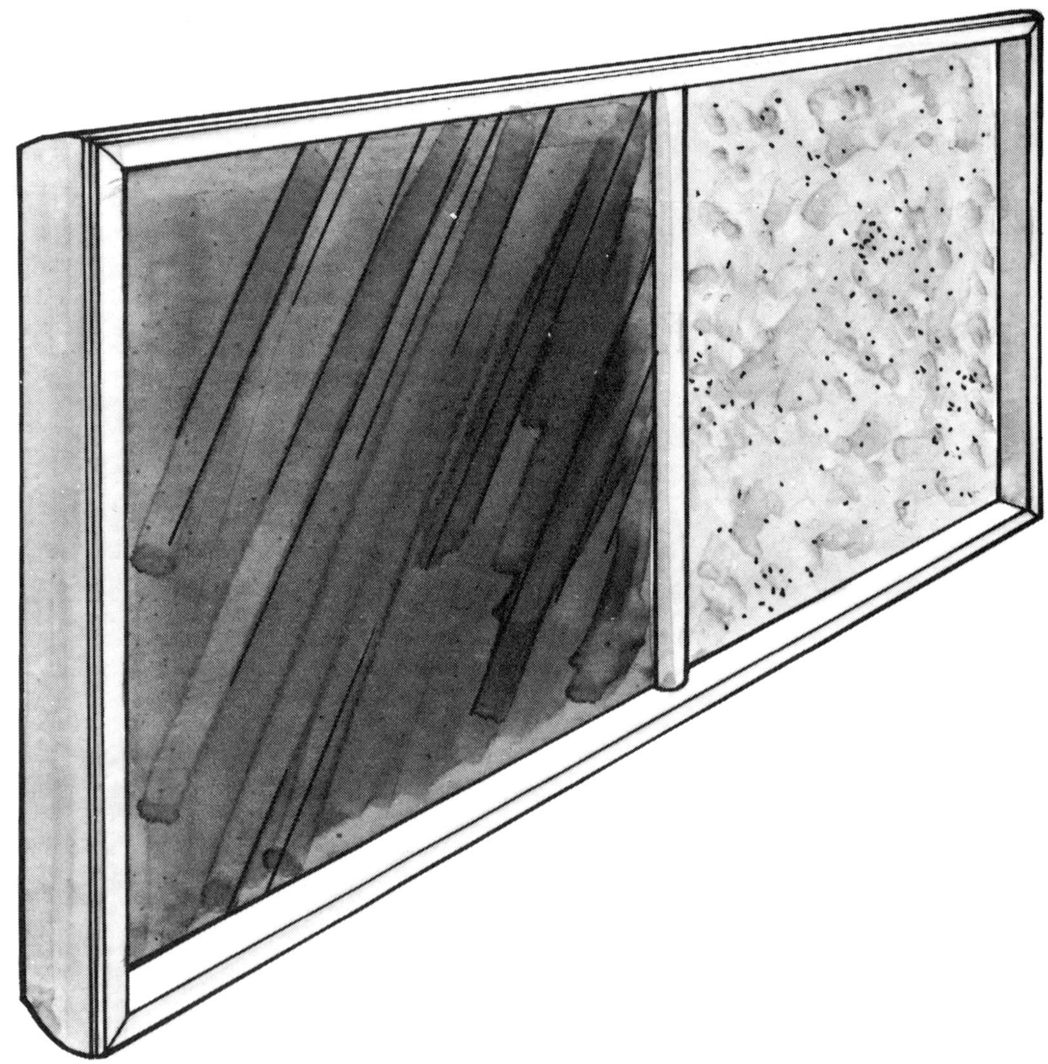

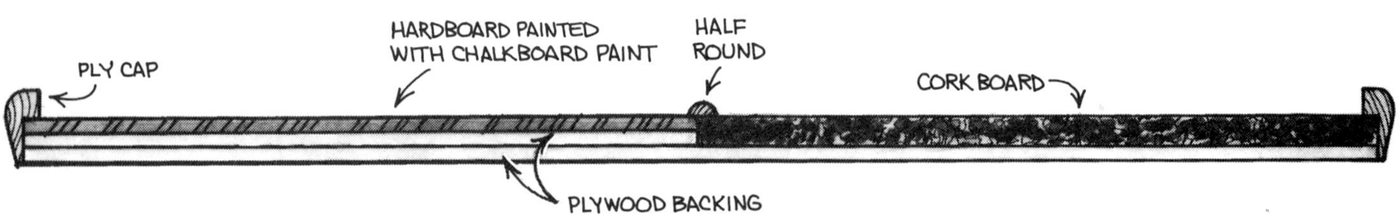

BOOKENDS

Tired of your books falling over? You can make them all upstanding editions just by building some bookends. In order for a bookend to be heavy enough to hold a series of books, it must be either very heavy itself or it must be attached to some kind of base for the first few books to set upon. This can be small metal plates of various types and sizes, which can usually be purchased at a metal fabrication shop, sheet metal shop or steel service center. Another method is to connect both bookends together with dowels. The ends then become a fixed distance apart making the size of your display permanent. The actual size of your bookends will depend upon the size of your books. Because bookends are usually under pressure, make sure they are glued and nailed so they won't collapse. The two framework samples shown can be even further strengthened by adding corner plates to the backside. It's also a good idea to add felt to the undersides of bookends to protect the surface on which they sit.

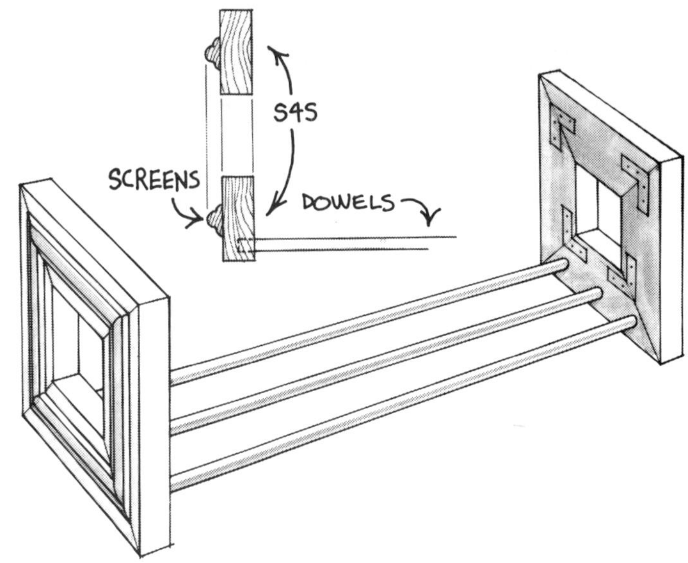

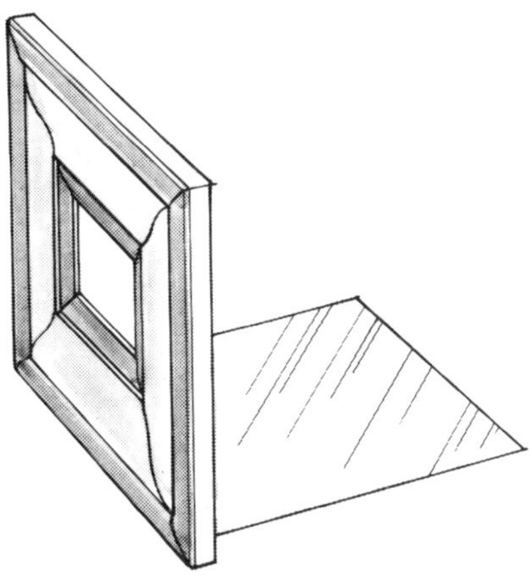

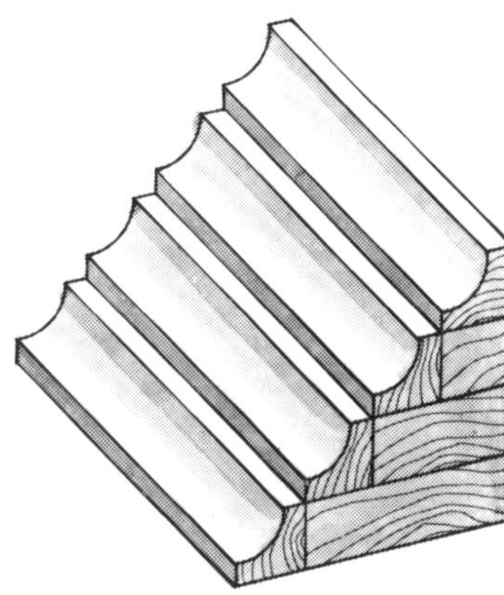

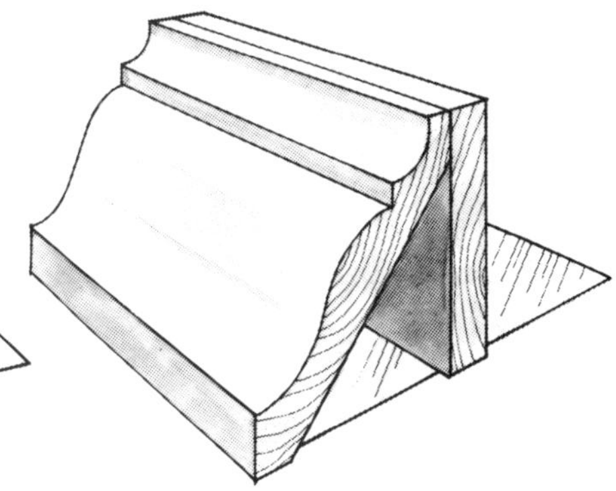

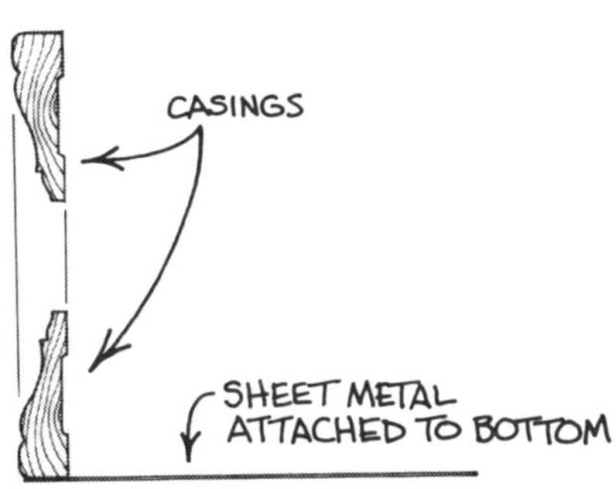

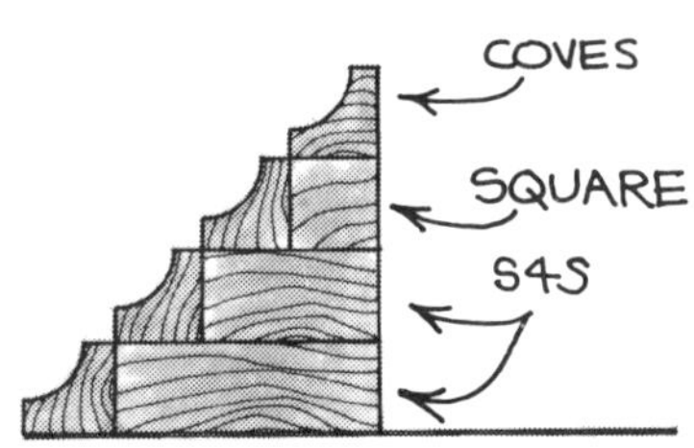

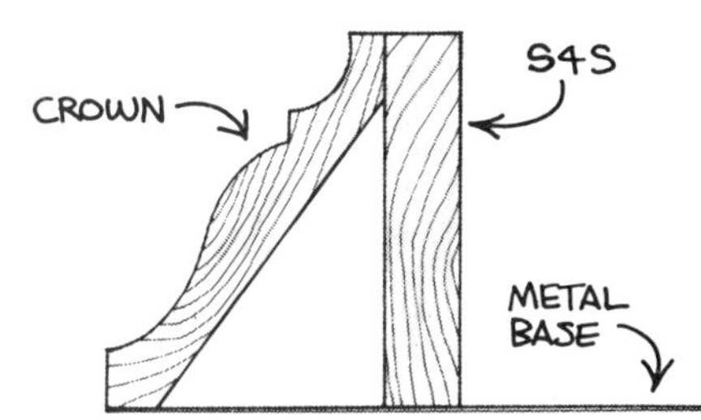

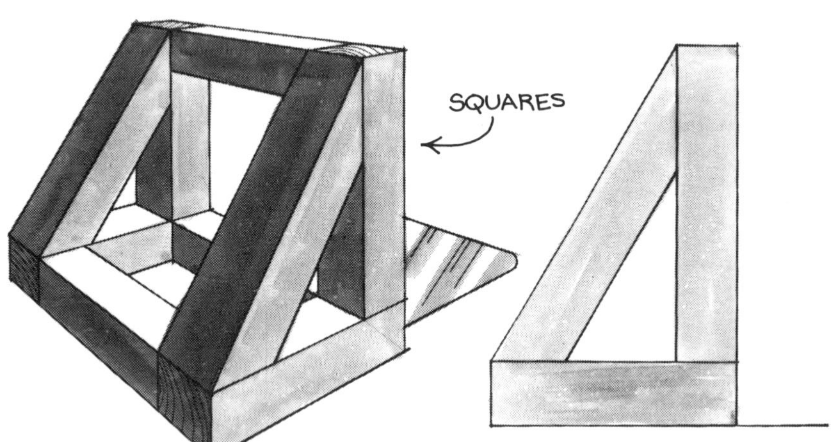

BOXES (Cigarette, Jewelry, Utility, Etc.)

These handy little boxes can be used for all sorts of goodies, such as jewelry, cigarettes, stamps, buttons, cookies or whatevers (whatever whatevers are). They all utilize a ¼″ plywood or hardboard for backing. Hardboard is fine for gluing but difficult for nailing. Your miter box is a must on these projects. The size of each box will depend on your needs. Remember, if you can't find the type of moulding shown here at your lumber yard, it is easy to substitute something you find available.

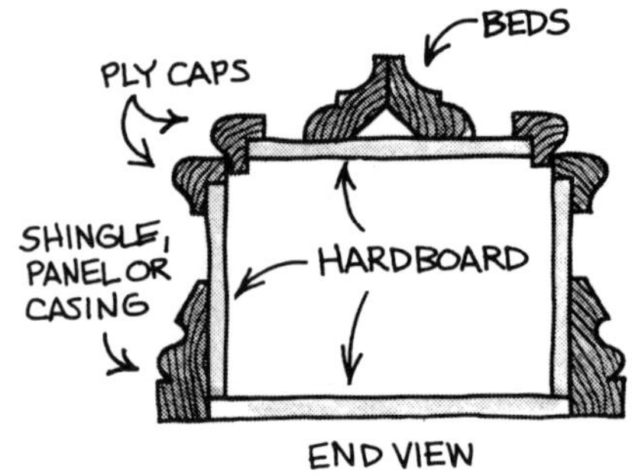

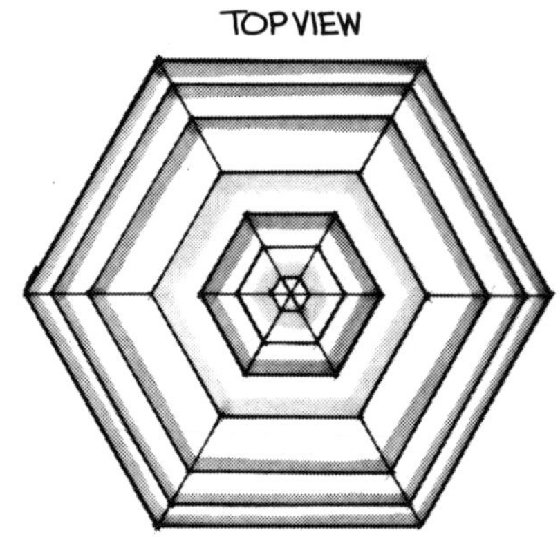

The hexagonal box shown here starts with two small bed mouldings back to back for the handle, ply cap mouldings are used at the edge of the lid and top of the sides, and shingle moulding for the bottom sides. Each angle will be 60° to make up the hexagon.

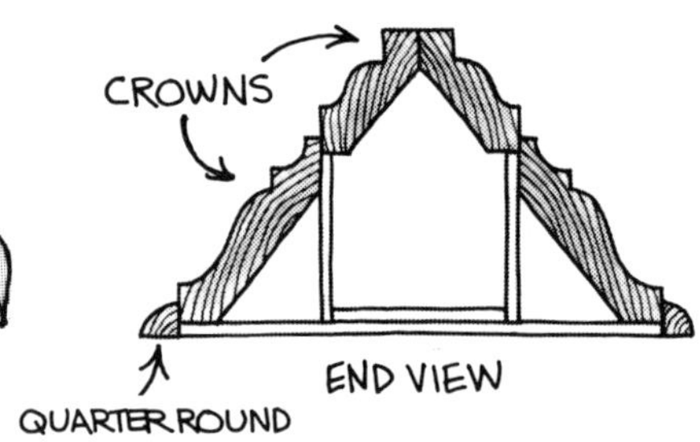

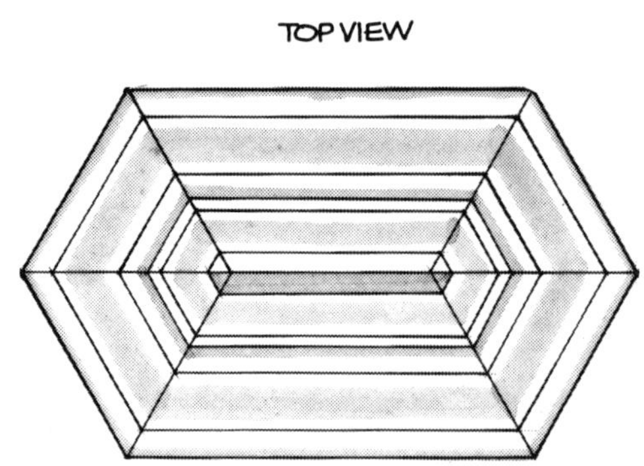

For variety, try a "stretched" hexagon. It uses two different sizes of crown moulds with a small quarter round to finish off the base.

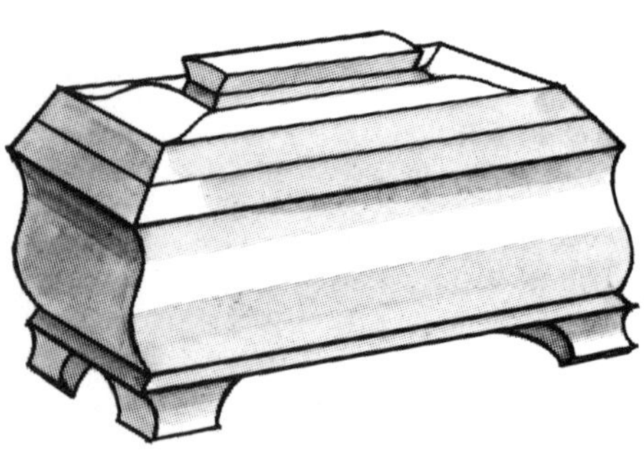

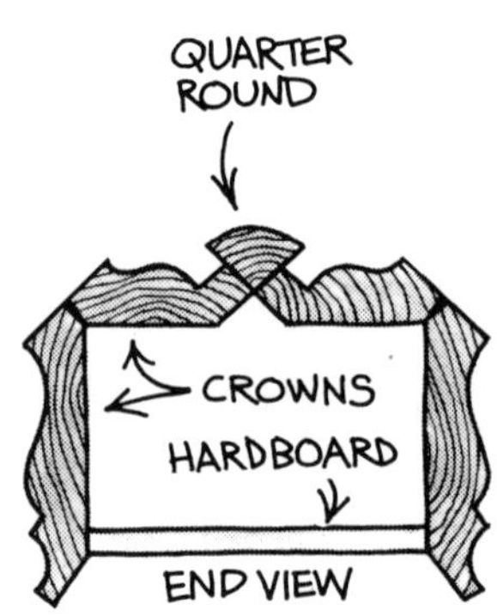

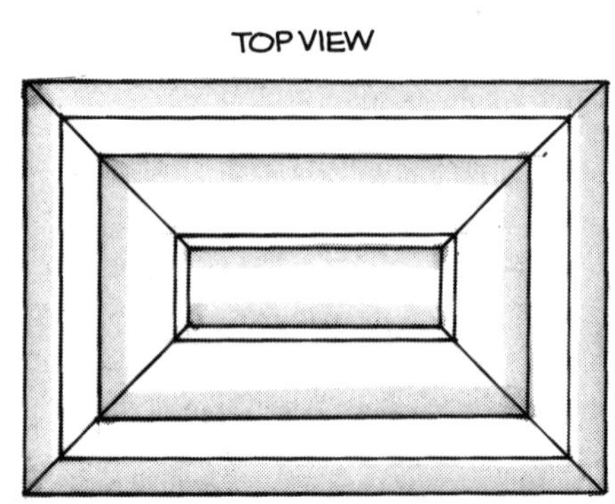

This one utilizes crown mould for the sides and top, and a piece of quarter round in the center for the handle.

BOXES (Cigarette, Jewelry, Utility, Etc.)

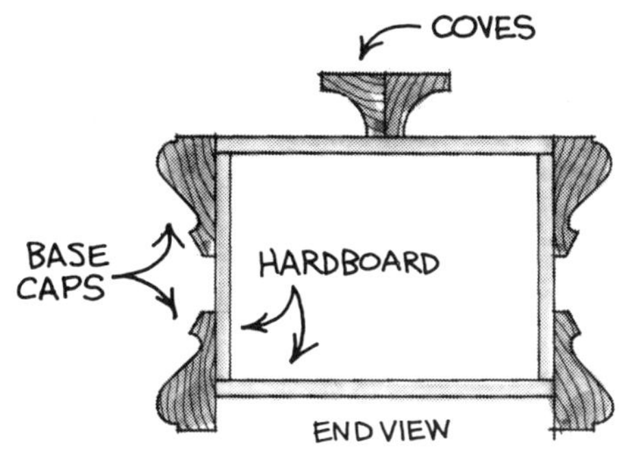

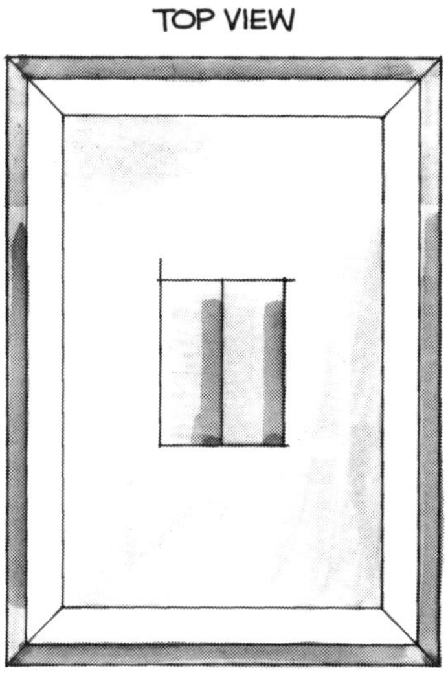

This ornate looking but simple-to-build box merely uses coves for a handle and base caps at the top and bottom of the sides.

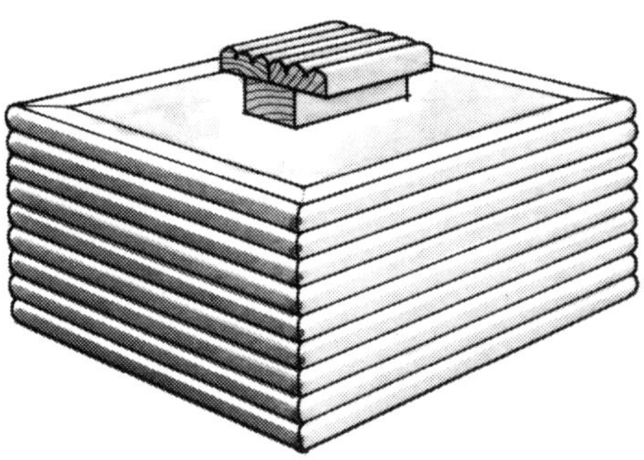

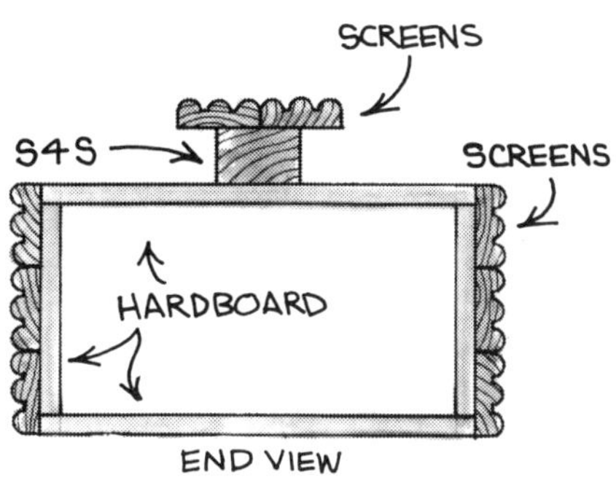

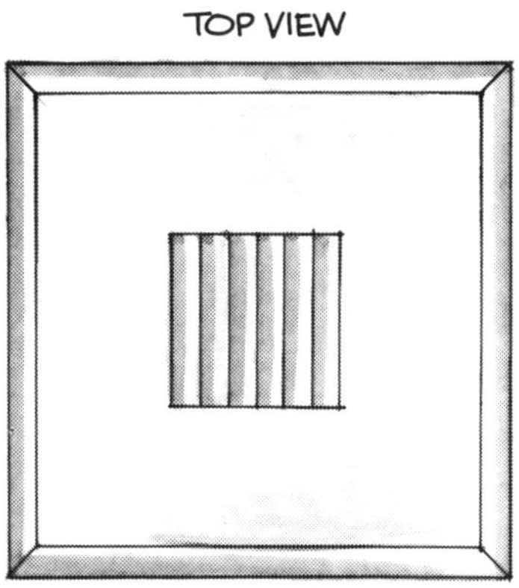

This is easier than it looks with screen bead for the sides and the handle. A small piece of S4S stock is used under the handle to provide finger room.

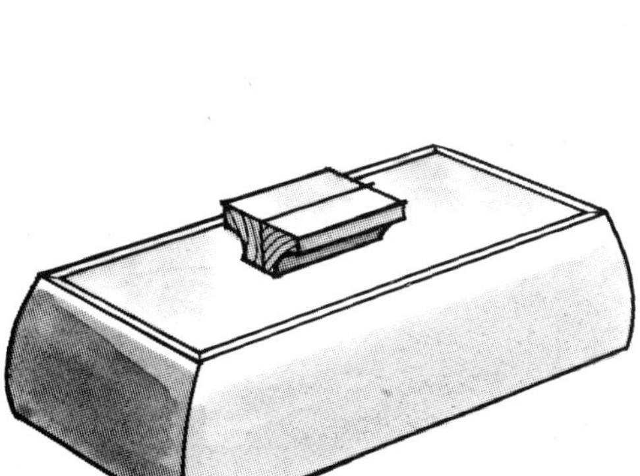

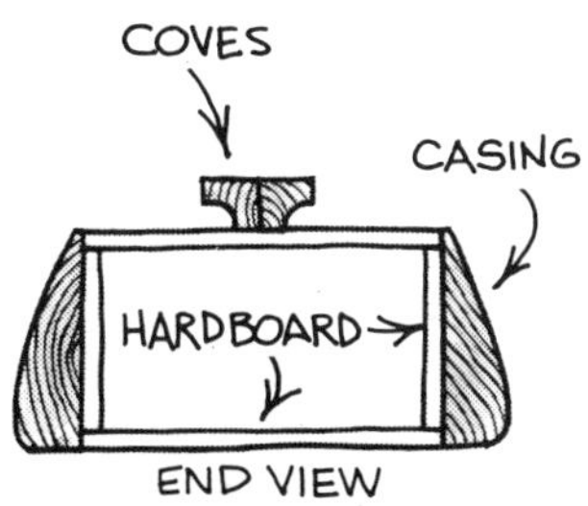

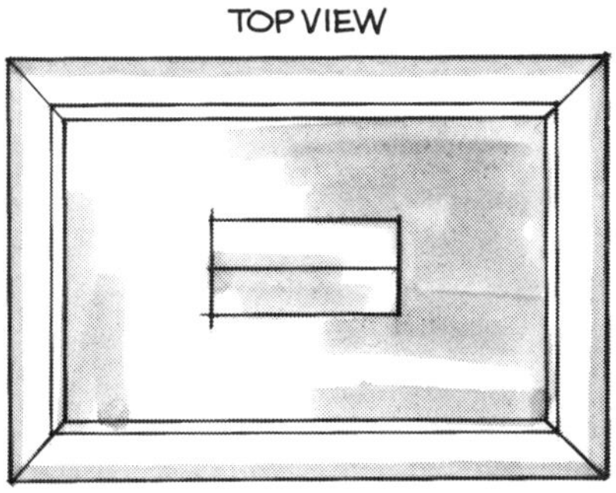

Here's the ultimate in simplicity. The sides are made of a large piece of base and the handle is two pieces of cove.

CABINET DOORS

The possibilities for customizing your kitchen or bathroom cabinets are endless. The variety of moulding patterns available, multiplied by the many ways they can be placed on the cabinets, plus stains and paint choices and the use of vinyl wallcovering, all add up to an unlimited assortment of designs. Wood mouldings can also be used as a framework to hold tile or stained glass. All these variations give you a unique opportunity to show off your customizing creativity. Now go to it. The choice is yours.

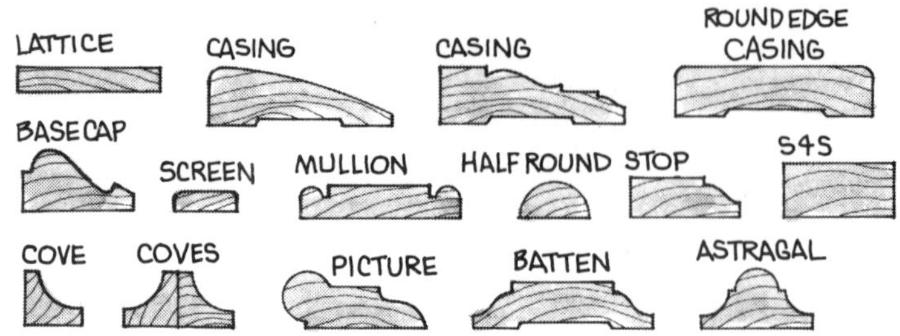

CEILINGS

If you're remodeling or just want to add some pizzazz or elegance around the edges of your ceiling, install a simple moulding pattern as shown. If you want to make it more elaborate, use a combination of patterns. The last design shown here looks fairly complicated but is actually quite easy to do. It's merely a crown resting against S4S stock with a base cap below. The dentil moulding is just a piece of lattice or screen moulding sawn into small pieces and spaced along the S4S and attached with glue and small brads.

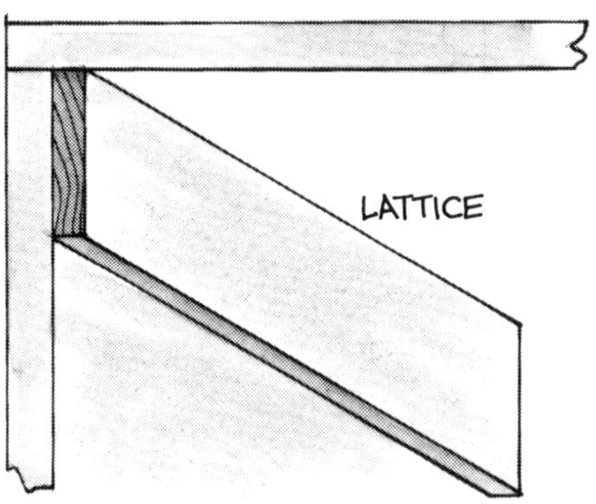

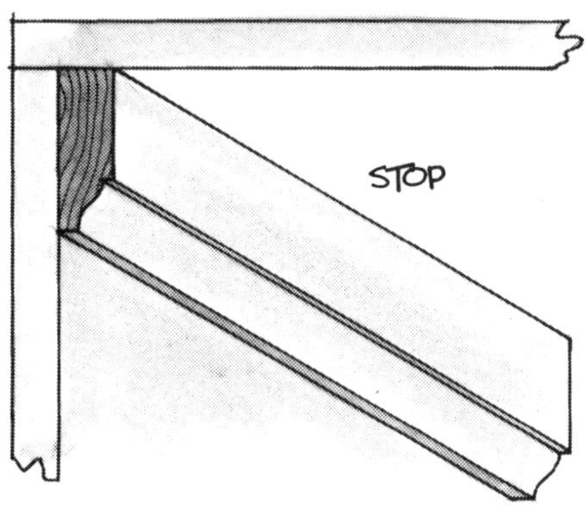

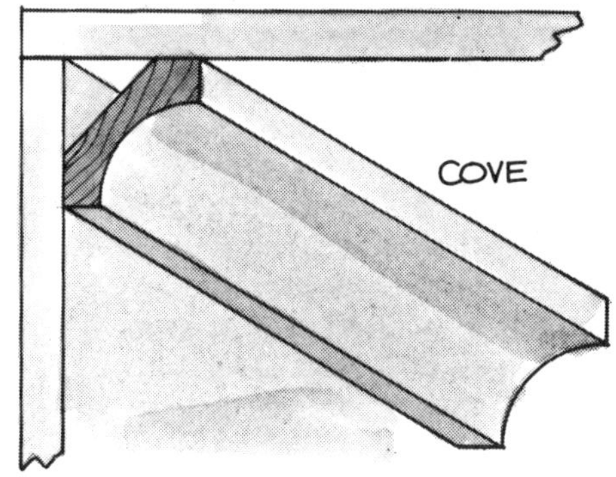

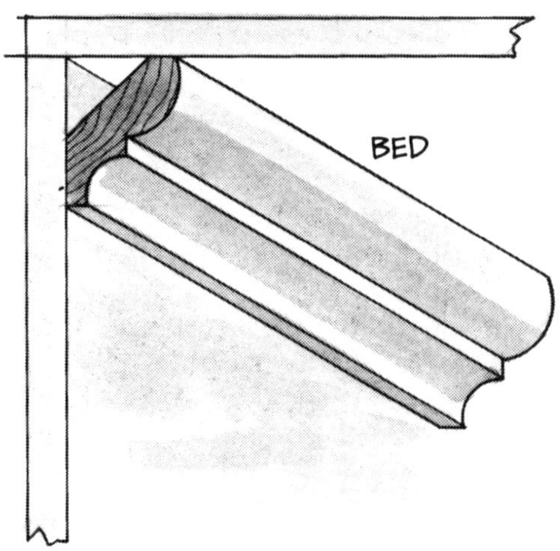

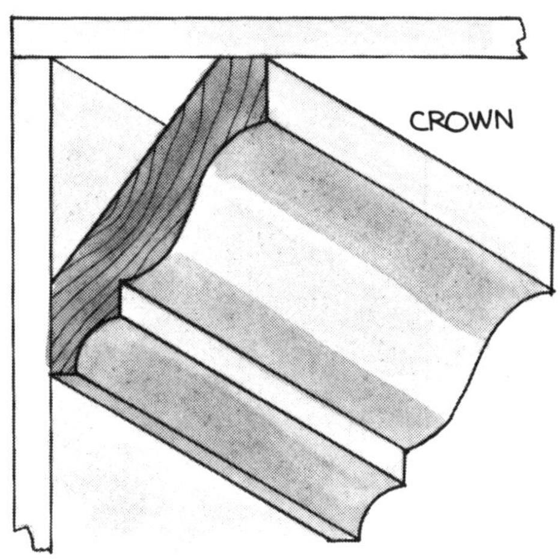

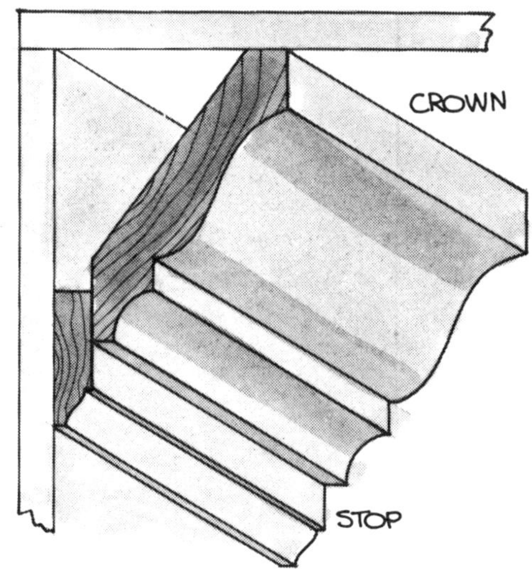

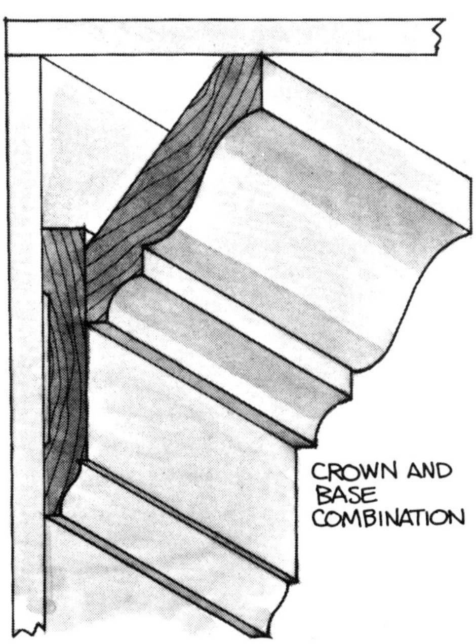

CEILINGS

If your ceiling is a little dull or has no character, add some interest with wood mouldings. There's a variety of patterns available, whether used alone or in combinations. Since you'll be viewing it from below and from many directions, it's usually best to choose a symmetrical moulding pattern. The layout of the overall design, however, depends on the shape of the room, the lighting arrangements and your own interest. Color and ceiling material can also add uniqueness. With just a little work and imagination you'll soon have a ceiling people can look up to.

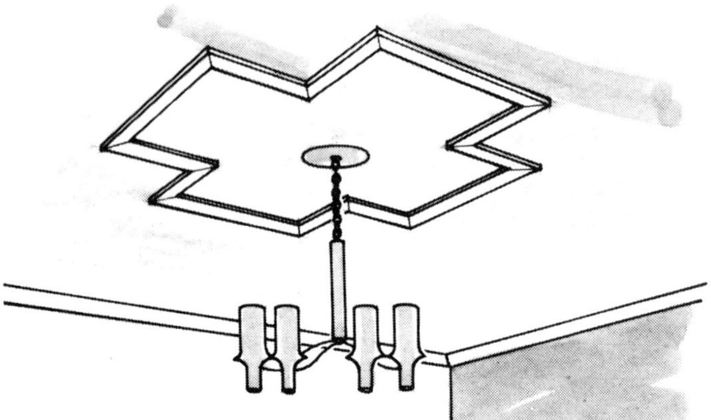

A series of 45° miters around a light fixture.

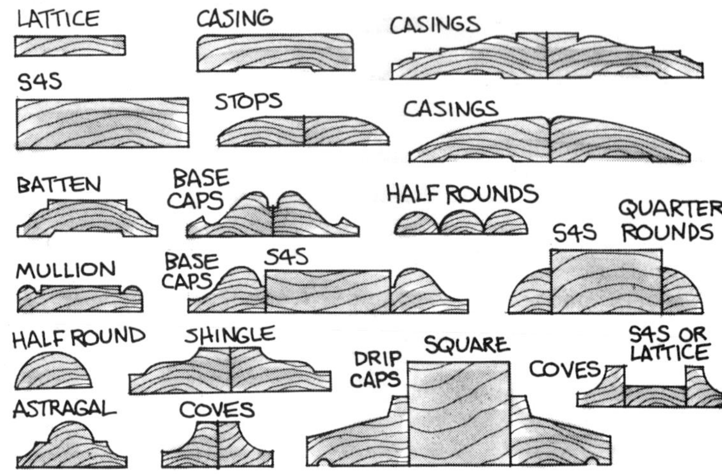

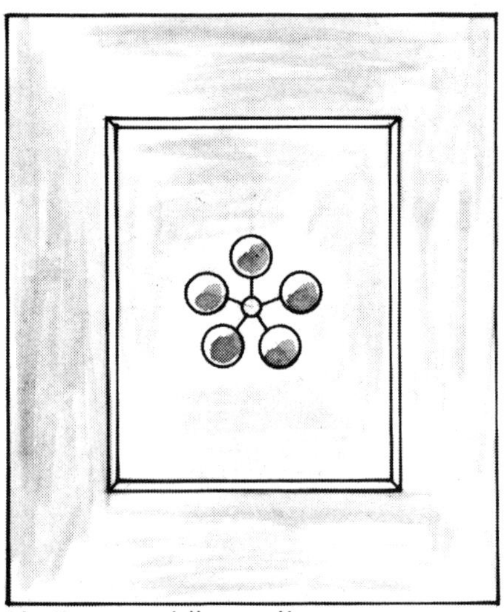

Various moulding patterns

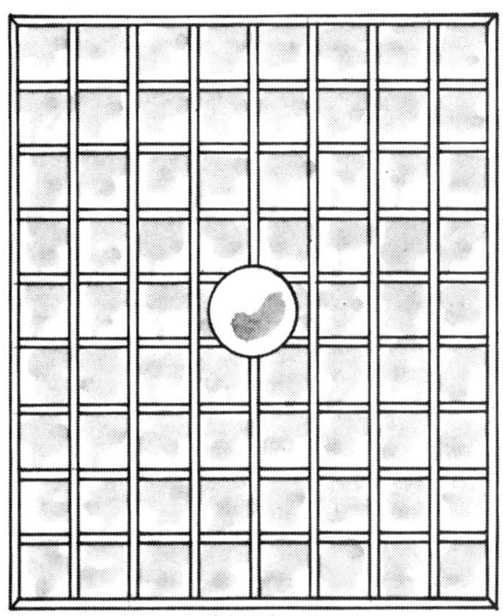

A grid pattern of lattice.
Can also be used over recessed lighting.

Various moulding patterns

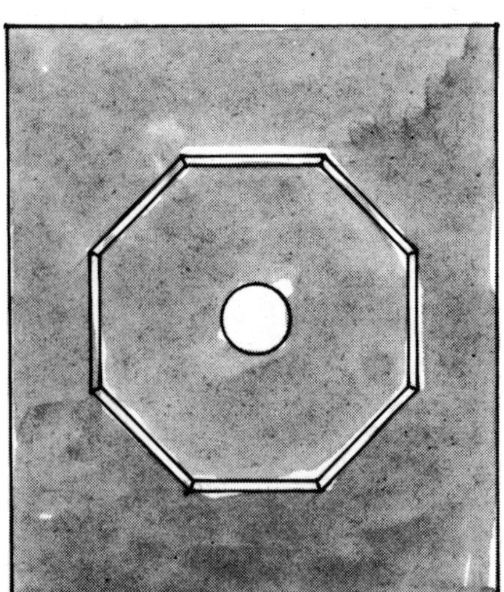

Octagon design using various moulding patterns

Lattice with multi-painted panels for variety

Various moulding pattern combinations

CEILINGS

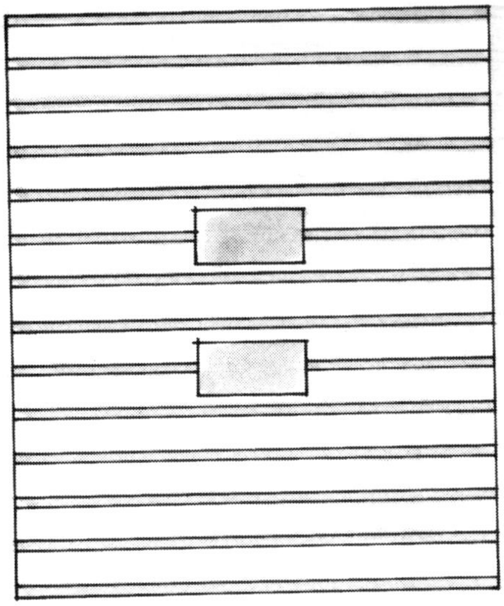

Lattice.

A beamed ceiling edged by bed or cove moulding. Base caps, quarter rounds, drip caps or stops can also be used.

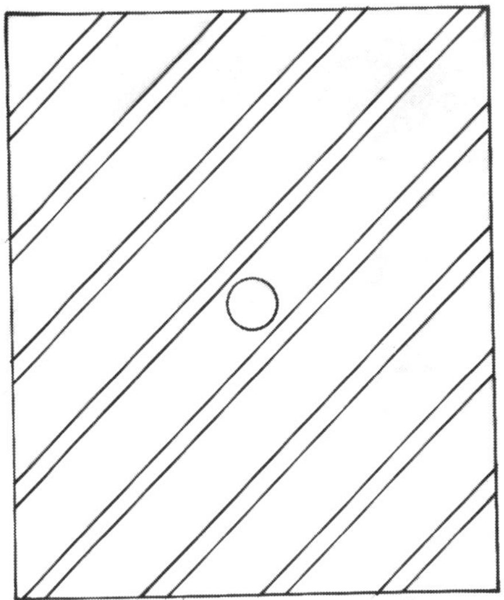

Various patterns used diagonally

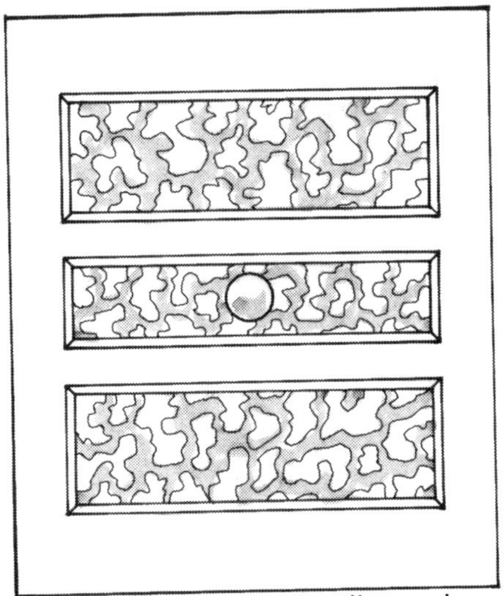

Framed panels of vinyl wall covering or wallpaper.

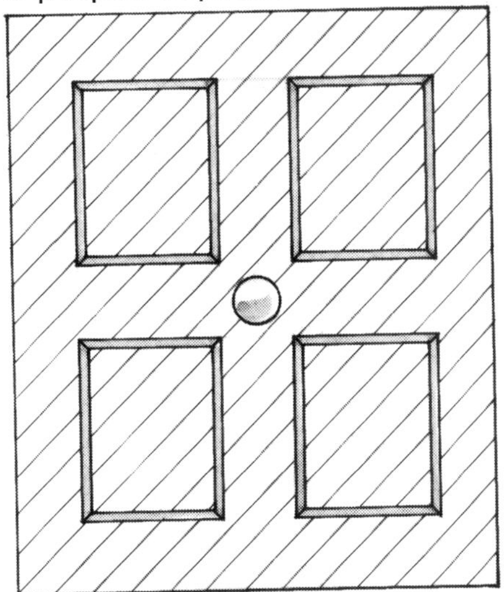

Moulding frames over a panel or lumber ceiling.

Various moulding patterns

Various patterns

Various patterns

Various patterns

CANDLE HOLDERS

The warm glow of candle light is even further enhanced by the warm richness of wood. These candle holders made from wood mouldings are simple to make, yet offer a variety of styles. Whether you use one candle or a cluster, there's a moulding pattern to fit the job. When drilling the holes, make sure they're perfectly vertical and are the correct size for the candles you want to use. Then stain or oil to the desired richness. One word of caution. If the candles are placed too close together when used in clusters, the heat may melt the adjoining candles. When constructing the clusters of rounds or squares, for instance, keep the candles as far apart as possible by using the largest moulding sizes available.

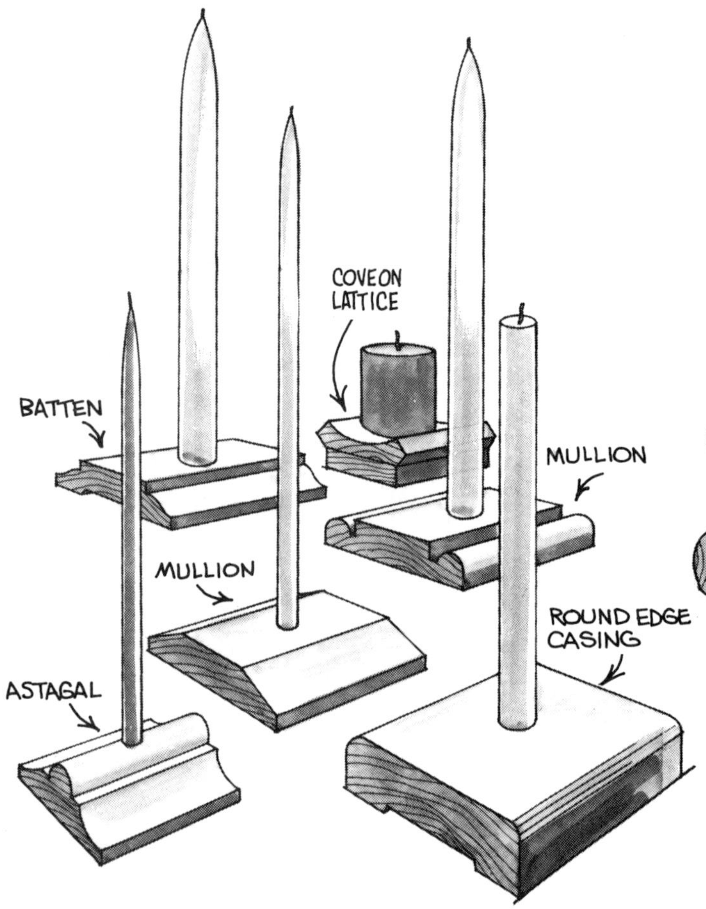

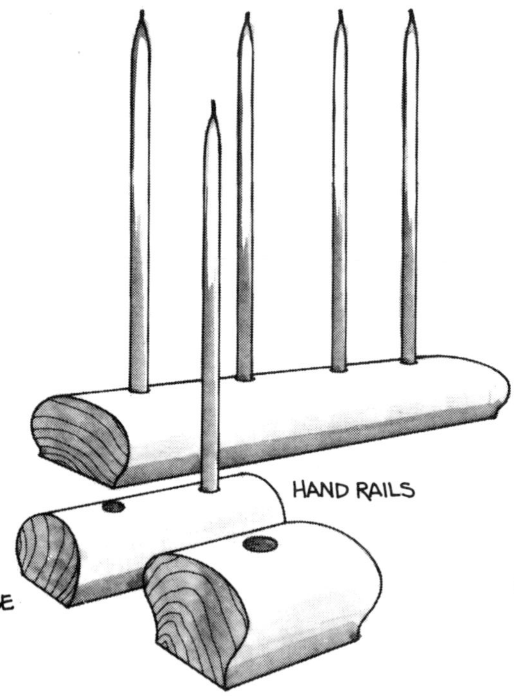

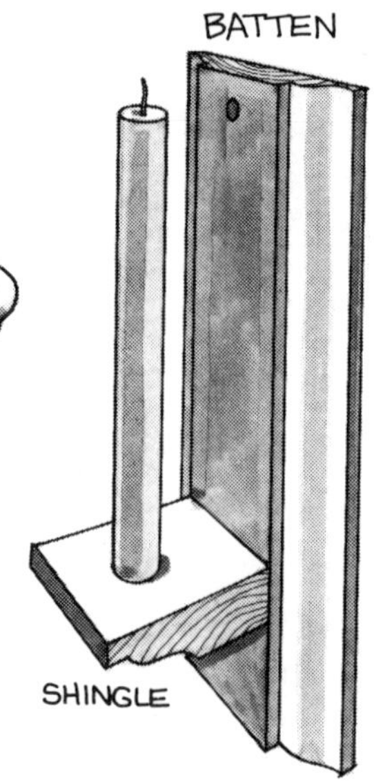

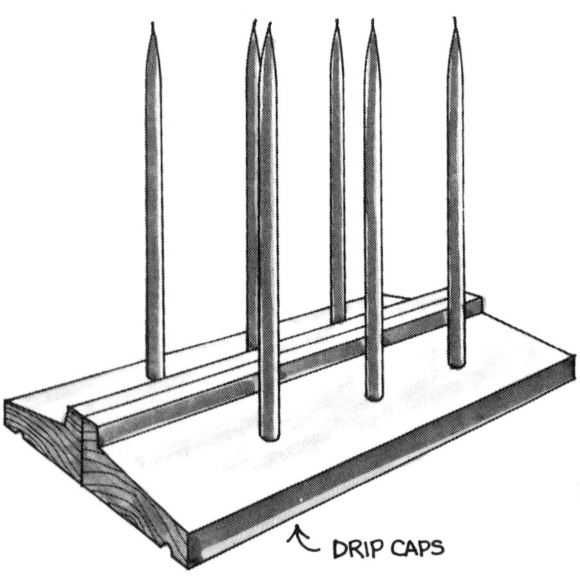

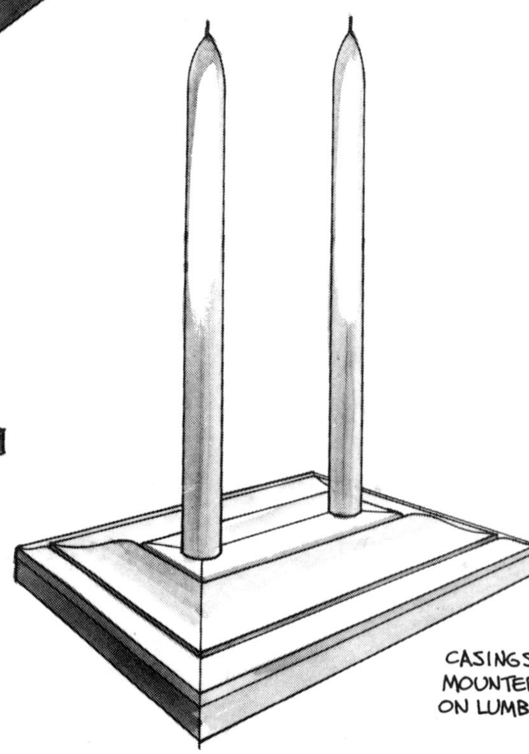

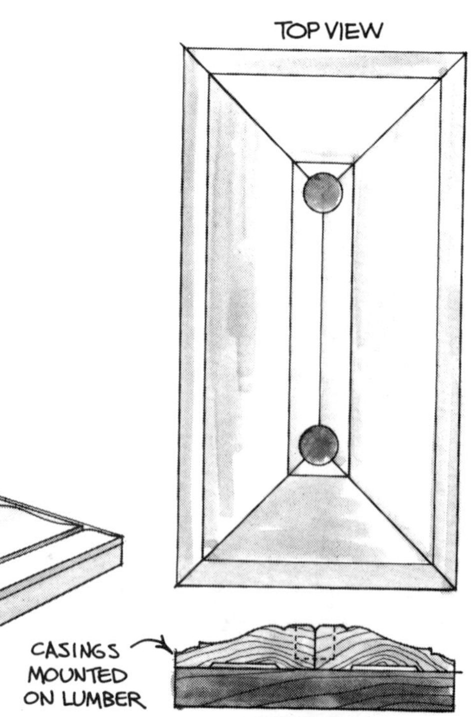

CANDLE HOLDERS

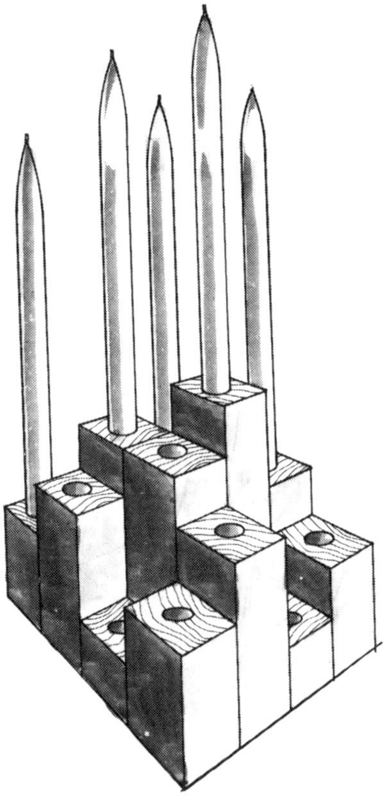

SQUARES CUT TO VARIOUS LENGTHS

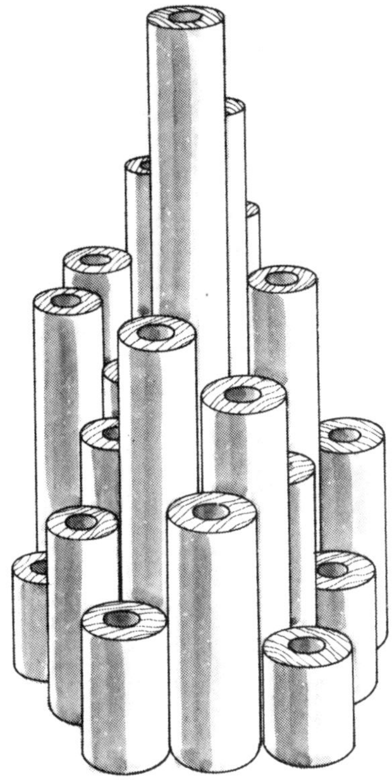

ROUNDS CUT TO VARIOUS LENGTHS.

ASSORTED ROUNDS CUT TO VARIOUS LENGTHS

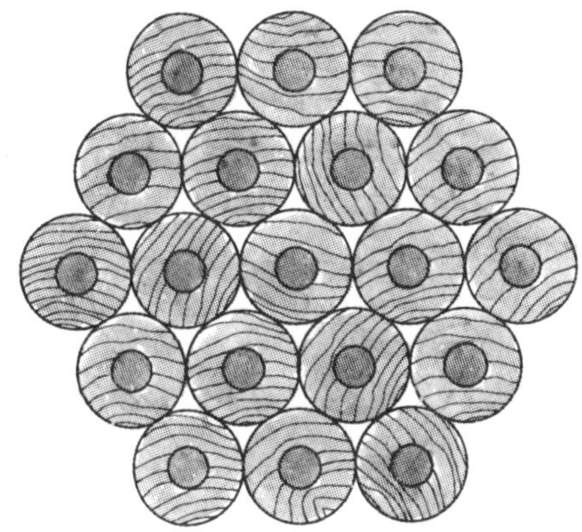

TOP VIEW

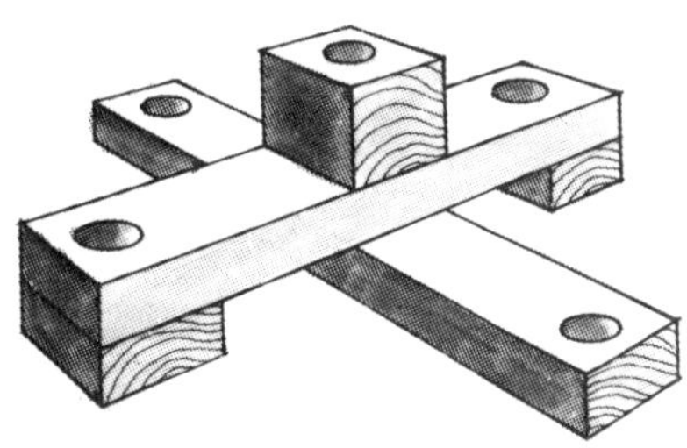

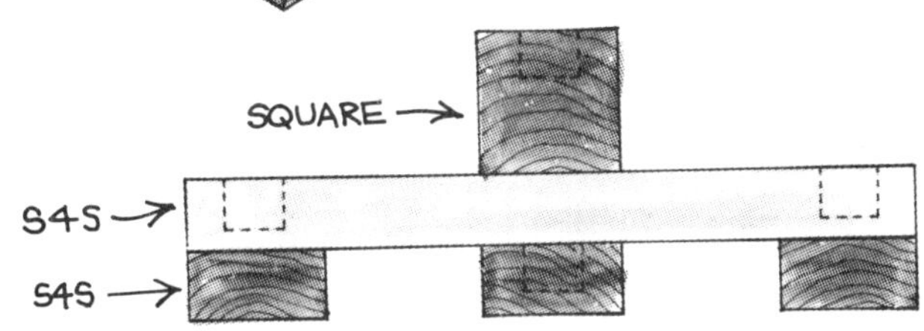

END VIEW

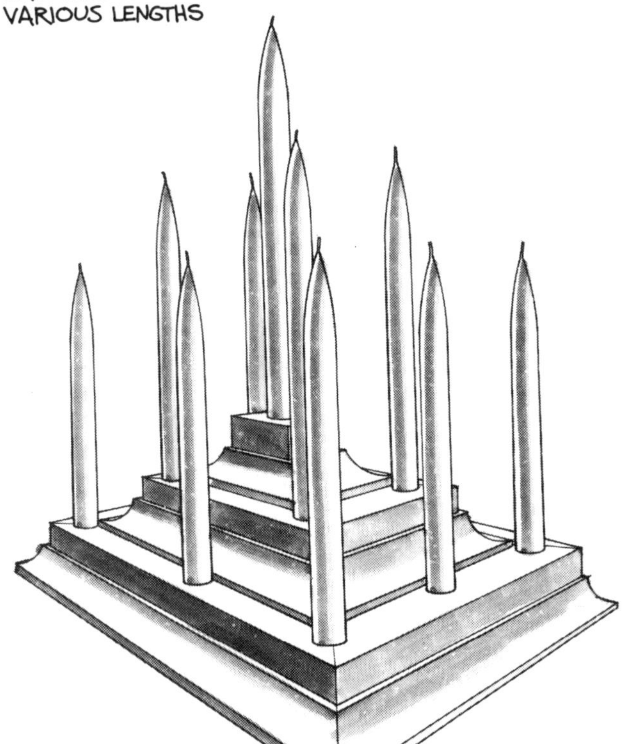

THE LENGTH WILL DEPEND ON WHETHER YOU WANT THE CANDLE HOLDER SQUARE OR RECTANGULAR

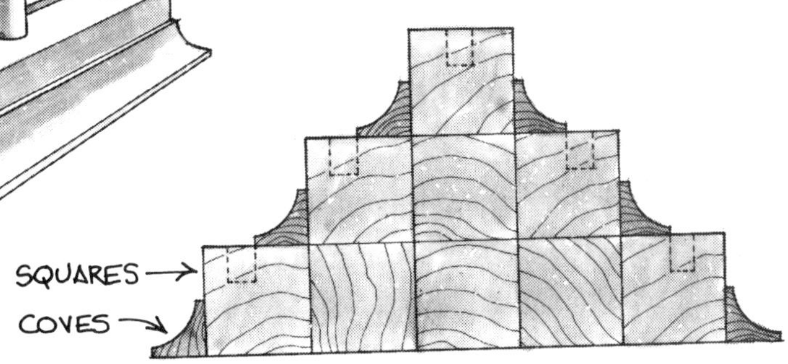

END VIEW

CHANDELIERS

If making your own chandelier causes you to hesitate because it involves working with lighting, wiring, electricity and all that stuff that's vaguely confusing, don't let that stop you. What's called for in these designs are ready-made sockets, which can be attached to the wood moulding chandeliers by either lattice supports, dowels or metal rods which come with lamp and lighting kits. So go ahead, light up that room with a chandelier of your own creativity.

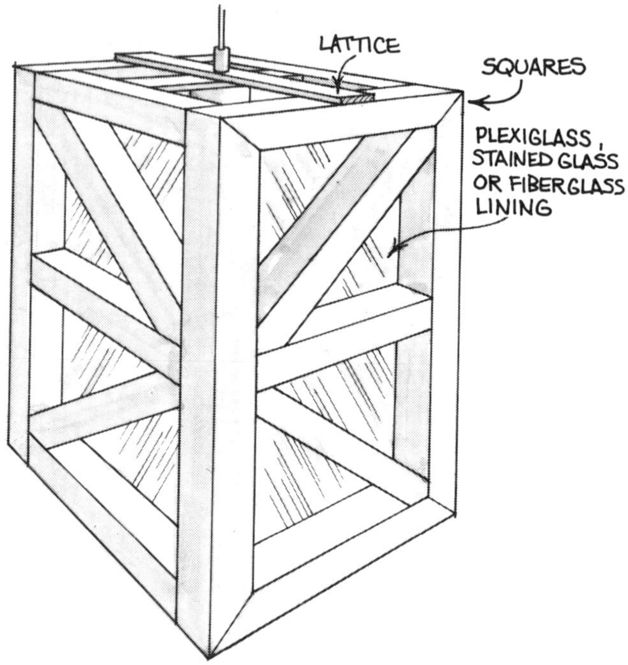

This rustic looking light is made up of moulding squares. First decide the size you want and then cut out the moulding to make the rectangle for one side of the lamp. It's then divided in half with a cross piece. Next, the diagonal pieces are fitted in place. Nail and glue together. Repeat the steps for the other side of the chandelier. With two sides built, it's easy to add the remaining cross pieces to form the whole outer structure. Next, line the inside with white fiberglass sheets or a cylindrical fiberglass lamp sleeve which may be purchased at most hobby shops or home improvement centers. The fiberglass is glued and tacked to the inside. Now either attach a lattice support bar to the top of the lamp or drill holes in the top cross member and insert a supporting dowel. Drill a hole in the support piece, string heavy duty round fixture wire through, secure with a knot and add a socket.

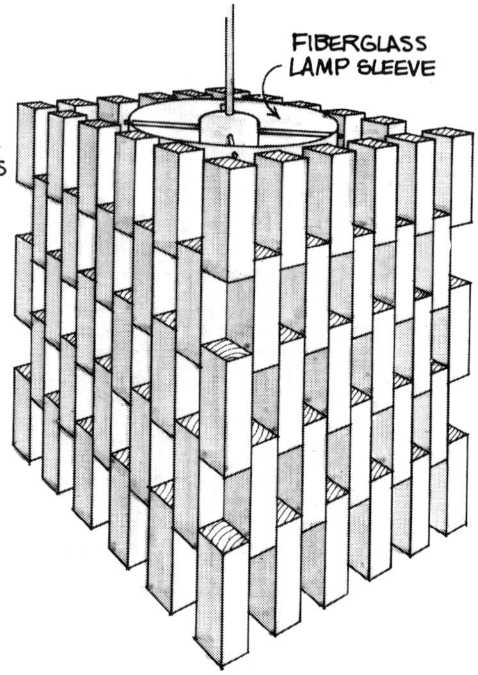

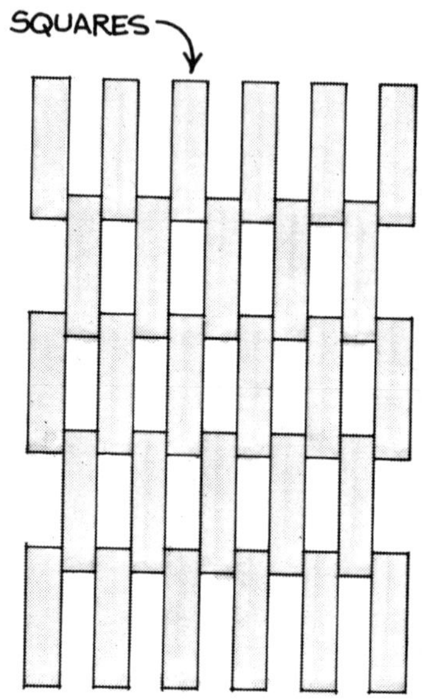

SIDE VIEW

This lamp is made of 100 short lengths of square moulding. Rather than cut each piece individually, it's best to clamp the squares together in groups to saw them. After all the pieces have been cut to size, work on one side of the lamp at a time. The pieces can either be attached to each other by gluing and nailing or by drilling a hole through a row at a time and inserting a dowel. When one side has been completed, do the opposite side of the lamp. The remaining two sides will have fewer pieces since the corners of the lamp have already been established by the first two sides.

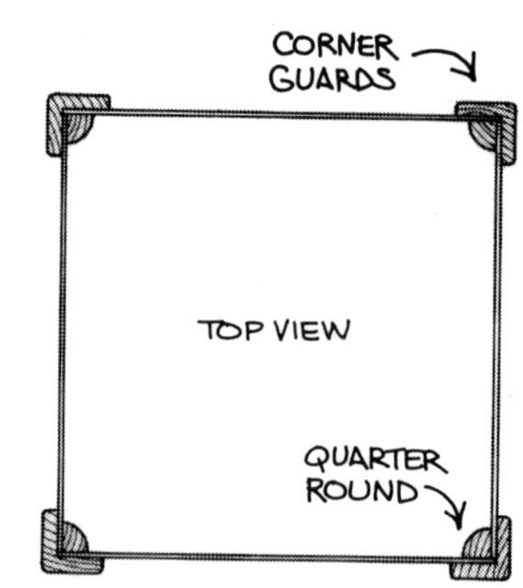

TOP VIEW

This colorful rectangular lamp is made of plastic, fiberglass or stained glass panels glued to corner guards and quarter rounds.

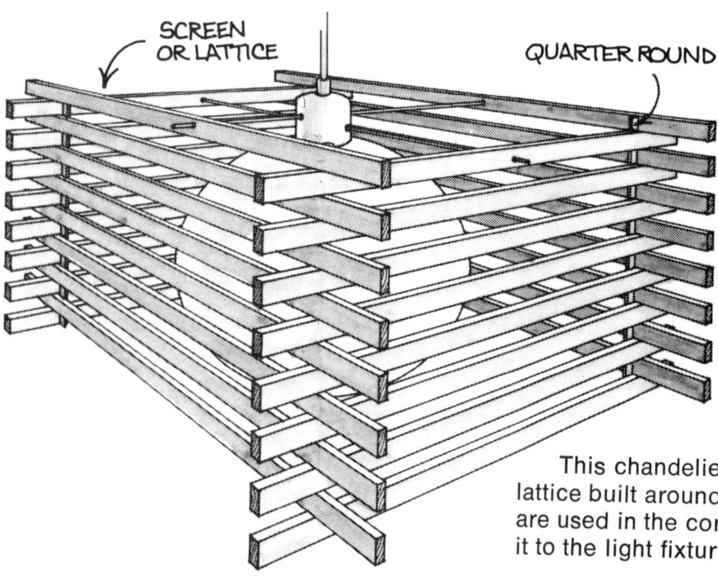

This chandelier is simply a fence of screen moulding or lattice built around a light fixture. Quarter rounds or squares are used in the corners to hold the structure together. Attach it to the light fixture in any of the ways previously mentioned.

CHANDELIERS

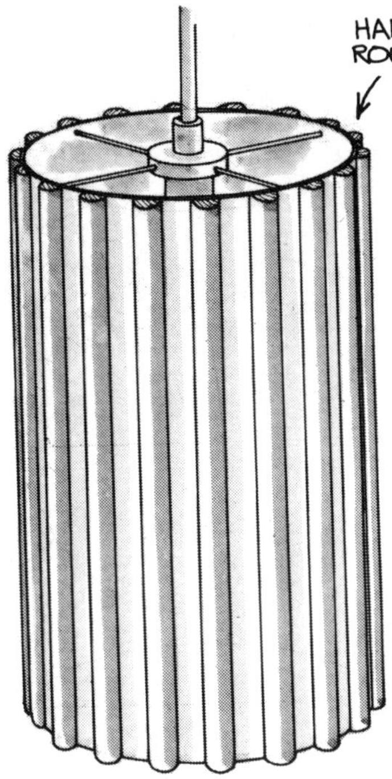

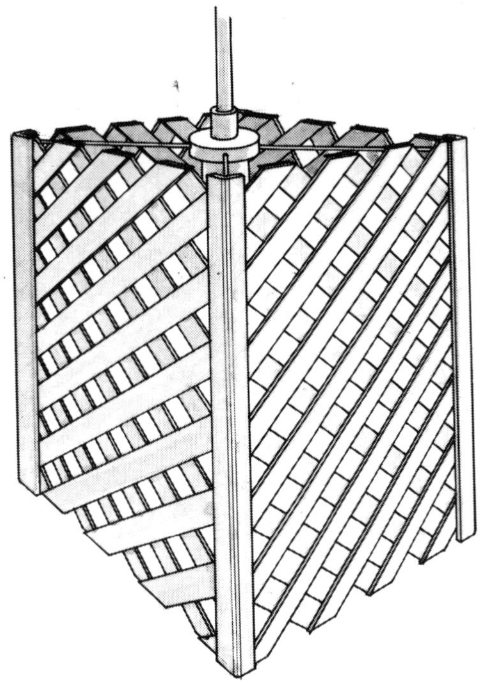

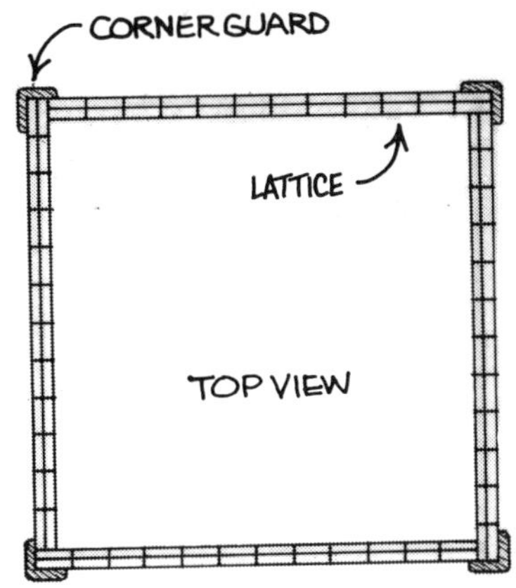

A simple plastic or fiberglass lamp sleeve with painted or stained half rounds glued to the outside make this an easy chandelier to build. Finish it off by adding a lighting kit or attach a socket as shown before.

Here's a chandelier for lattice lovers. Simple make four sides using a diagonal grid pattern of lattice and connect the sides with corner guards. Again, use fiberglass sheets or a fiberglass sleeve. Attach the light fixture as outlined before or buy one of the many lamp kits that are now available. The fixtures from an existing chandelier could also be used.

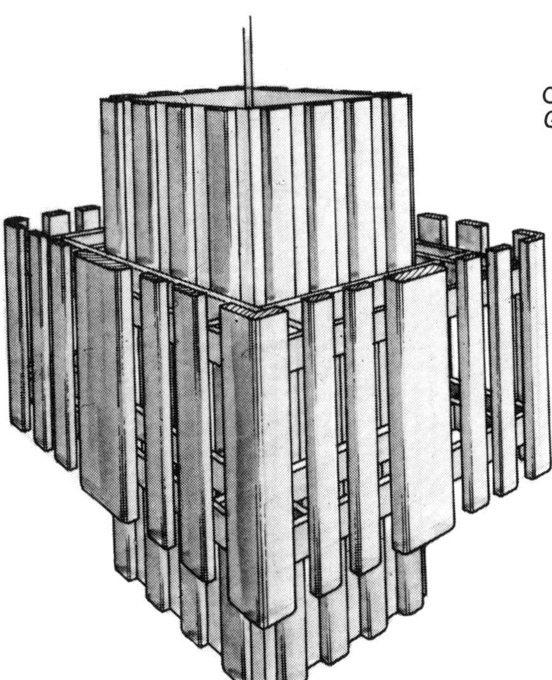

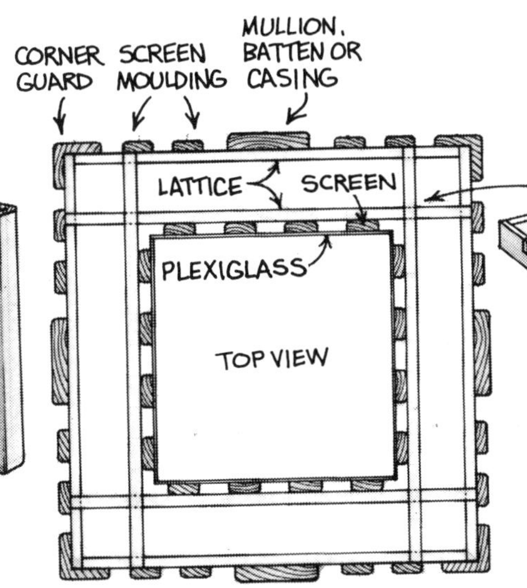

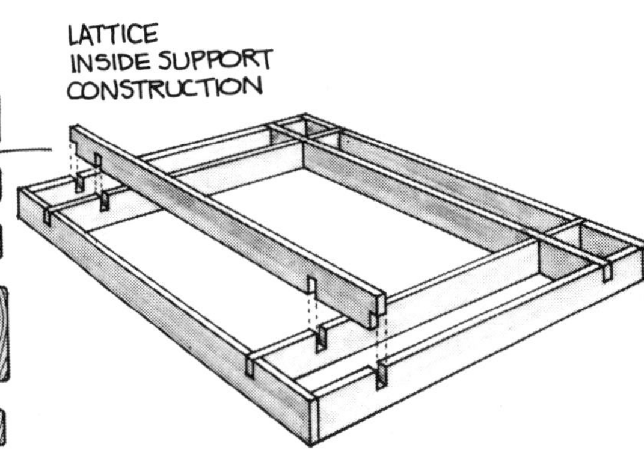

This chandelier takes a little more time to build but is worth the effort. It consists of a tall, narrow inside structure of screen moulding glued to panels of plexiglass, plastic or fiberglass. The shorter outside structure is also made of screen mouldings, with corner guards at each corner and mullion or round edge casing in the center. Holding the two structures together are two inside supports constructed of lattice which is fitted together as shown in the illustration. To put the whole thing together, it's best to first cut and notch the inside supports to the proper size. Make sure they fit together but don't do the gluing yet. Now attach the outside pieces of the chandelier to the outside braces of the supports. Then attach the inside pieces of the chandelier to the inside braces of the support. Slip the two structures together and glue and tack in place. Add the fiberglass panels next, with the lighting fixture as the final step.

CHRISTMAS TREE ORNAMENTS

Homemade ornaments add a special touch to your Christmas tree. Everybody wants to get in on the act of creating these items. Whether it's the designing, cutting, gluing, painting or hanging them on the tree, it's fun for the whole family. The first step is to take short pieces of the mouldings you want and glue them together to form the desired shapes. Let the glue dry overnight before sawing. When cutting them out, make sure you have a good sharp saw and a miter box to hold the moulding steady. They should be cut at least 3/8″ thick. Any less than that and they tend to break. After sanding off the rough spots, paint or stain them and attach a small eye screw for hanging. Have a Merry Christmas.

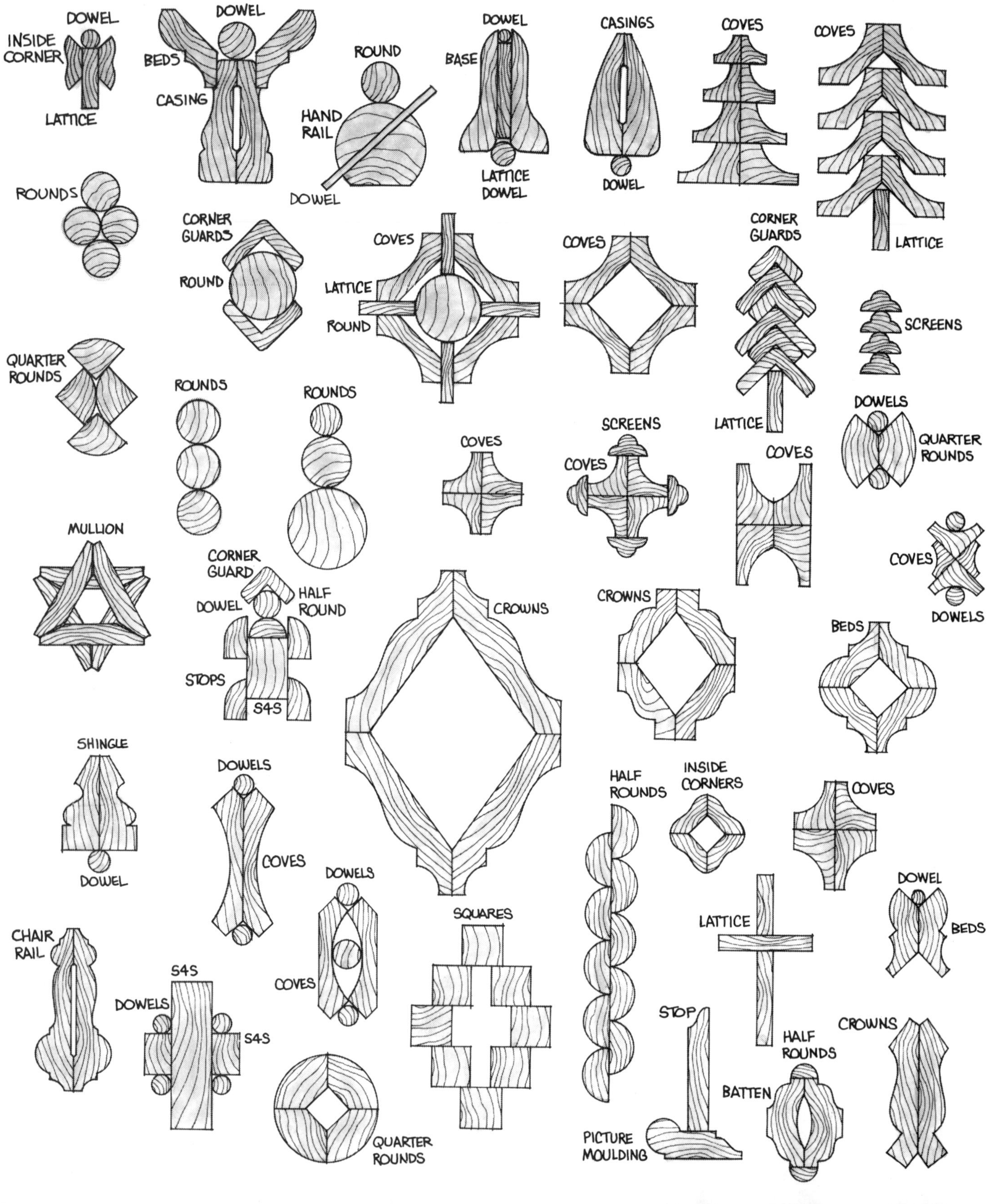

CLOCKS

If you're a clock watcher, build a clock worth watching. These handsome examples are simply made of mouldings cut to frame the clock faces which are made of plywood, lumber, paneling or plastic laminate. The clock's numerical symbols are made of decorative tacks or moulding rounds or half rounds. Drill a hole through the center of the face to attach the hands. The clock workings themselves are attached to the back of the clock. These workings can be purchased at many hobby shops or you may want to use the working parts of an old existing clock.

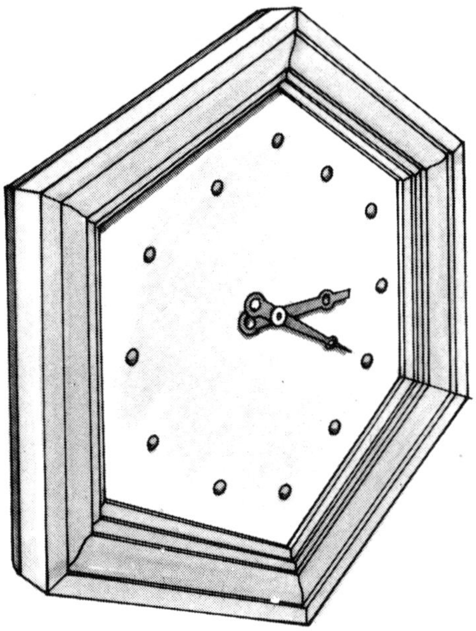

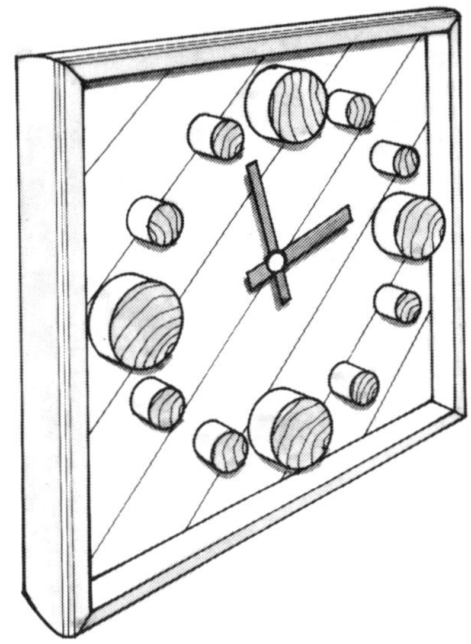

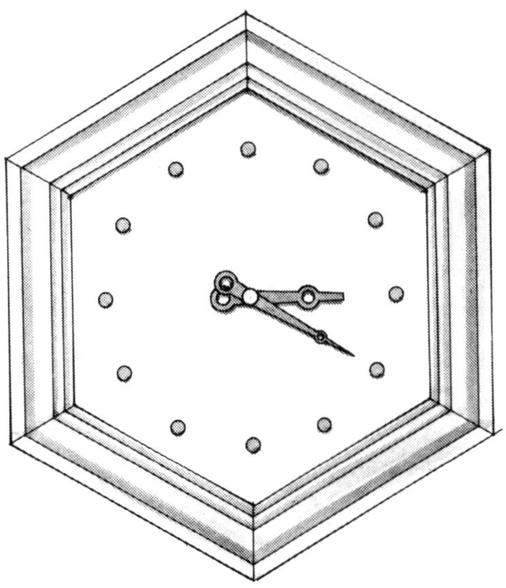

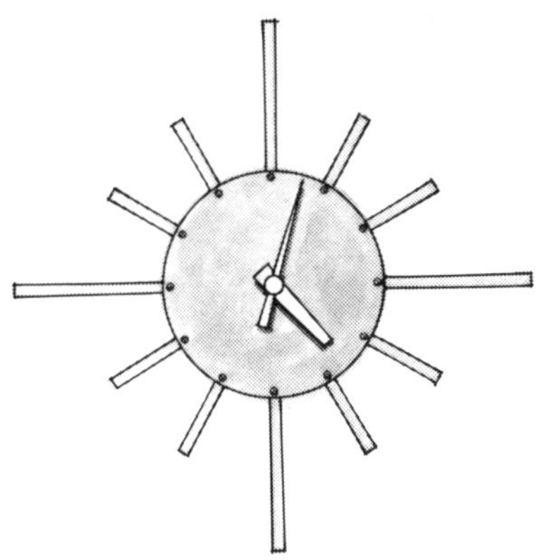

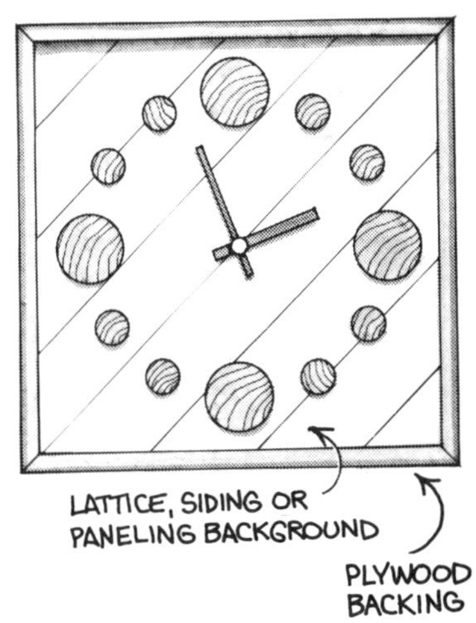

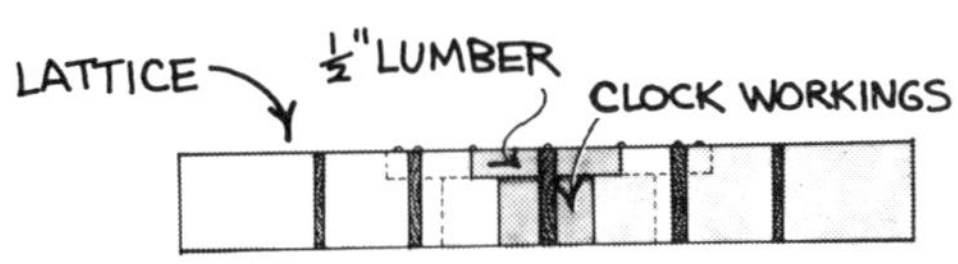

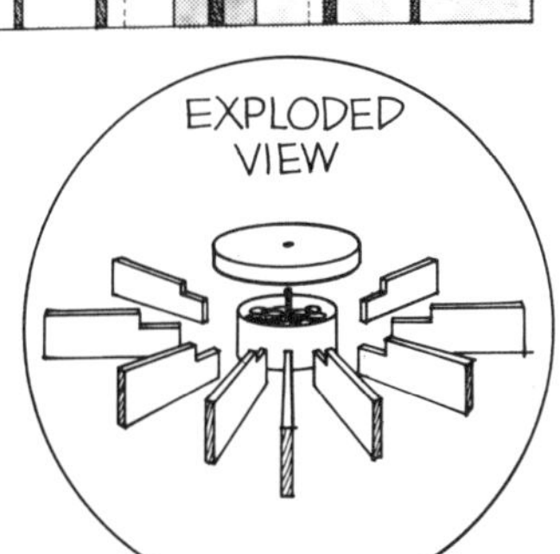

COAT RACKS

Here's a group of coat racks built to go anywhere . . . in the hallway, the kid's room, the rec room, the basement, the back porch, the office . . . anywhere you want to hang your coat. These ideas are flexible enough to put together a small rack for hanging one coat, or a large rack that runs the complete length of a hallway with space for dozens of coats. It's up to you. Whether it's small, large, contemporary or traditional, wood moulding has the size and pattern you want.

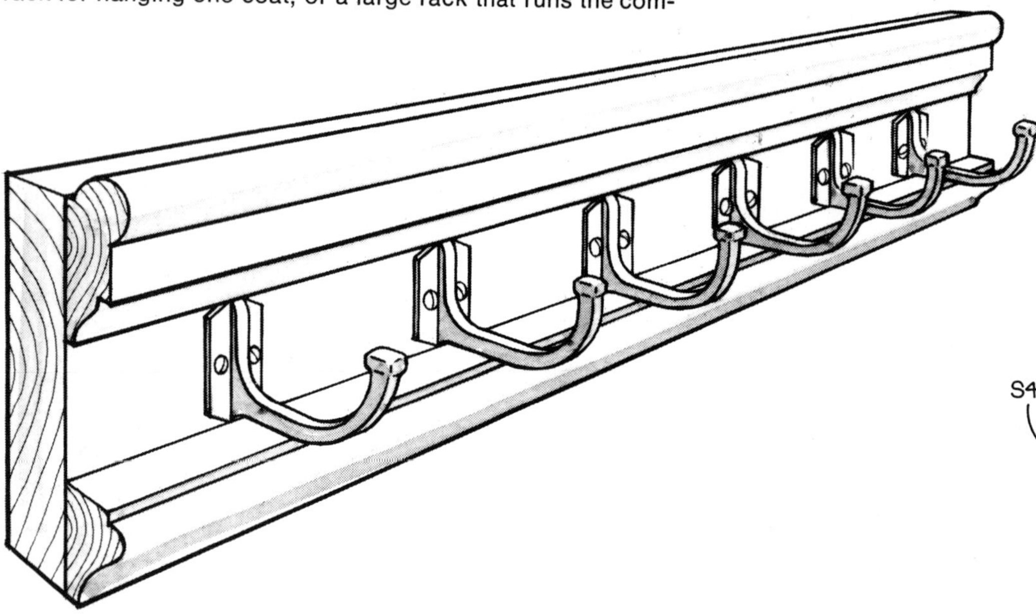

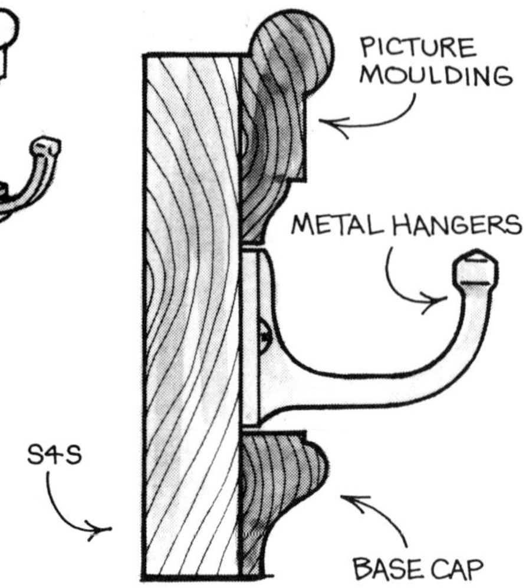

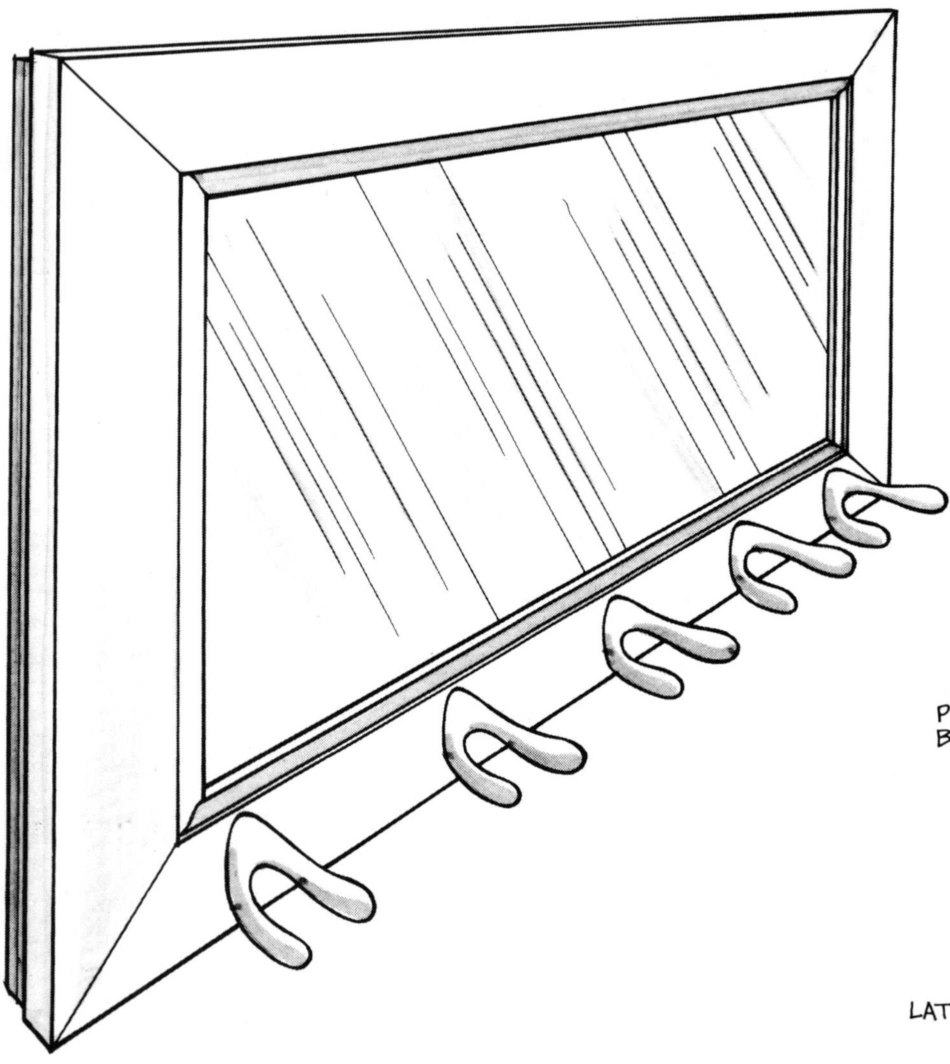

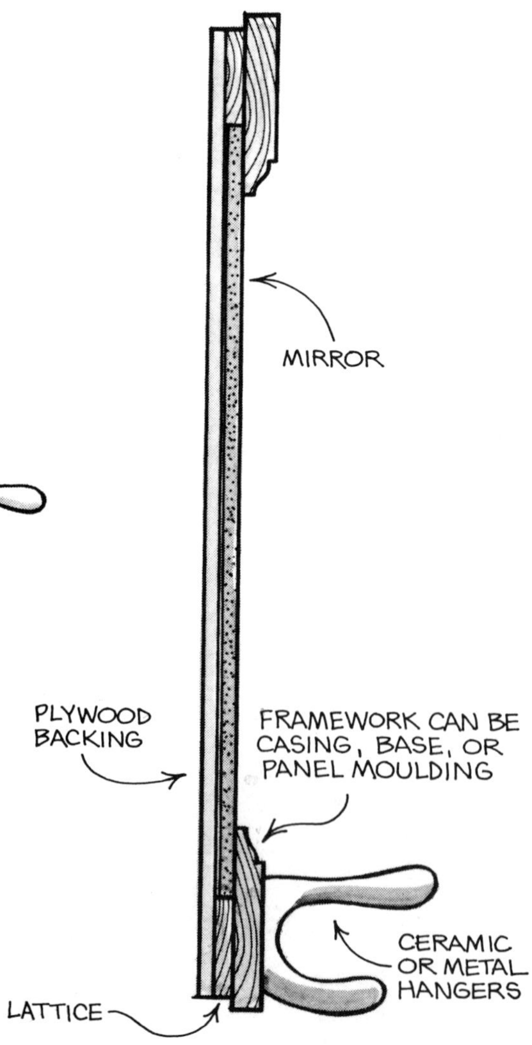

COAT RACKS

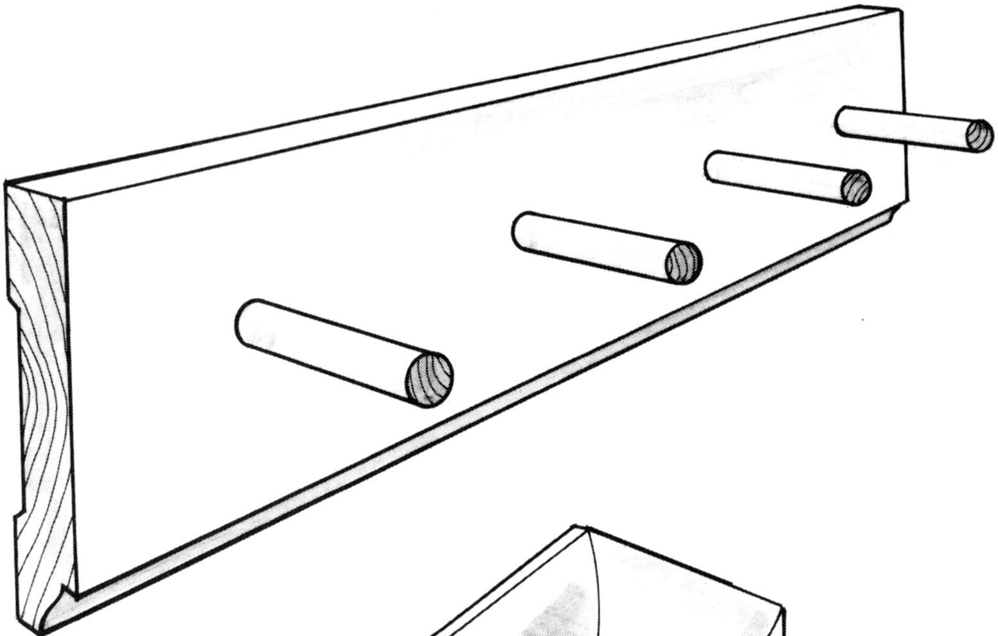

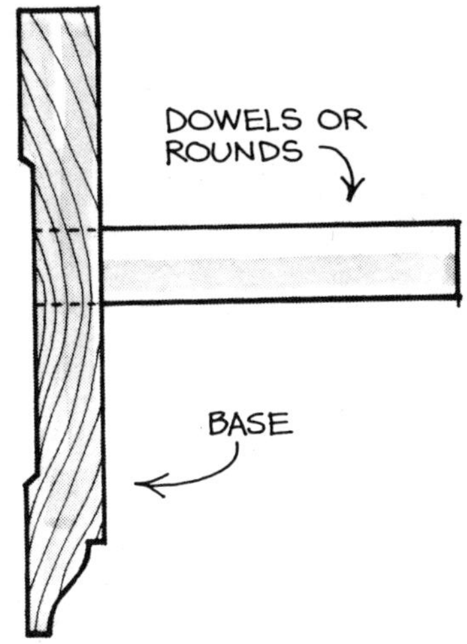
DOWELS OR ROUNDS
BASE

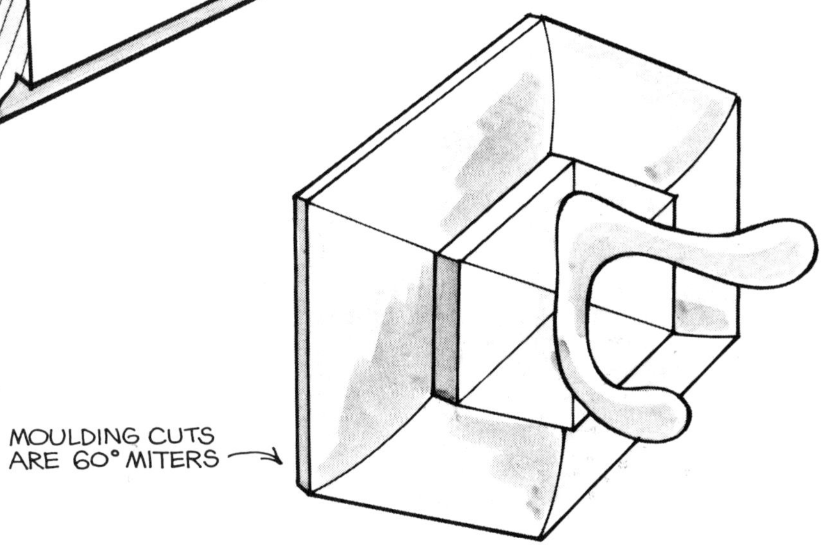
MOULDING CUTS ARE 60° MITERS

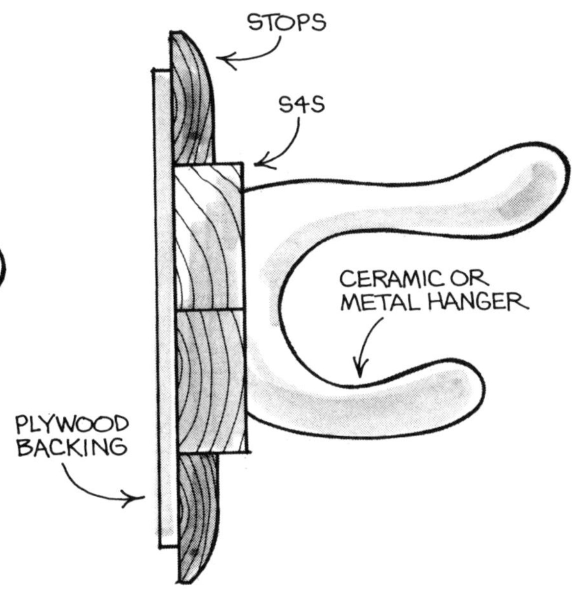
STOPS
S4S
CERAMIC OR METAL HANGER
PLYWOOD BACKING

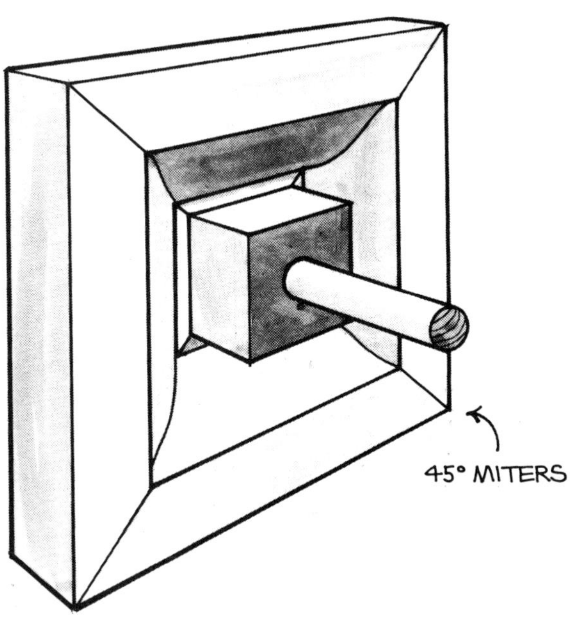
45° MITERS

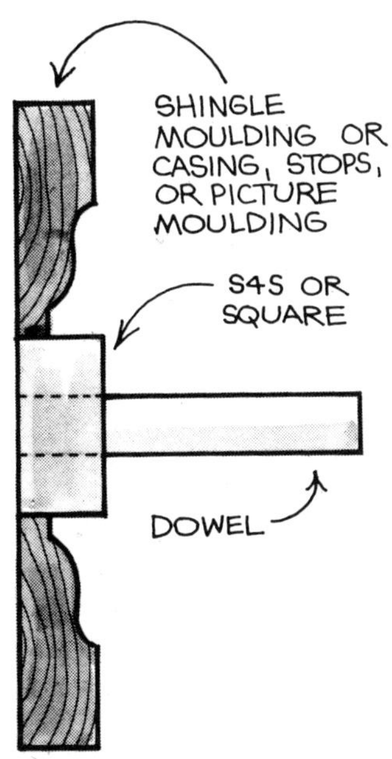
SHINGLE MOULDING OR CASING, STOPS, OR PICTURE MOULDING
S4S OR SQUARE
DOWEL

COASTERS

No matter how small a job is, it can still be made so it's individually yours. Even something as simple as a set of coasters can have a touch of quality merely by being made from wood mouldings. Careful sawing is important here. When gluing, clamp and leave overnight. Staining or oiling them will add to their good looks.

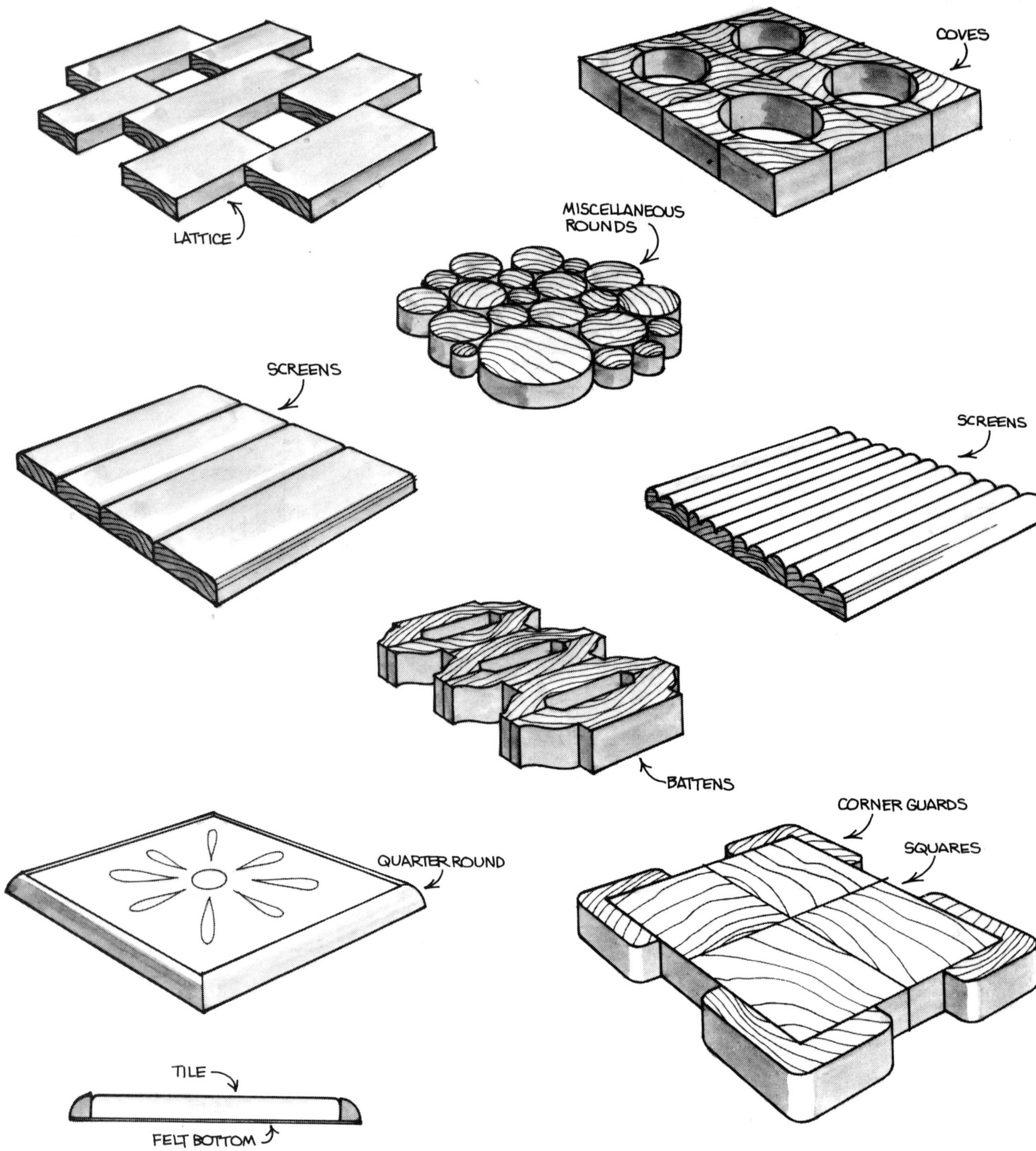

COUNTER FRONTS/BAR FRONTS

Does the counter that sticks out in your kitchen, or the kitchen island, or possibly the bar in your game room look a little dull? Add visual impact by giving it a wood moulding facelift. It's easy to do. There's such an endless variety of patterns and designs to choose from. Probably your only problem will be deciding which one to use. Once you make that decision, just be sure you do your measuring and mitering carefully. Then glue and nail in place and paint or stain it to match your decor.

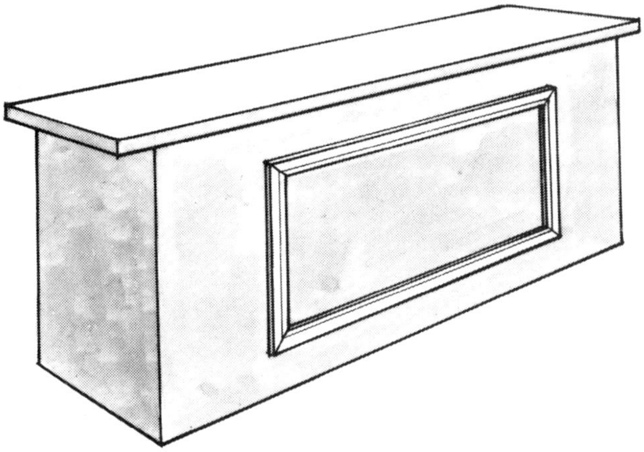

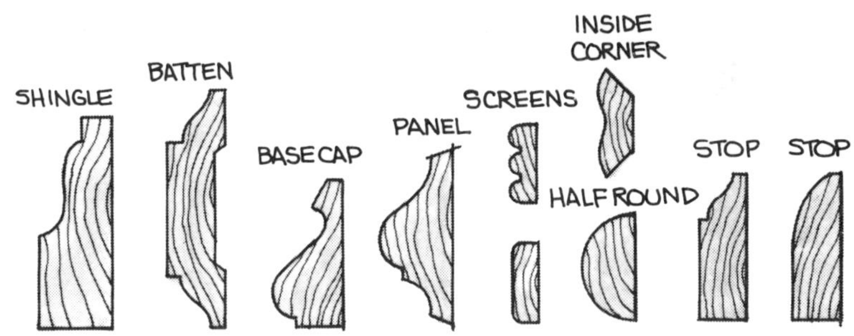

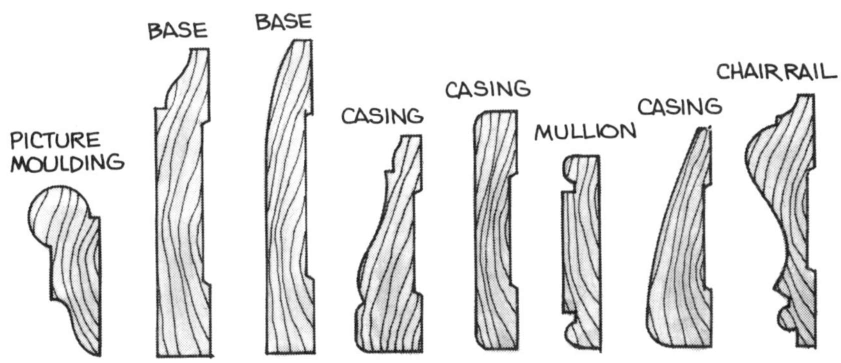

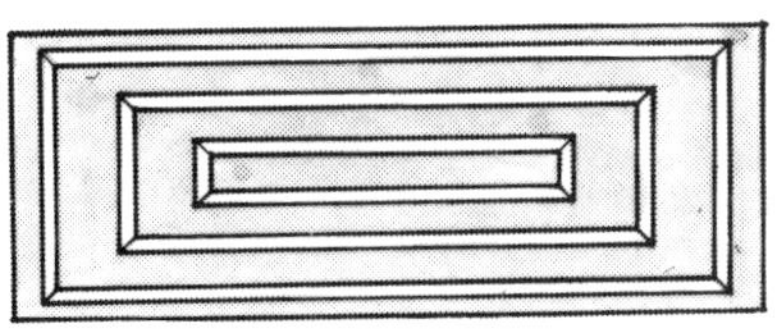

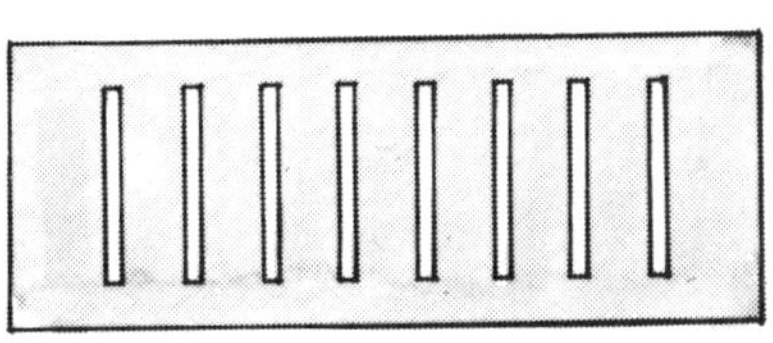

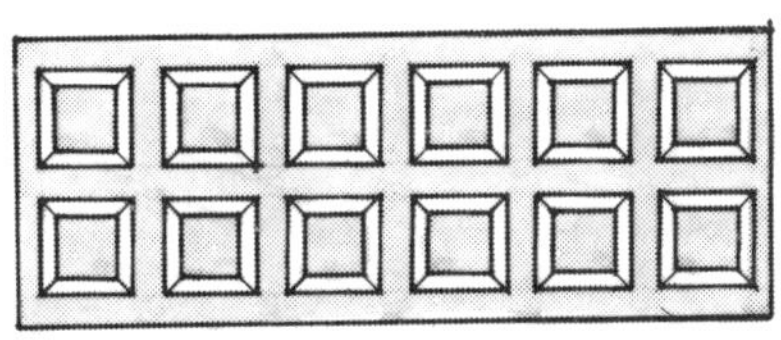

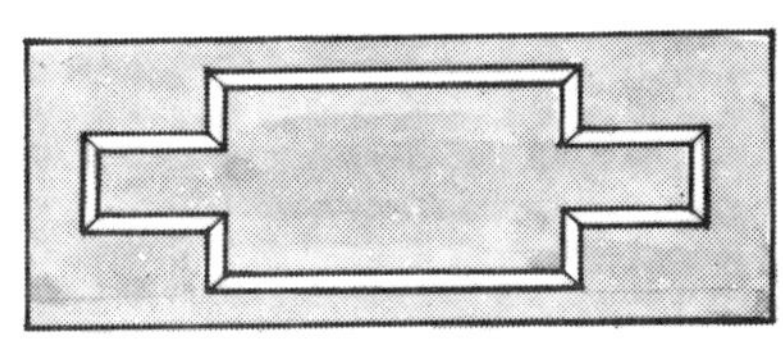

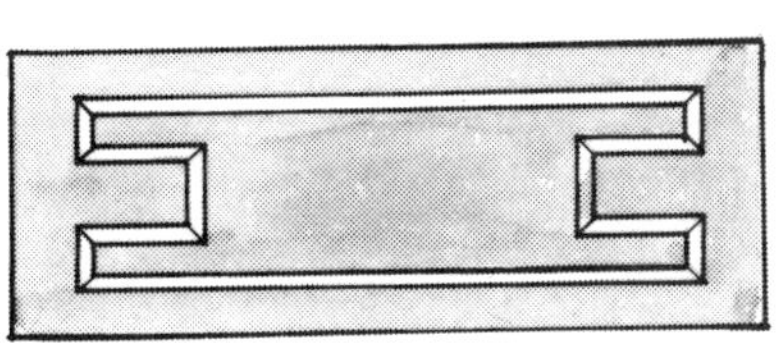

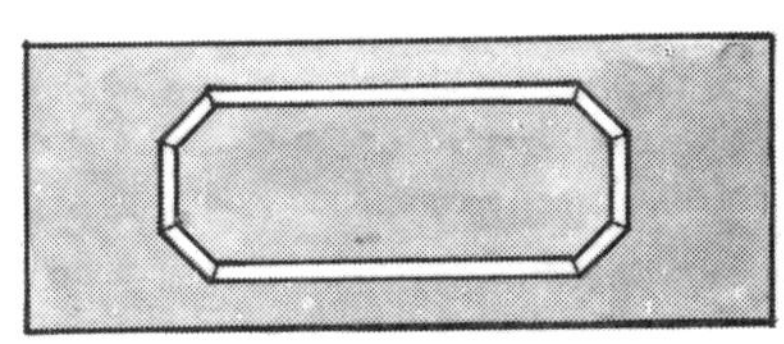

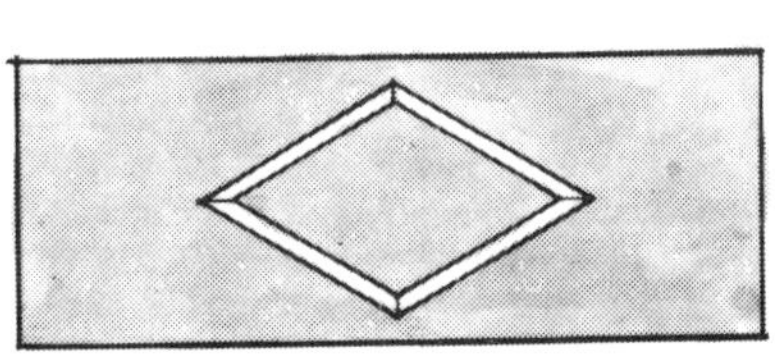

COFFEE TABLES

Building these coffee tables is much easier than it looks. Each one is simple in its own way. Go ahead. Try one. There's probably a spot in your home or office just waiting for a coffee table. The length and overall size of each table will depend on where and how it will be used. If it's to be used in front of a couch, it's usually long and low. About 16 inches high is normal. If it's to be used as an end table or to hold a lamp or plant, it can be slightly taller. Anywhere from 20 to 24 inches is normal.

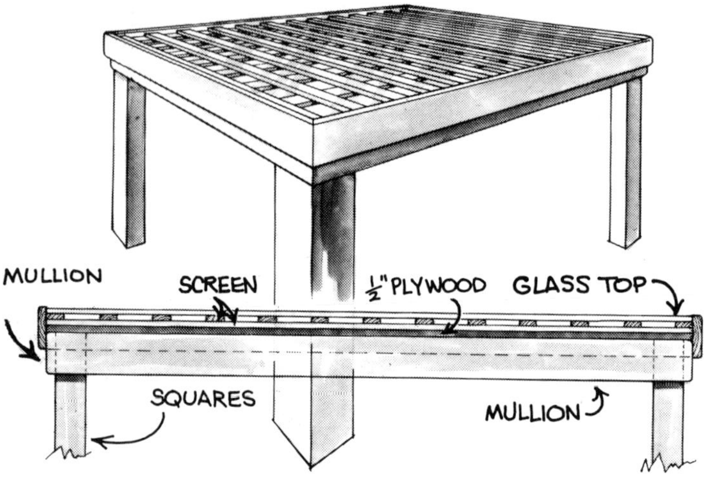

This table is just a grid pattern of screen mouldings sitting on a plywood base and covered by a glass top. These four layers then sit on top of a framework of mullion casing with legs of square mouldings attached to each inside corner. The plywood base board is also nailed to the legs for added stability. Metal corner braces will also help. An outer layer of mullion or shingle moulding is then attached around the whole structure to hide all the rough edges and to give it a clean finished look. Be sure to miter the corners.

This table is merely a plywood box with notched leg openings. All four corners, the top and bottom edges and all around the openings are framed with screen mouldings. A piece of quarter round is added to each corner to give it a smoother look. This type of table works well in a family or game room and looks great when painted in bright contrasting colors.

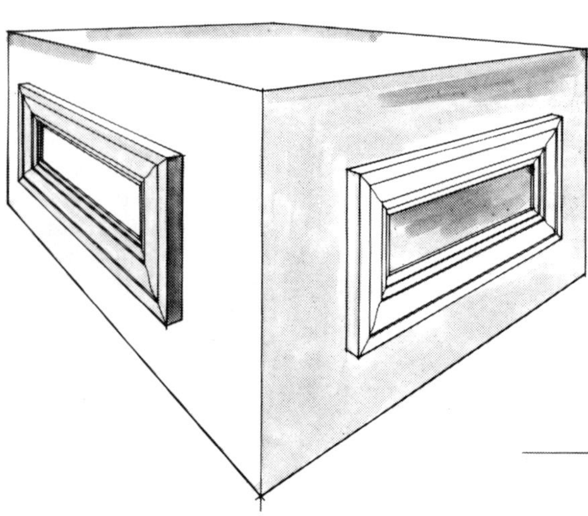

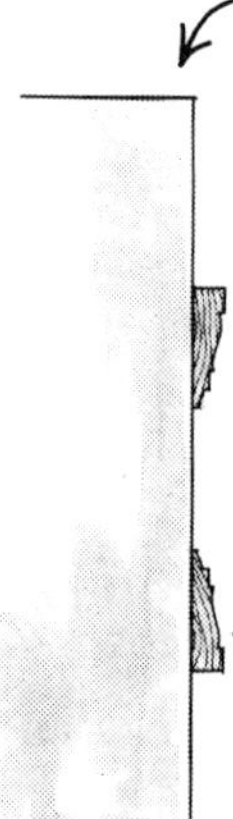

This is another plywood box, but without the leg openings. A smooth seamless look is achieved by covering it with plastic laminate. A decorative framework of wood moulding is then applied to the sides. Either casing, battens, mullions, chair rails or base caps can be used. These tables work well when made into cubes and used either individually or in groups. They can also be easily moved around when entertaining.

COFFEE TABLES

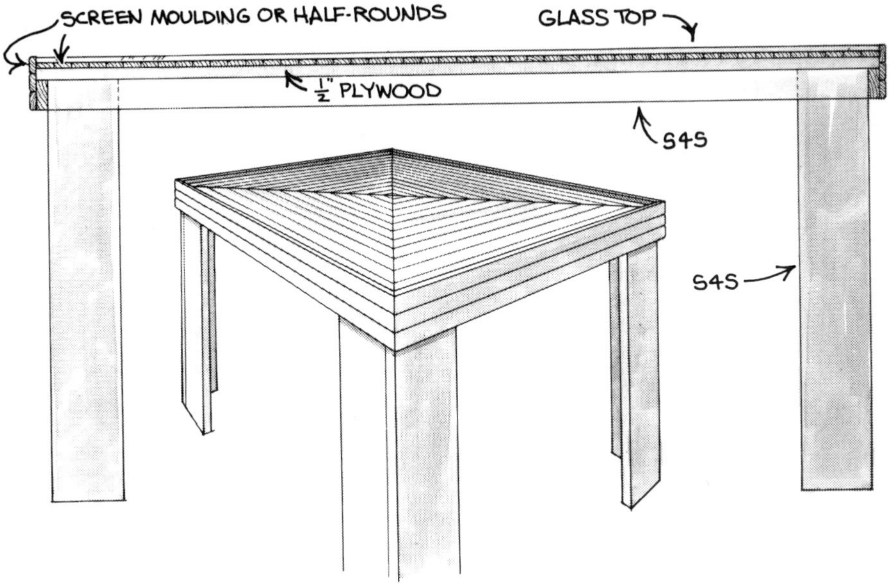

The top of this table is made up of a continuous series of mitered screen mouldings or half rounds. They are glued to a plywood base which sits on top of an S4S frame. Legs made with two pieces of S4S are then nailed to the inside corners of the S4S framework. A glass top is then set over the whole table. An outer frame made from a group of screen or half rounds is then attached all around the table. Be sure to miter the corners accurately.

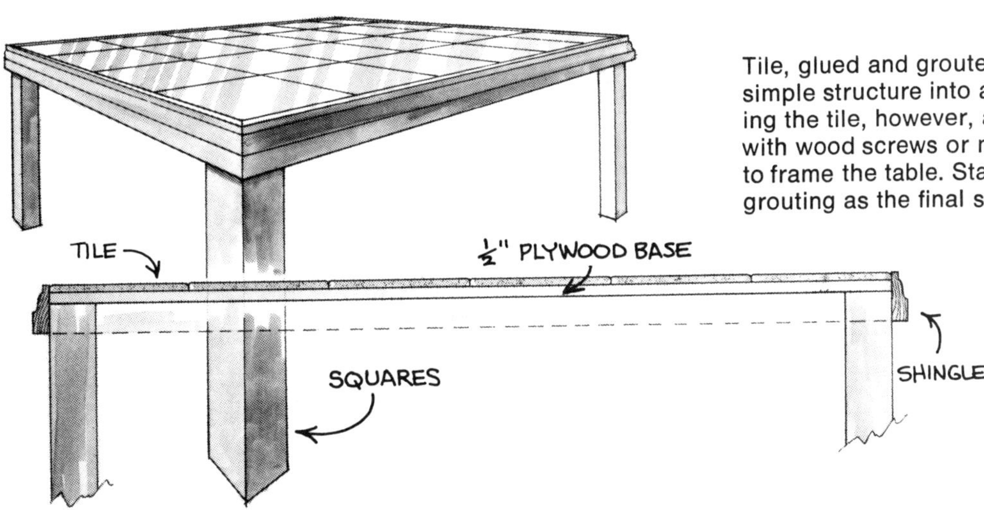

Tile, glued and grouted to a plywood base, transforms this simple structure into a handsome coffee table. Before applying the tile, however, attach the legs of square moulding with wood screws or nails. Shingle moulding is then used to frame the table. Stain or paint it next, with the tiling and grouting as the final step.

This elegant little square is just a plywood box, except the top and bottom extend out past all four sides. How far they extend will depend on what size crown moulding is used to frame the top and bottom. Accurate measurements here are a must. Also the accuracy of the corner miters are very important. Don't worry though, it's not as hard as it sounds. The finished job will make it all worth while. The top of this table can be plywood or plywood covered with tile or plastic laminate. Add a strip of quarter round on the edge to give it a smooth finished look.

CUP RACKS

Here's a couple of simple coffee mug racks that can be used in your kitchen, game room, bar, office, conference room, lunch room, or wherever you hang your mug. They can be built to handle from two to twenty cups. The smaller design shown here can be hung either vertically or horizontally, depending on the space you have available. When drilling holes for the dowels, it's best to angle them slightly so the cups are less likely to slip off. A drill stand will be helpful in keeping the angles uniform, but even if you don't have one, you'll still be able to drill a good set of holes. It just takes a little practice and a steady hand. When finished, attach the rack to the wall with screws.

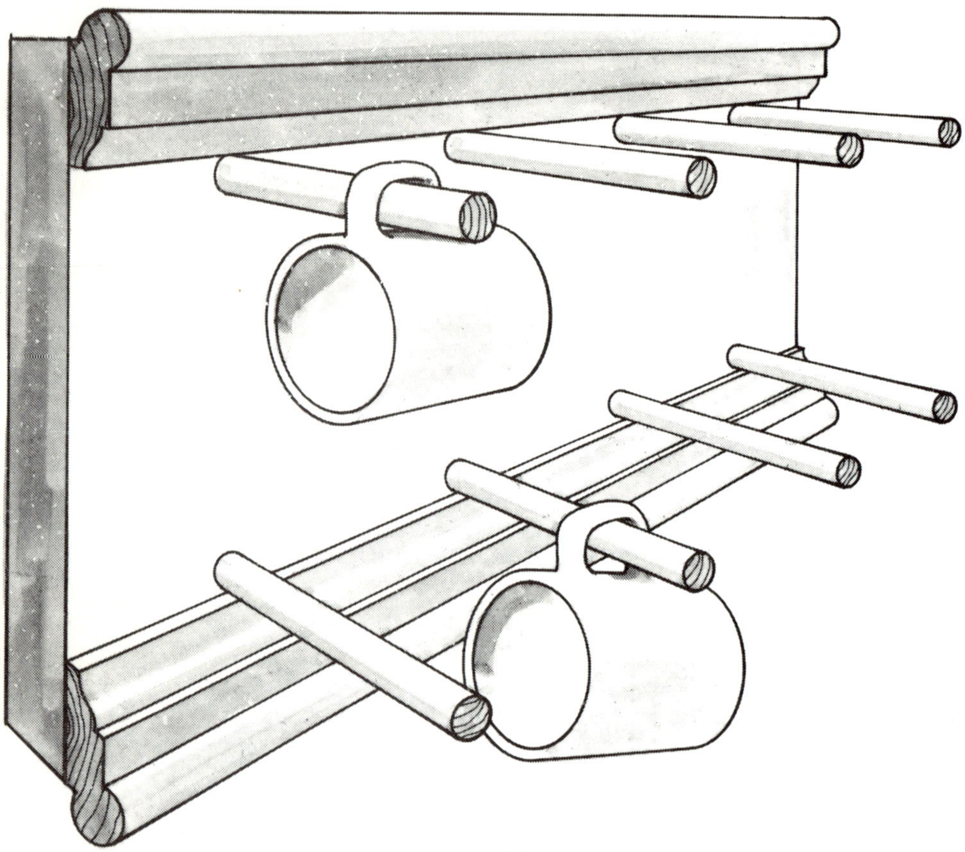

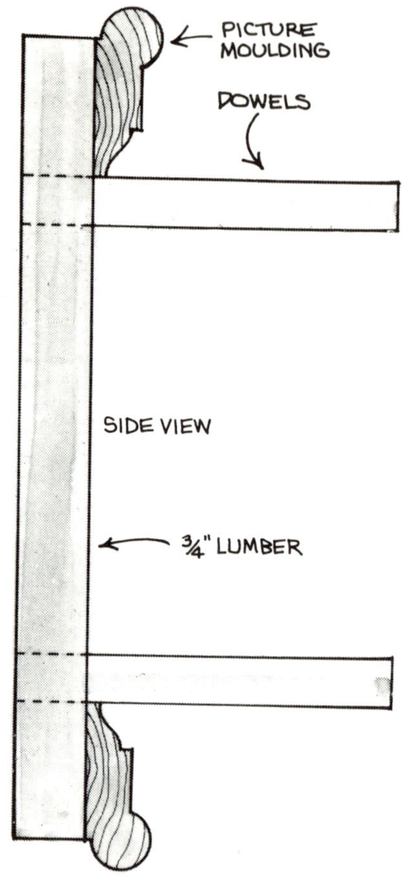

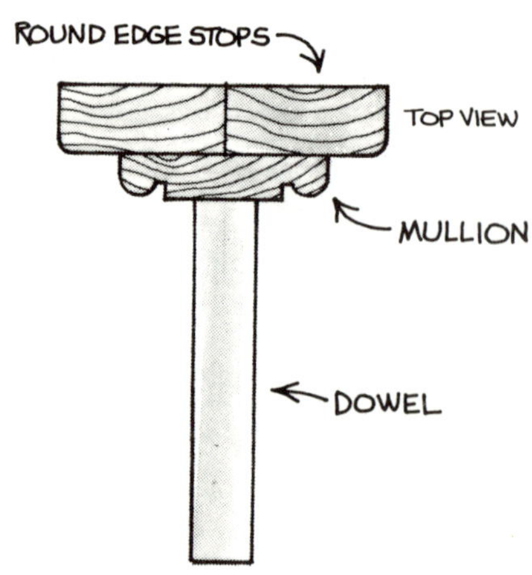

CURTAINS (Doorway, Window, Divider)

Uniqueness and charm will make this curtain design a hit in your home, wherever it's used. Whether it's hanging in a doorway, covering a window or used as a room divider, it's sure to be noticed. It can be made from ply caps, inside corner guards, outside corner guards, battens or coves. The pieces are cut to size, then glued face to face with just enough space for the cord to fit through. Knots in the cords will hold the pieces in place. The size of the pieces can either be equal or can be a mixture of varying lengths. The finish can be subdued by staining or painting them all the same, or it can be a mixture of wild colors, depending on the decor of the room.

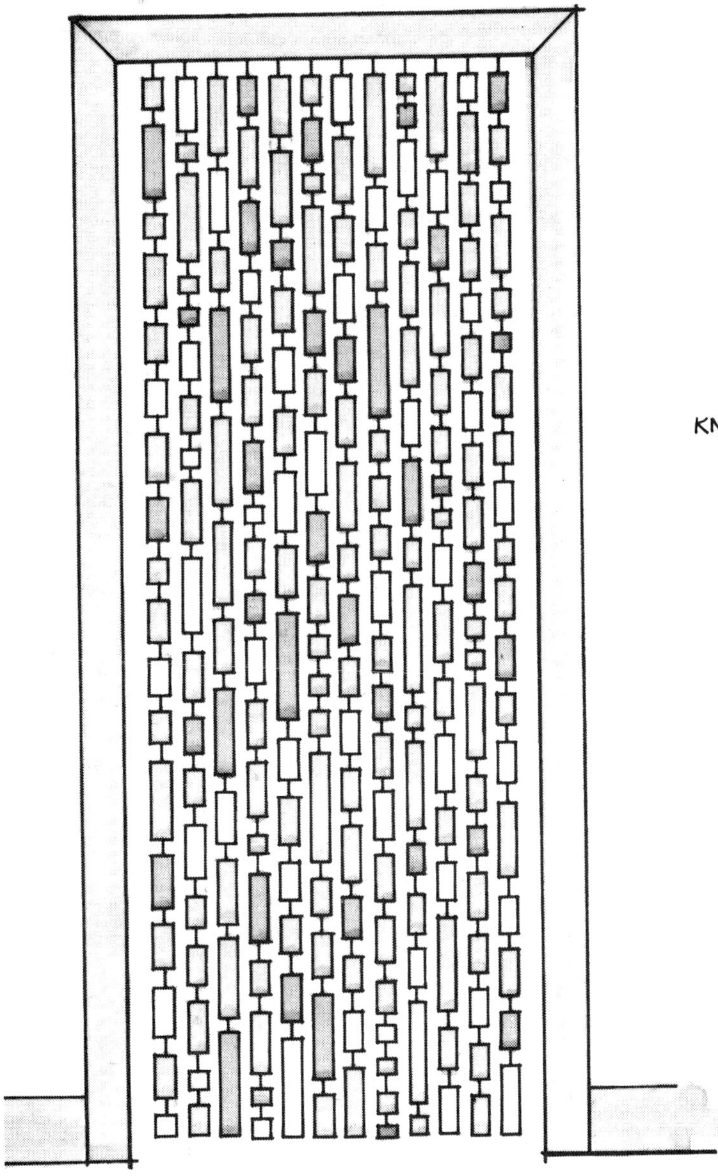

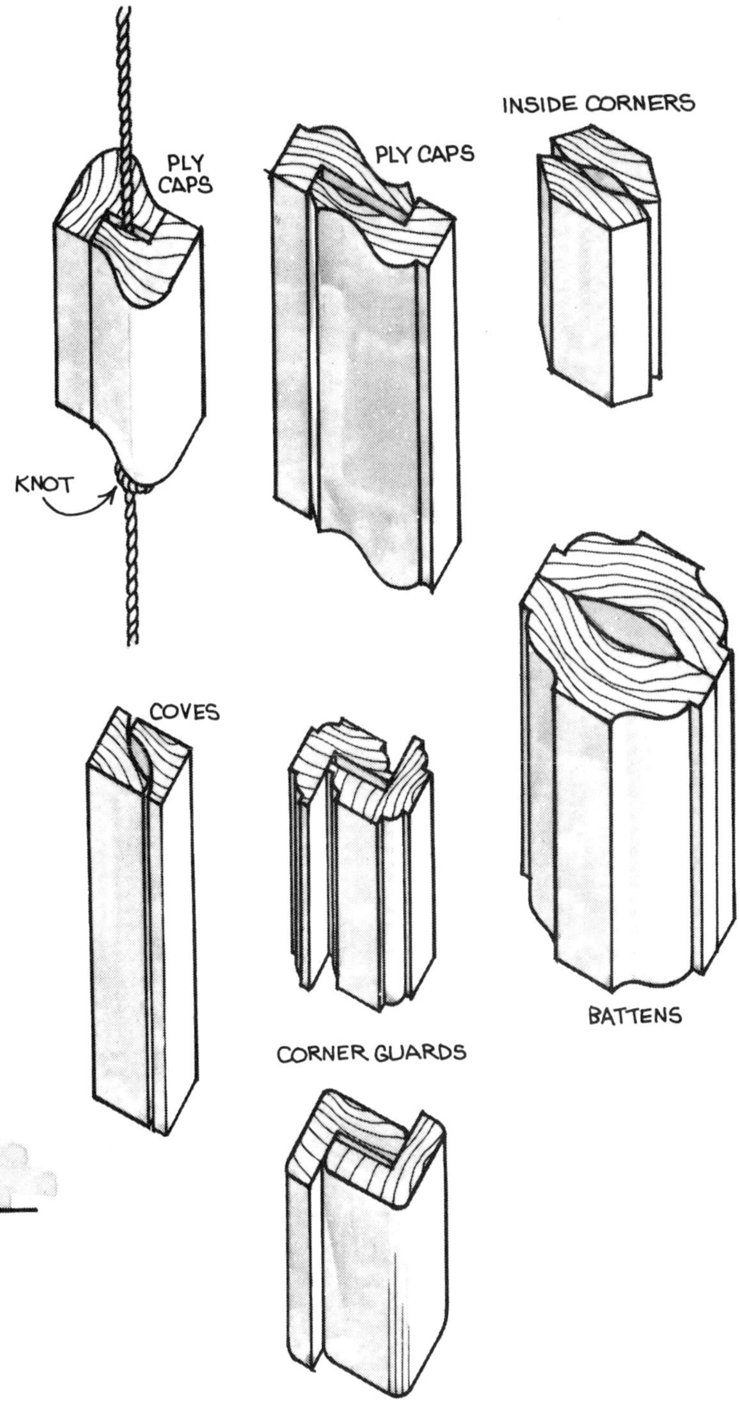

DECORATIVE ROOM TRIM

Trace a bold pattern in wood moulding, then add color with tile. Moulding gives that important finished look to today's clean-lined architecture. Here the handrail, designed with mouldings, sets the decorative theme that continues around the door and window. Against the wall, the handrail becomes a ledge to hold a changing art display. Then the same design trims the built-in cupboards. The decorative railing is easily constructed with cove moulding and finish grade lumber. Tile serves as colorful inserts. Simply mount them on a strip of lumber. Next trim the floor and ceiling with a simple contemporary moulding, then stain or paint and you're done.

DESK ORGANIZERS

Can't find a pencil when you need it? Need an envelope, fast? Wonder where the gas bill disappeared? Sounds like your desk is a mess. What it needs is a little organization and here's a simple way to do it. They're quick and easy to build and can be as small or large as you want, depending on your needs and your desk or table size.

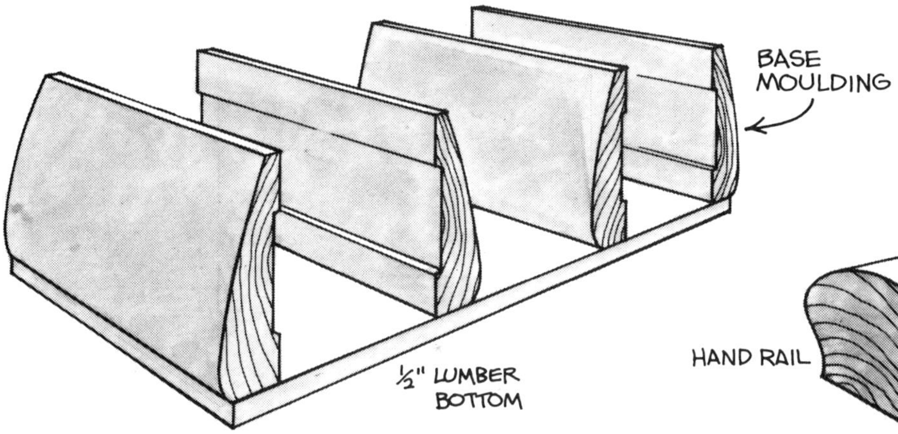

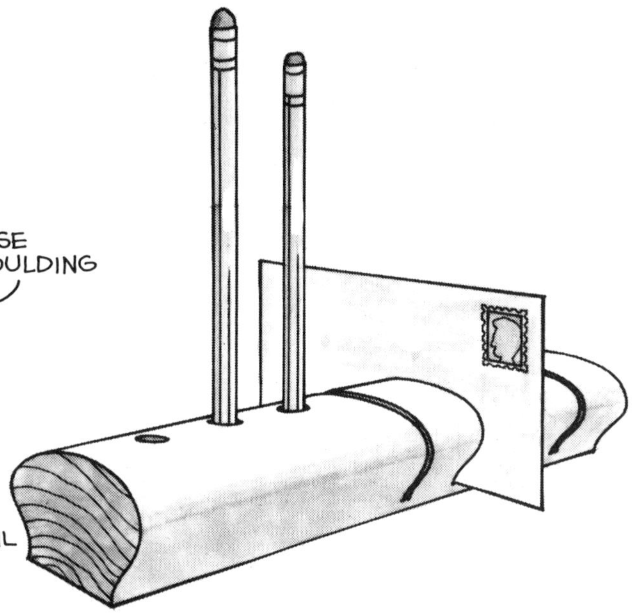

This dandy little organizer is just a group of uprights sitting on S4S stock or a piece of lumber. The uprights are made from either casings or base mouldings. How many to use and how far apart to put them again depends on your needs.

A single piece of hand rail makes a super simple organizer. Drill as many holes as you need, with the size depending on what kind of pens and pencils you use. The width of the envelope slots will depend on whether you stick a single envelope in it or a group of them. One wider slot for your address book might be a good idea, too. Cut the depth of the slots about three fourths of the way through the hand rail.

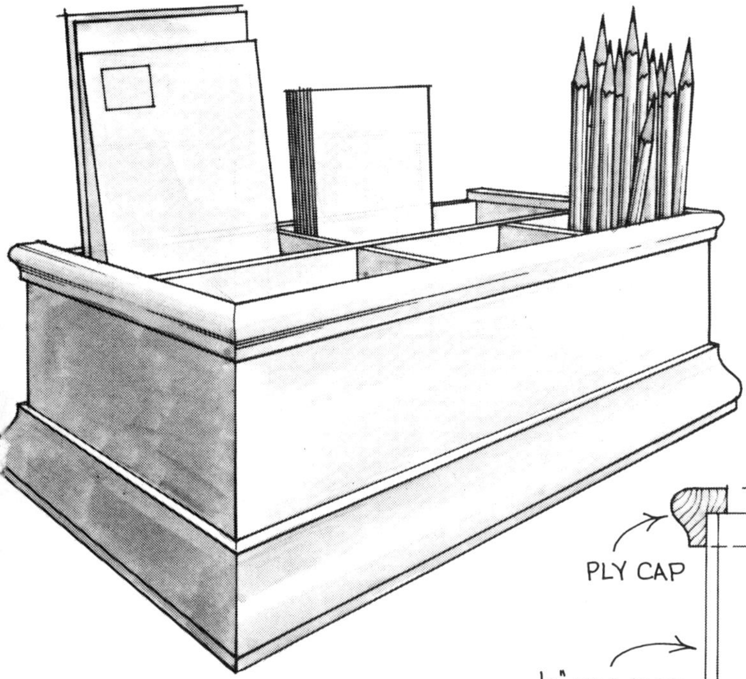

This one is just a box subdivided into sections. The upper front and side edges are framed with ply caps and the bottom and side edges are framed with base caps.

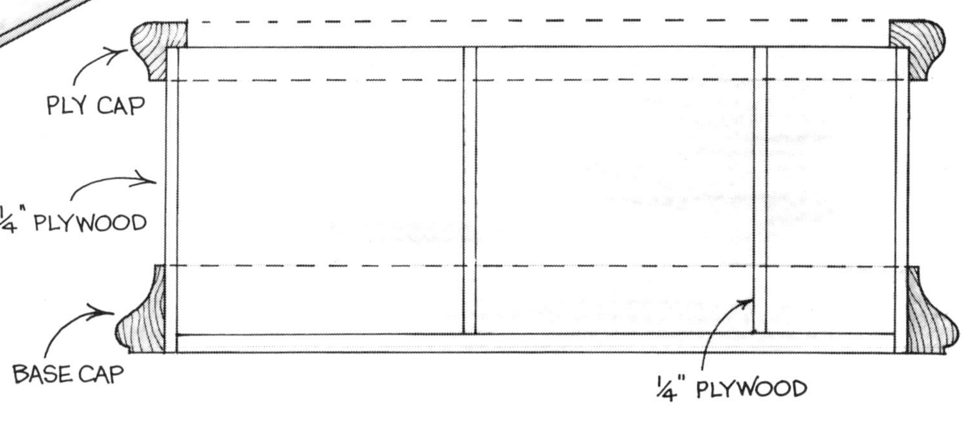

DOORS

Is the door that greets your welcome visitors as handsome as you'd like it to be? If not, why not give it a face lift? A little wood moulding will give it a whole new personality. The variety of patterns shown here are but suggestions. Use one pattern or a combination of patterns. Remember, when mitering, good accurate measurements are important. It's best to cut all the pieces to size before attaching them to the door. It's also a good idea to take the door off its hinges and lay it on saw horses to do the work. When all the pieces are cut and positioned it's time to glue and tack them in place. Since most outside doors swing in, do not position any moulding clear to the edge of the door or it will not close properly. Now paint or stain your new door to give it that final touch of class.

DOORS

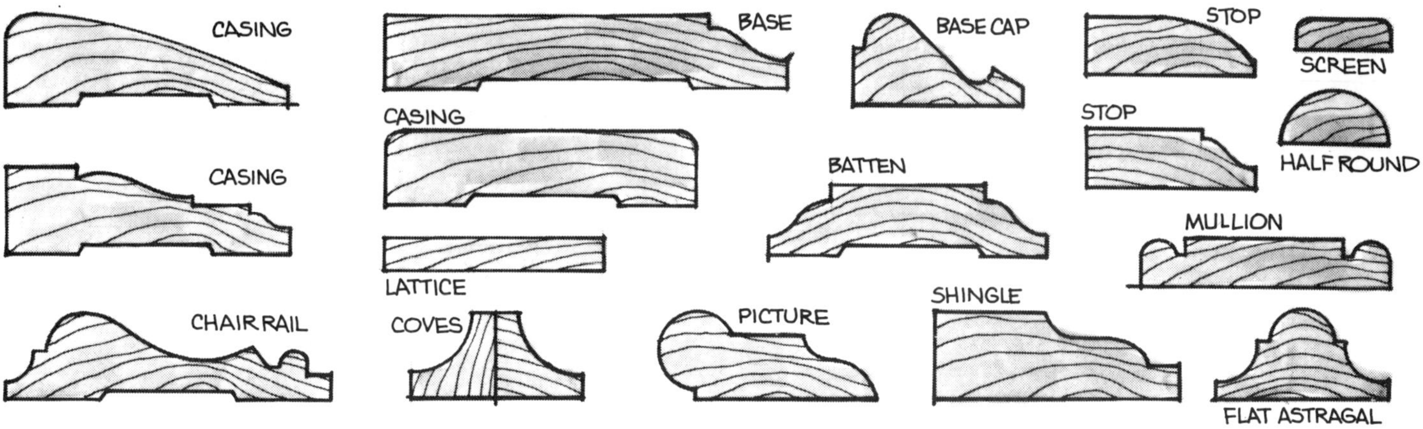
CASING
BASE
BASE CAP
STOP
SCREEN
CASING
STOP
HALF ROUND
CASING
BATTEN
MULLION
LATTICE
CHAIR RAIL
COVES
PICTURE
SHINGLE
FLAT ASTRAGAL

DISPLAY RAILS

Here's a good way to display paintings, photos, charts or trophies at your home or office. Most of the shelf samples here are made of S4S stock or lumber. Its size and whether it's used vertically or horizontally will depend on what you want to display. The front edge is capped with base caps, ply caps, panel moulding, picture moulding or stop moulding to give it a decorative appearance and to keep the objects from slipping off. Most of the shelves also have a support underneath made of coves, quarter rounds, stops, base shoes or base caps. The shelves are attached to the wall studs with recessed lag or flathead screws. Drill holes about half way through the shelves to recess the screws. After the shelves are attached to the walls, the drilled holes are then covered with the front capping. If the shelf is extremely large, it might be better to attach it with decorative support brackets.

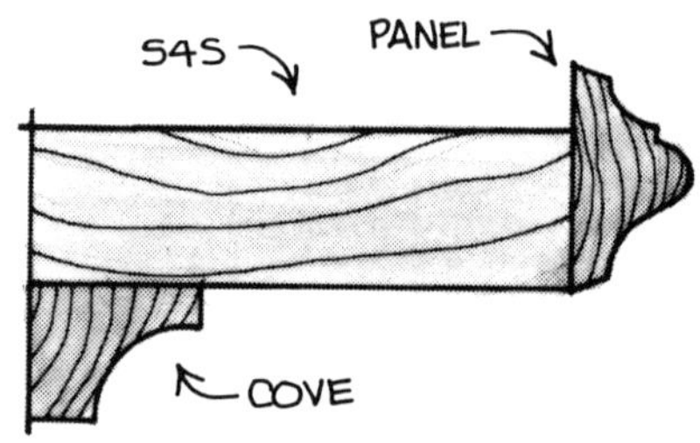

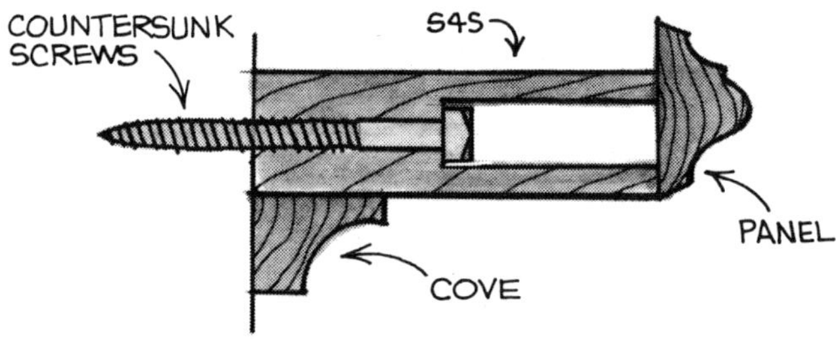

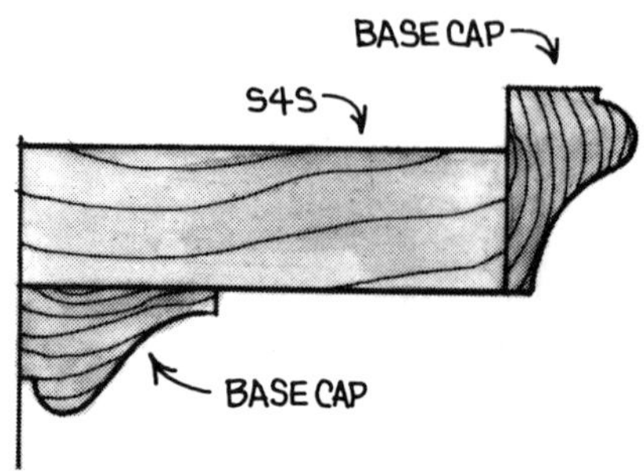

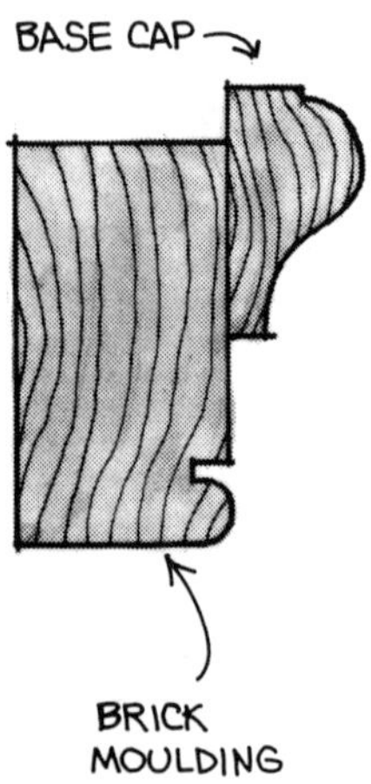

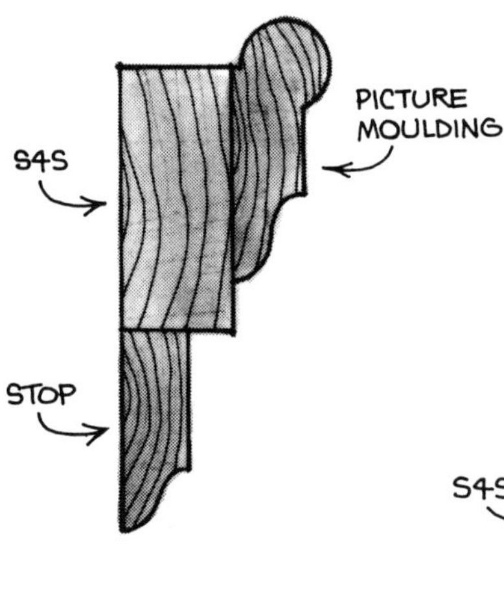

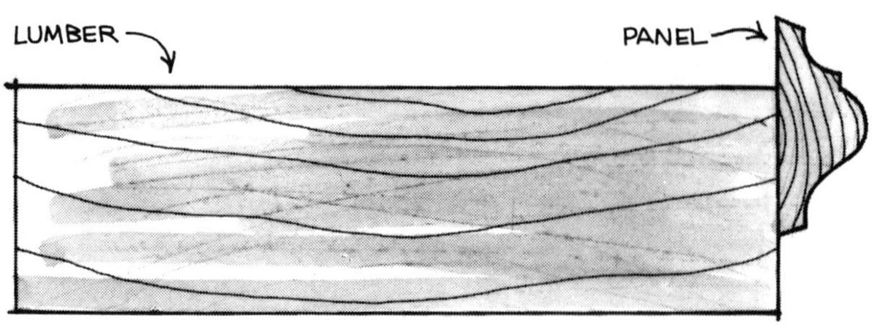

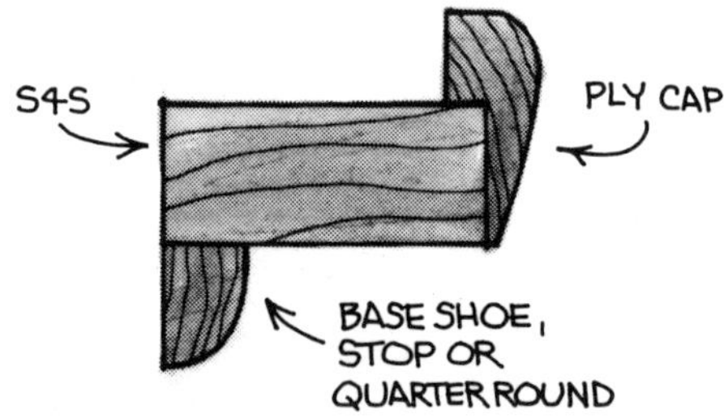

DRAWER DIVIDERS

Do most of the drawers in your house look like junk drawers? Some drawers are supposed to look that way, but there are some that could stand a little organizing. That's where drawer dividers come in handy. If you have tools, hardware, silverware, cosmetics, sewing supplies, fishing stuff, drawing equipment, or anything that is easier to find and use when it's separated, drawer dividers can be helpful. Just cut the moulding pattern that works best for your needs to fit the drawer width or length. Space them the way you want and glue in place.

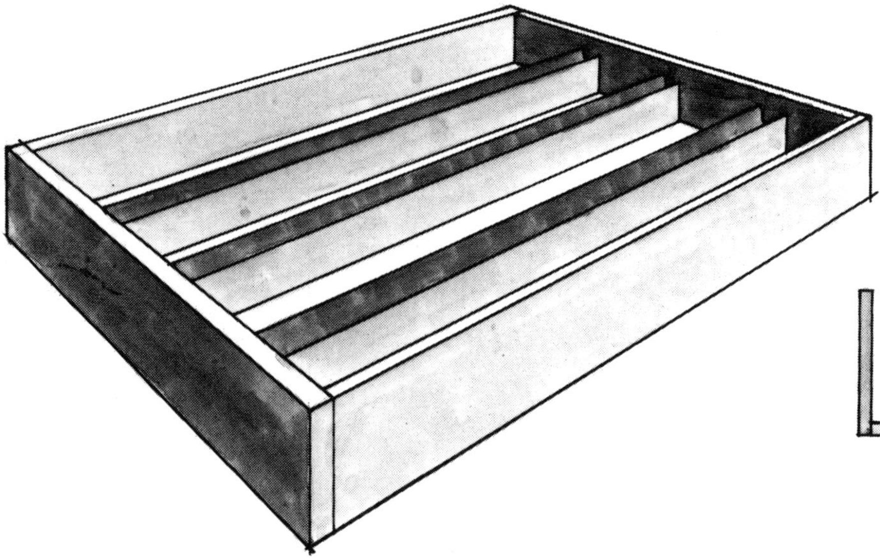

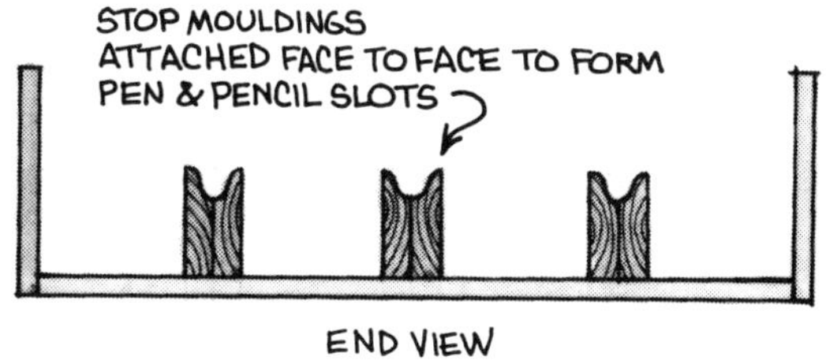

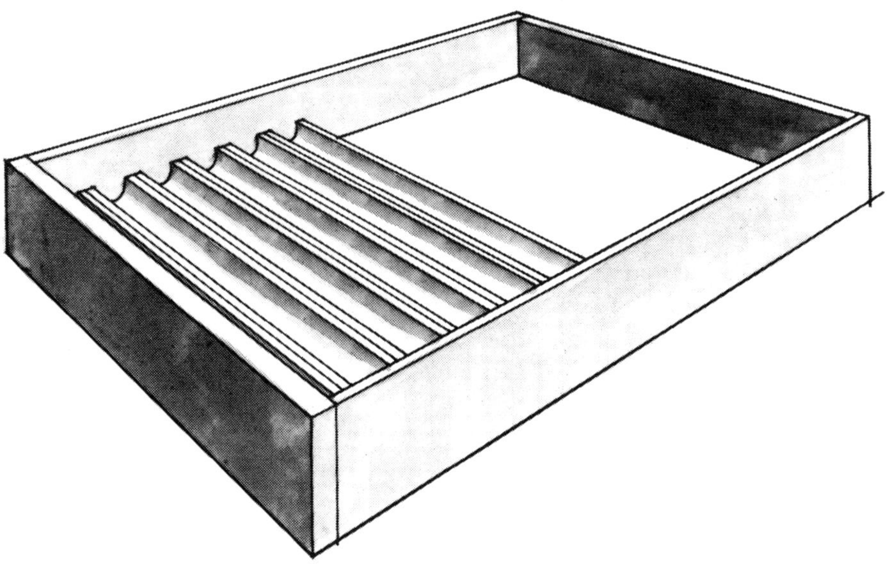

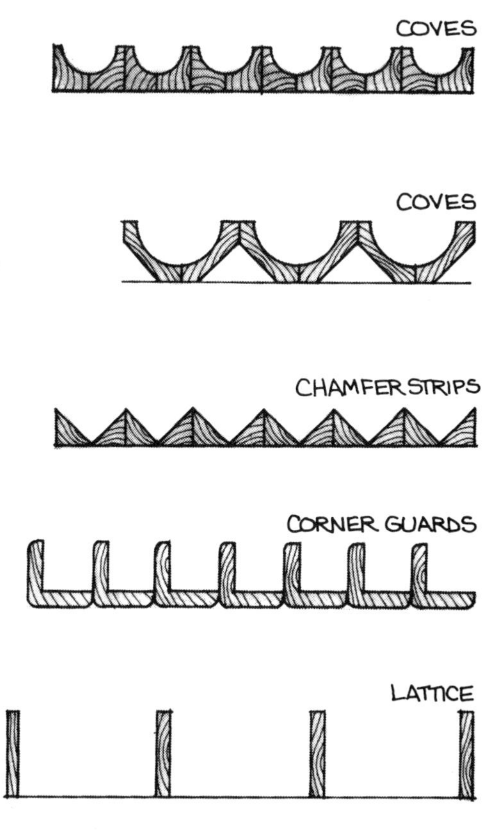

DRAWER PULLS & HANDLES

Make your drawer and cabinet door pulls uniquely your own by designing them with wood moulding. Besides the many patterns available, variations can be gained simply by changing the length or the placement of the handle. When choosing a moulding to use in your kitchen, bathroom, bedroom, or shop, keep in mind that the strength of the drawer handle must be in keeping with the size of the drawer and the weight of its contents. A small piece of ply cap moulding, for instance, would prove inadequate for use on a large drawer filled with heavy hardware, while a large hand rail drawer pull would be unnecessarily bulky if used on a small jewelry drawer. Wood screws work best for attaching the handles. An added note: Drawer pulls do not always have to run horizontally. They can also be attached vertically up the middle.

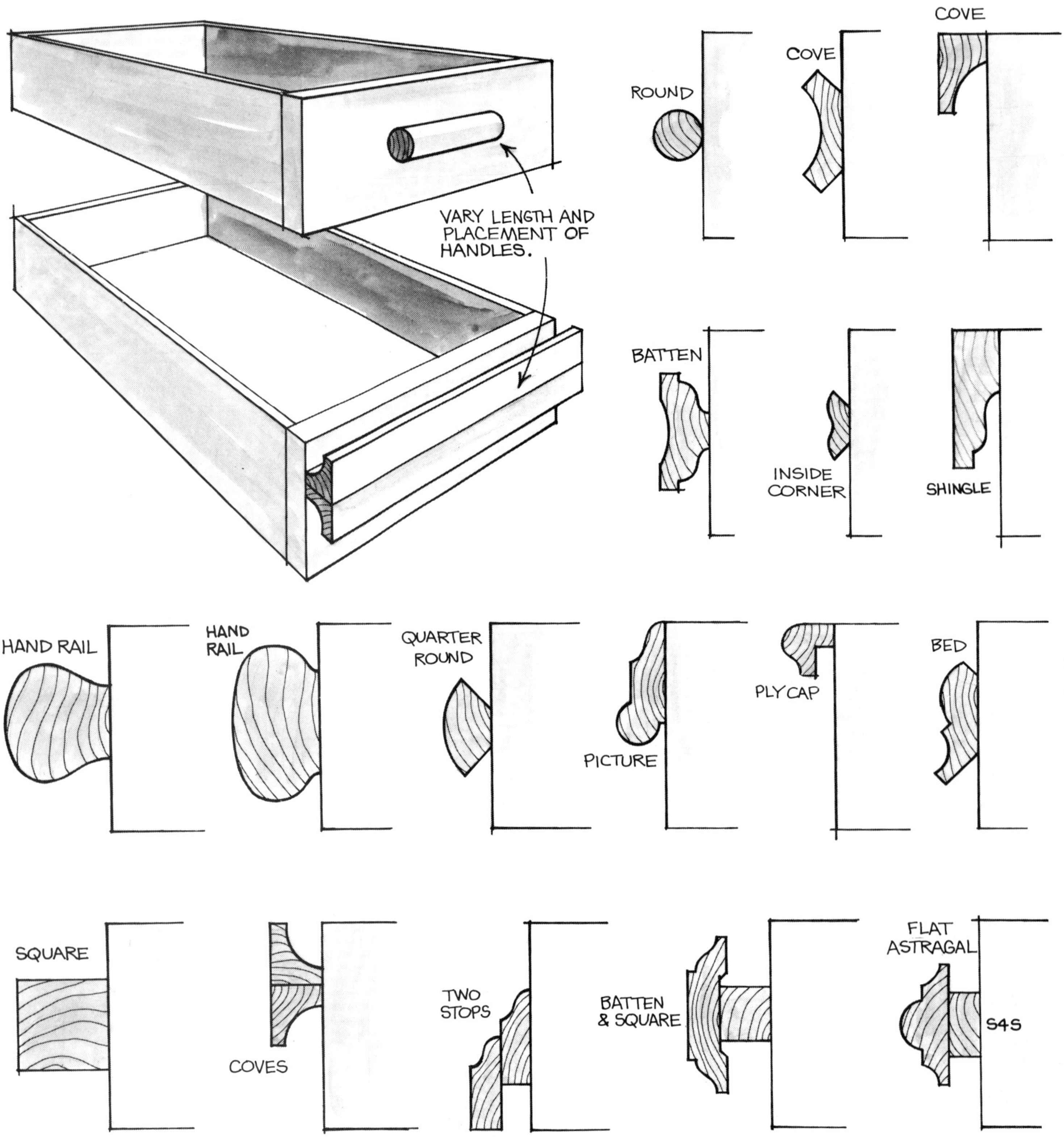

DRAWING BOARD TROUGH

This may not sound like a great idea to everyone, but to the artist, engineer, draftsman or architect its not only clever, but useful. No more will those wandering pencils roll off the table. The eraser that always seems to be in hiding will be right at your fingertips. That misplaced ruler is there when you need it. All it takes to build this simple little catch-all is a piece of lattice and cove moulding. Simply glue and nail the cove to the lattice and then attach the lattice to the bottom underside of the drawing board. A piece of stop moulding or mullion with quarter round would also work.

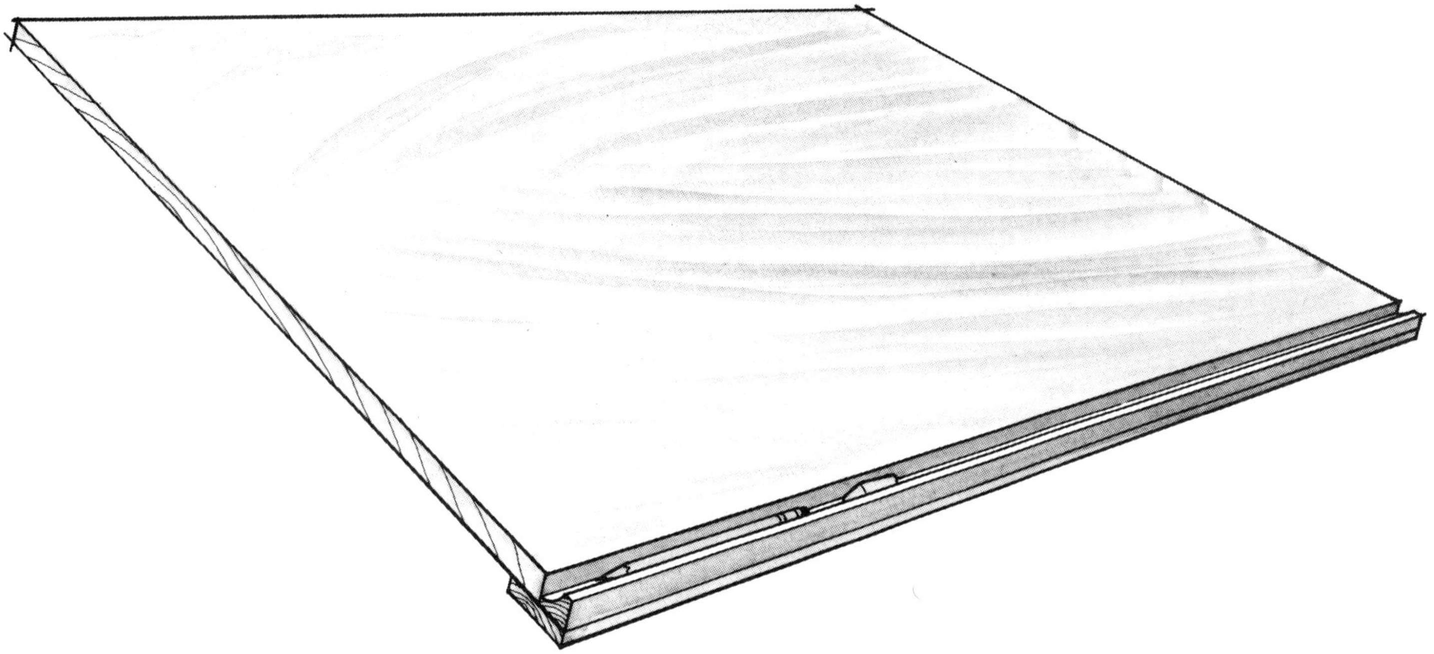

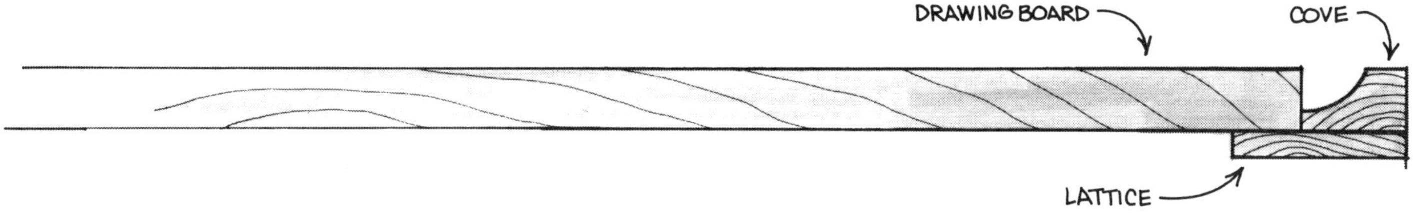

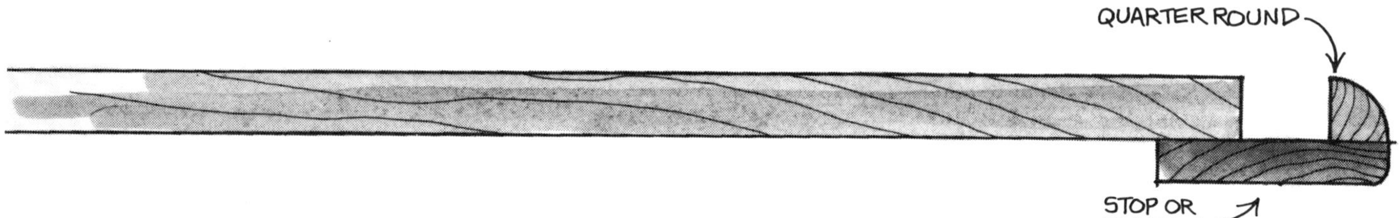

FACIAL TISSUE BOXES

There's not too much to say about facial tissue boxes. Most look like commercial paper boxes, which is what they are. While many are beautifully designed, they still lack the richness and permanence of wood. The three designs shown here are just wooden boxes that slip over the existing tissue box. Each has that distinctive third-dimensional design that can only be achieved with wood mouldings. Be sure to measure an actual box of tissue before you start cutting the mouldings. The new box must slip over the old dispenser with just a little room to spare. When working with pieces this small, accurate measurements and good mitering is a must. Also use small brads when nailing and gluing. Larger nails might split the wood.

Two of the boxes here have tissue slots that are first drilled and then cut with a saber saw. After cutting the slots, sand to smooth out the rough edges. On the third box the slot is merely a space between the two top pieces. After your new tissue dispenser has been assembled, take advantage of the wood by staining or oiling it to a beautiful finish.

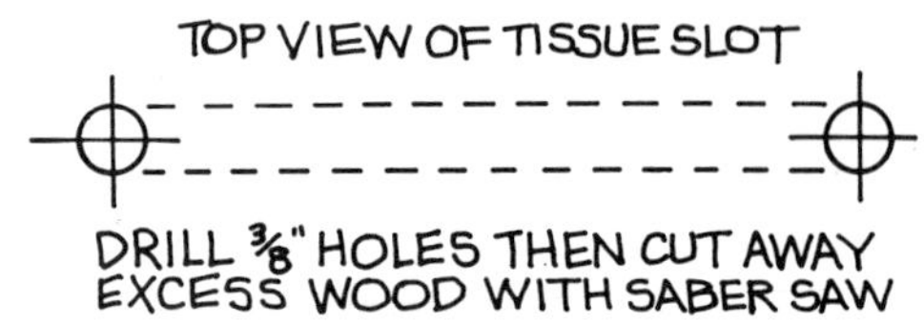

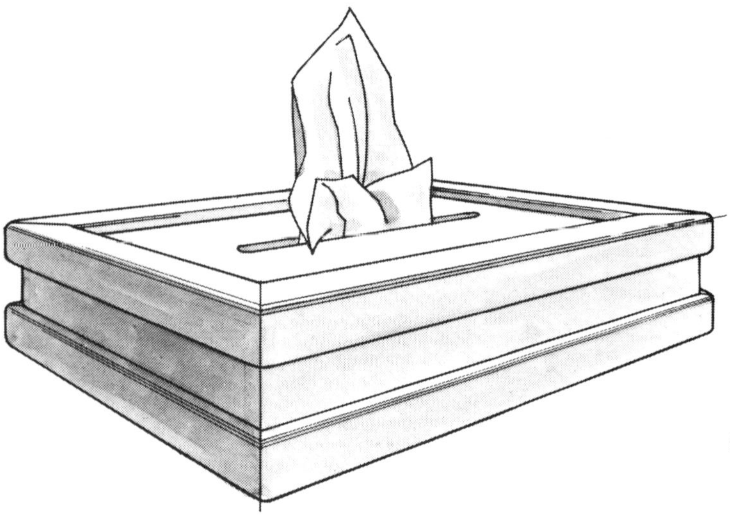

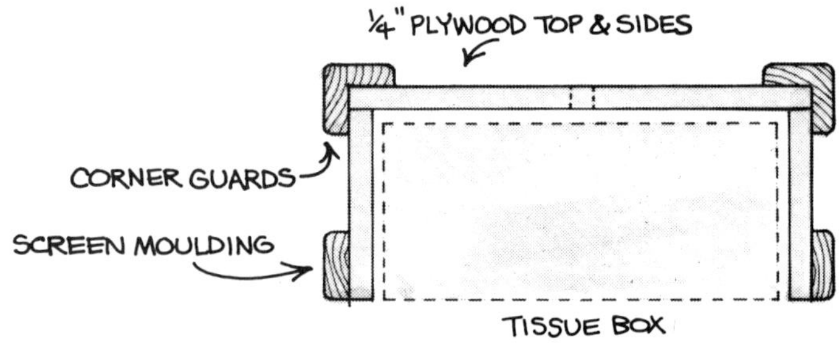

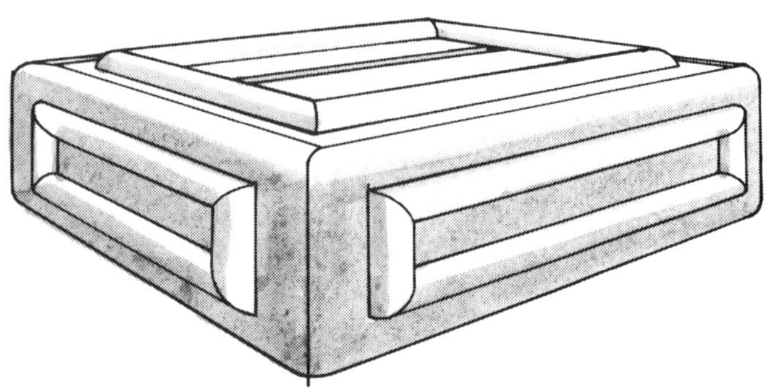

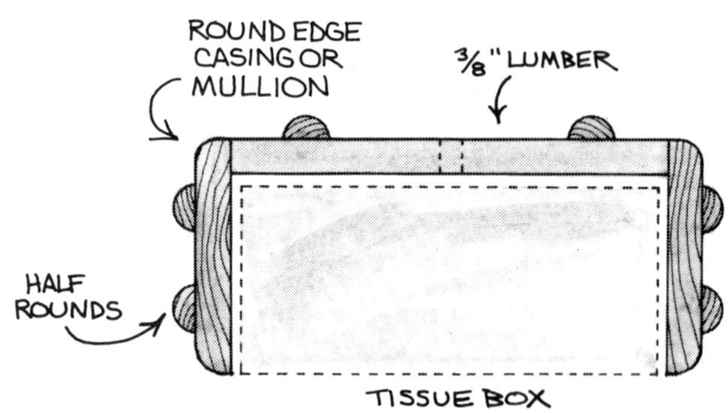

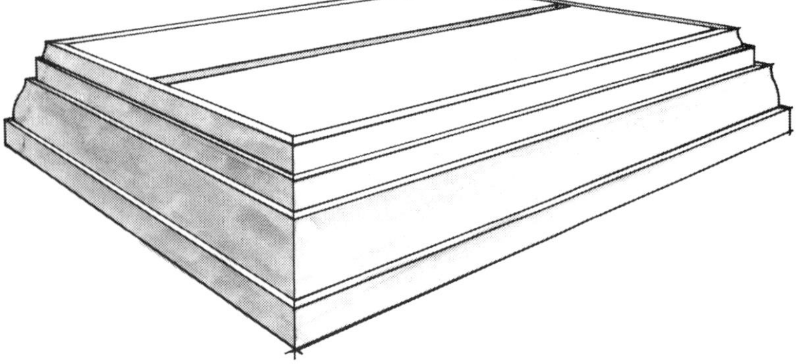

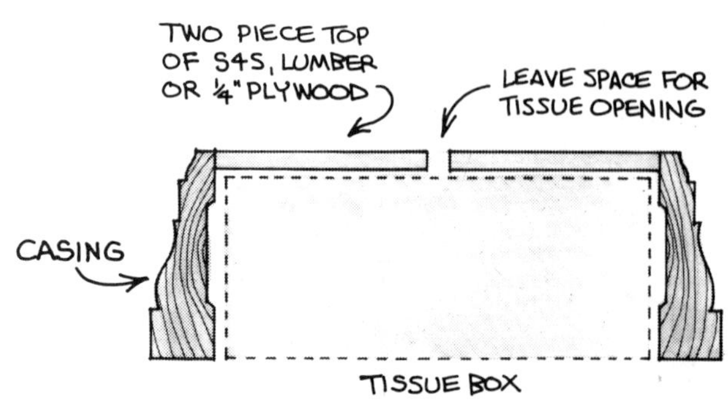

FISHING ROD RACK

Do you have to take time before each fishing trip to untangle your lines and hooks? Now you can keep those rods separated by hanging them on this rod rack. It's just a piece of shingle moulding and round edge stop that sits on a piece of cove, or if you want to make it even simpler, you can make both top pieces from the same stop moulding. Notches for hanging the rods are cut about half way through. Depending on how many fishing rods you have, you can make the racks as long or short as you want. Stain it to a warm richness, drill holes for screws and hang it in your den or office.

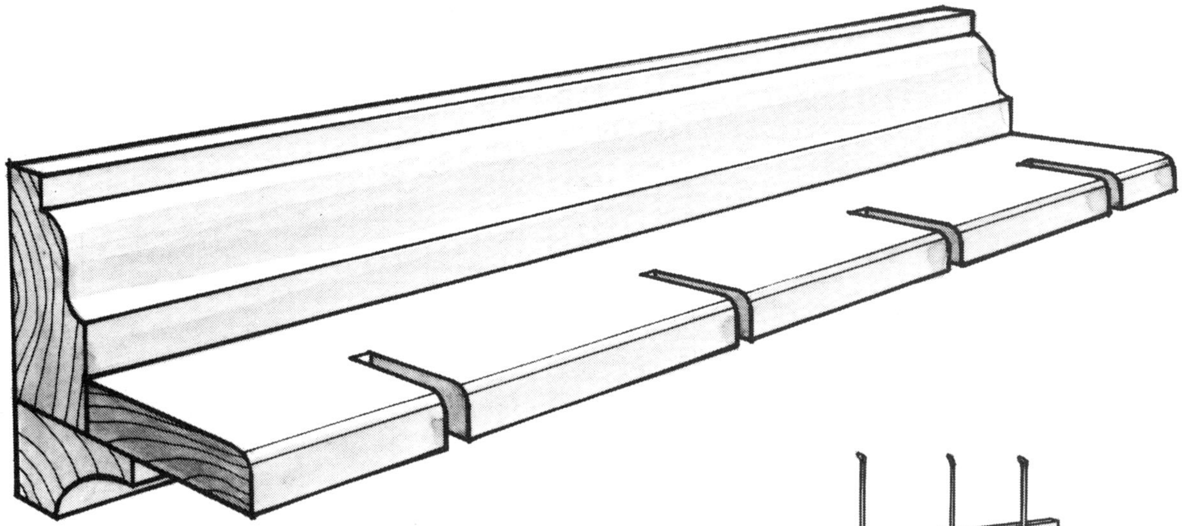

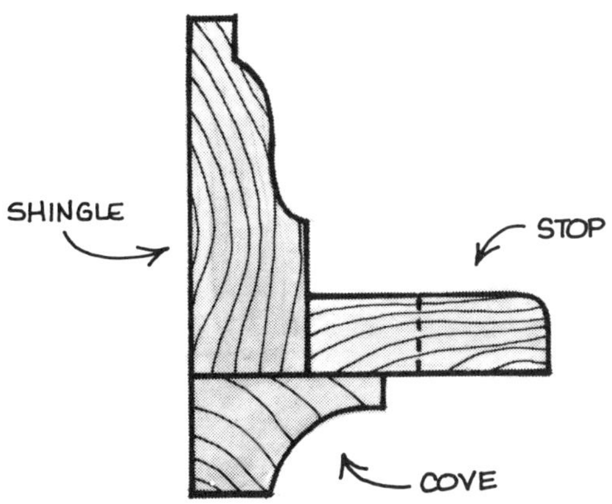

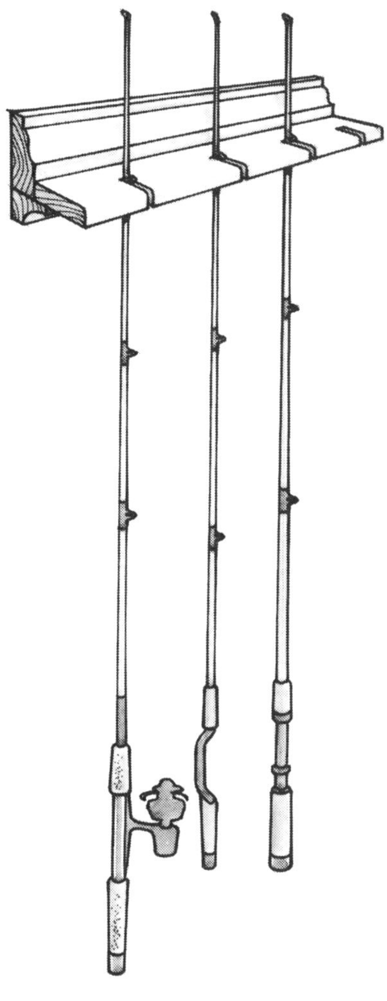

FIREPLACES

Most people feel that fireplaces are permanent untouchable monuments of masonry. To a point, that's true. The inside structure, the part that makes the fireplace function properly, shouldn't be tampered with. But the outside! That's a different story. You can change the face of your fireplace any way you want. Old mantels can come off and new ones can go on. Give it the personality you desire. Traditional or contemporary. Wood mouldings can make the changes amazingly simple. At first glance, some of these ideas may seem a little complex, but after close examination you'll see how uncomplicated they really are.

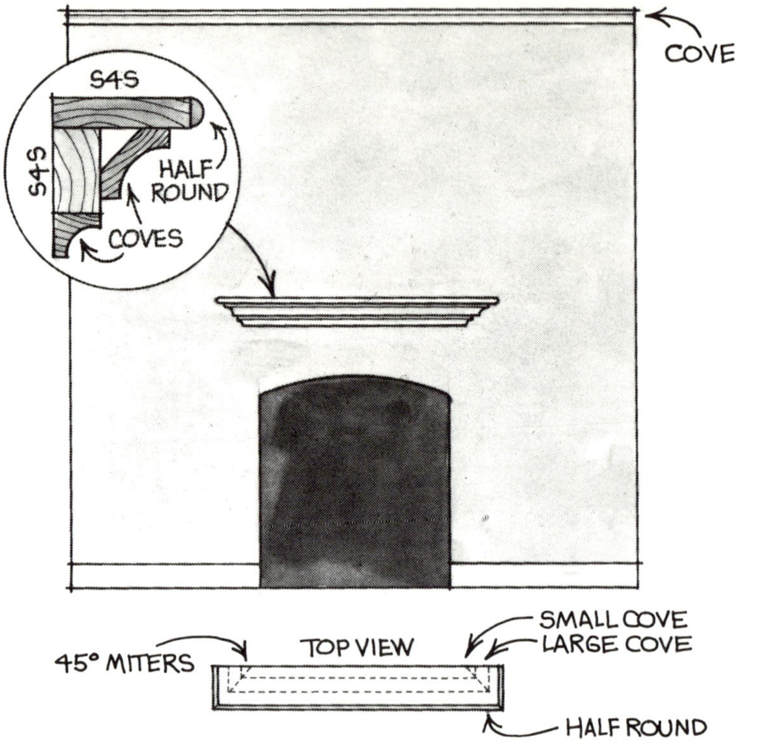

This contemporary fireplace is merely a plastered wall with a simple mantel made of mitered coves topped off with S4S which is capped on all three sides with half rounds.

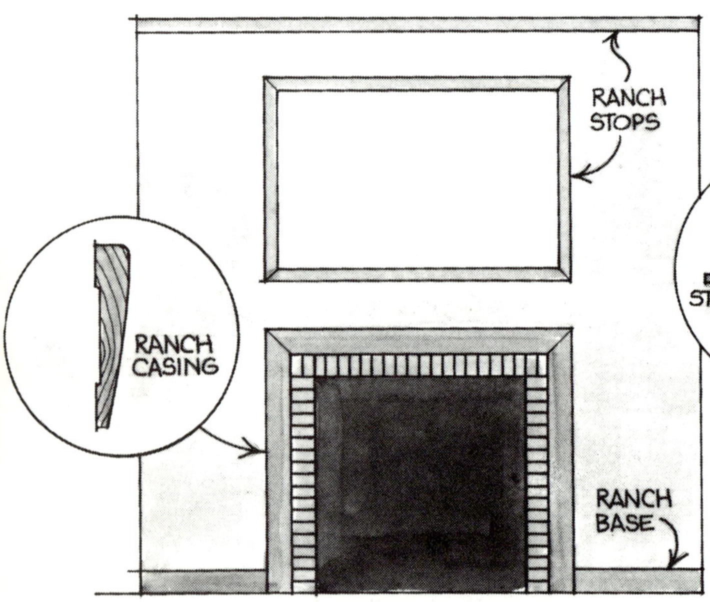

Here's an uncomplicated design where the fireplace is just framed with ranch casing. A matching framework of ranch-stops above the fireplace repeats the design.

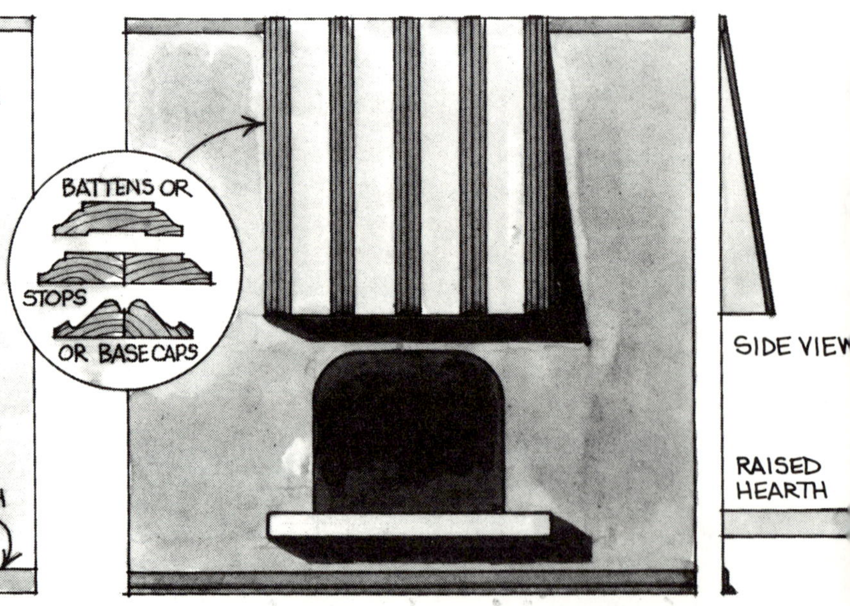

This model has a hood structure built out at an angle above the fireplace. It can either be painted or paneled and then wood moulding strips are repeated in vertical rows across the front. This particular fireplace has a raised hearth.

FIREPLACES

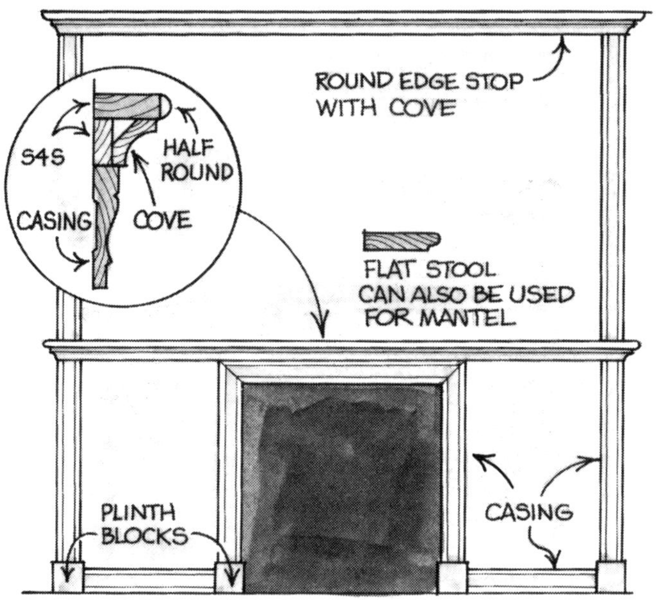

Here the fireplace wall is framed with casing and is repeated around the fireplace itself. Above the fireplace casing, a narrow mantel of cove and S4S stock capped with half round, extends to the outer casings and is mitered. The top of the mantel could also be built with a flat stool moulding. Next to the ceiling, a round edge stop with cove, repeats the mantel design. Casing is also used as base moulding at the floor line. The upright casings all sit on plinth blocks. These are thicker wood blocks used where the upright and horizontal mouldings join together.

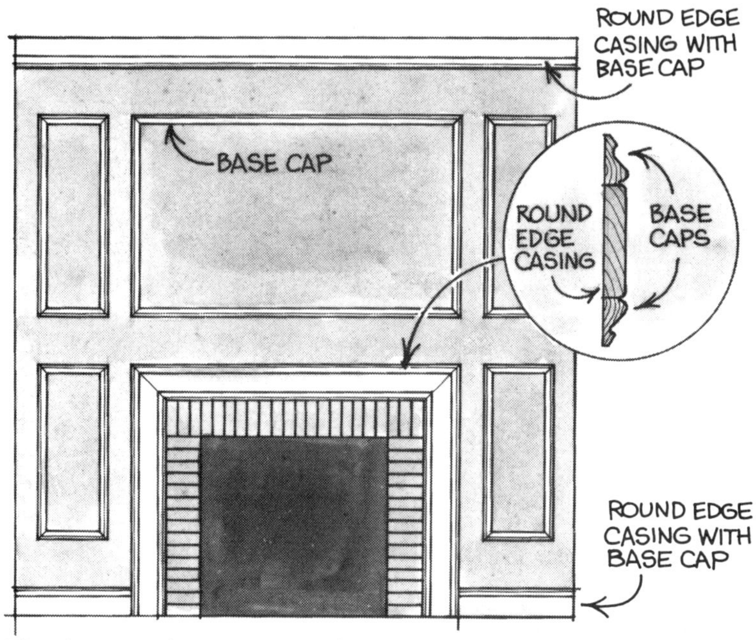

This fireplace has a framework of round edge casing between two base caps. The wall design is made up of base cap frames. Along the ceiling, round edge casing is repeated with base cap below it. Along the floor the same thing is done, except with the base cap above.

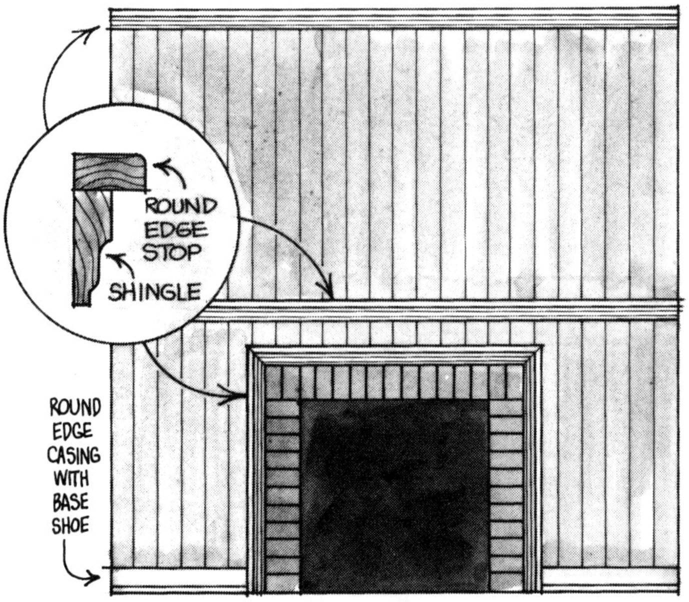

The wall pictured here is rough sawn lumber which contrasts handsomely with the simple mantel which is made with a piece of round edge stop and shingle moulding. The same patterns are repeated around the fireplace and along the ceiling. The base along the floor is made up of round edge casing and a base shoe.

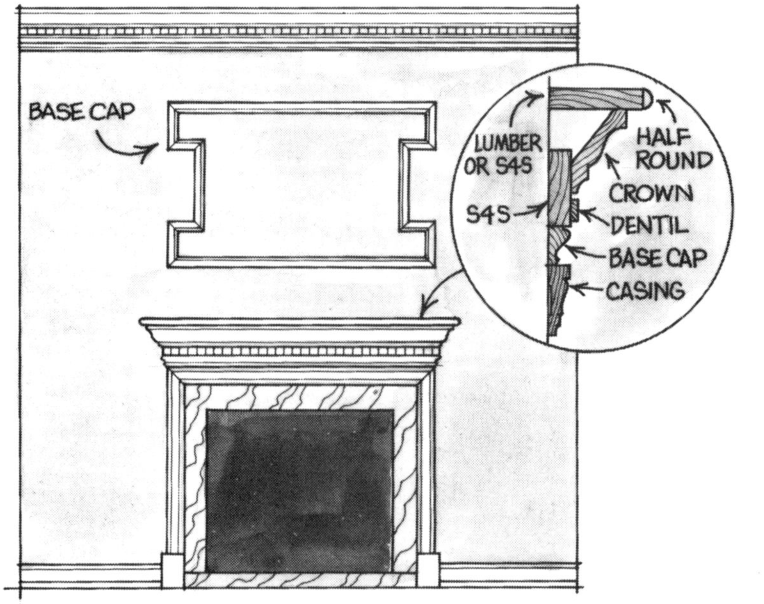

To take advantage of the traditional look of this marble fireplace, a more ornate style of mantel was designed. It is, in fact, the same ceiling design as shown on page 15, except the casing around the fireplace has been added, and in this case, the ends have been mitered. The same design is repeated along the ceiling but without the S4S shelf on top and the fireplace casing on the bottom. Traditional base with a base shoe is used along the floor line. The wall design above the fireplace uses base cap. The fireplace casing sits on plinth blocks to add to its traditional look.

FURNITURE TRIM

Add a new face to that old chest of drawers or buy a new unfinished set and refinish it with wood mouldings in whatever style you want.

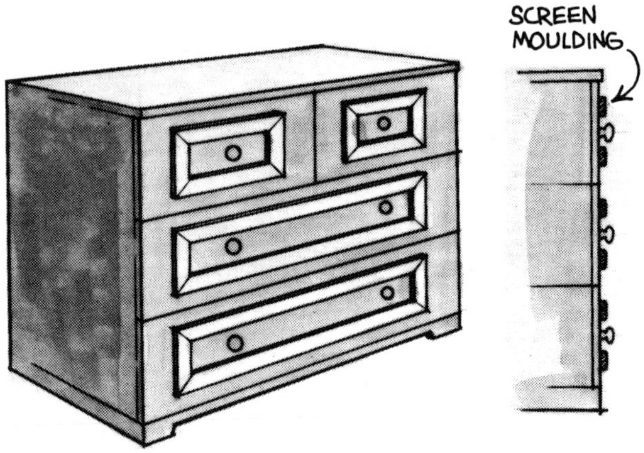

Simple frames of screen, stops or lattice moulding makes this design a quick and easy furniture face lift. A bright colored paint job and decorative ceramic knobs will make it a happy addition to the kid's room.

Rosettes made from casings gives this dresser a handsome look. Add wooden circular knobs to the center, then stain as desired. For additional rosette patterns, see page 88.

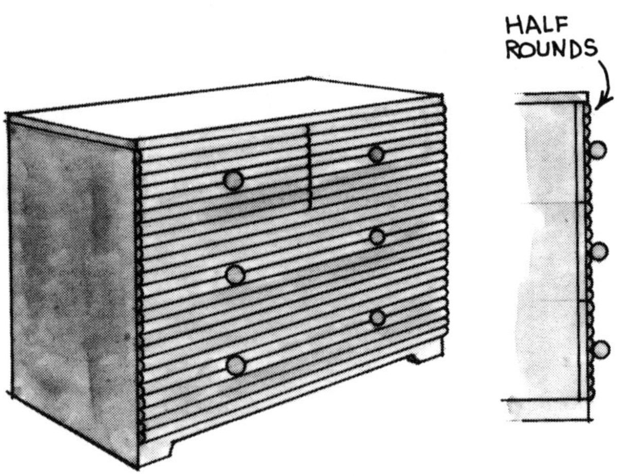

Rows of half rounds give this dresser a long low look.

Shingle mouldings frame each drawer and is repeated around the base to give this set a handsome traditional appearance. Ornate drawer pulls further enhances its charm.

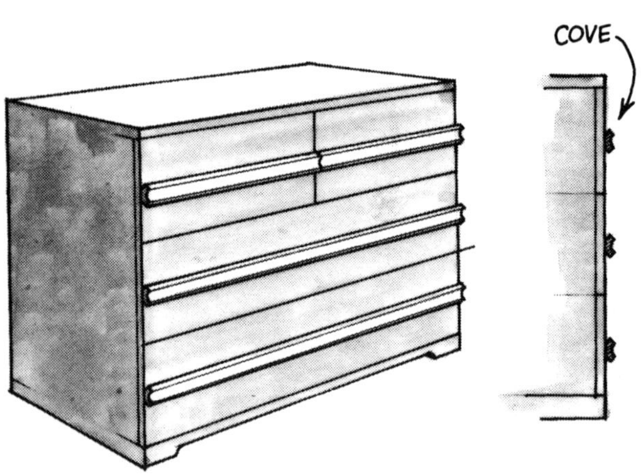

This clean contemporary style is achieved merely by attaching strips of cove moulding to the center of each drawer. For other drawer pull ideas, see page 38.

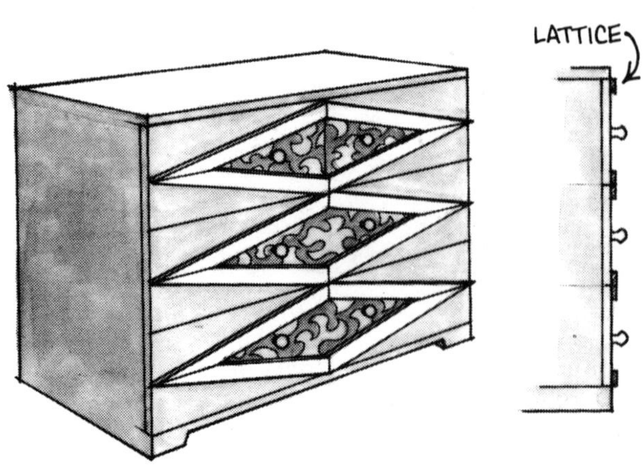

Diamond patterns of lattice or screen moulding takes the blandness out of this dresser. To make it even livelier, add vinyl wall covering within the shapes.

GAMES

This checker set is made of squares which sit on a plywood base and are then framed with base cap moulding. Use contrasting stains on each alternating square to give it a rich checkerboard appearance. Sand before staining. It's also best to do the staining before you glue the squares in place. Cut out 16 smaller squares for one player's pieces and 16 more pieces from rounds for his opponent.

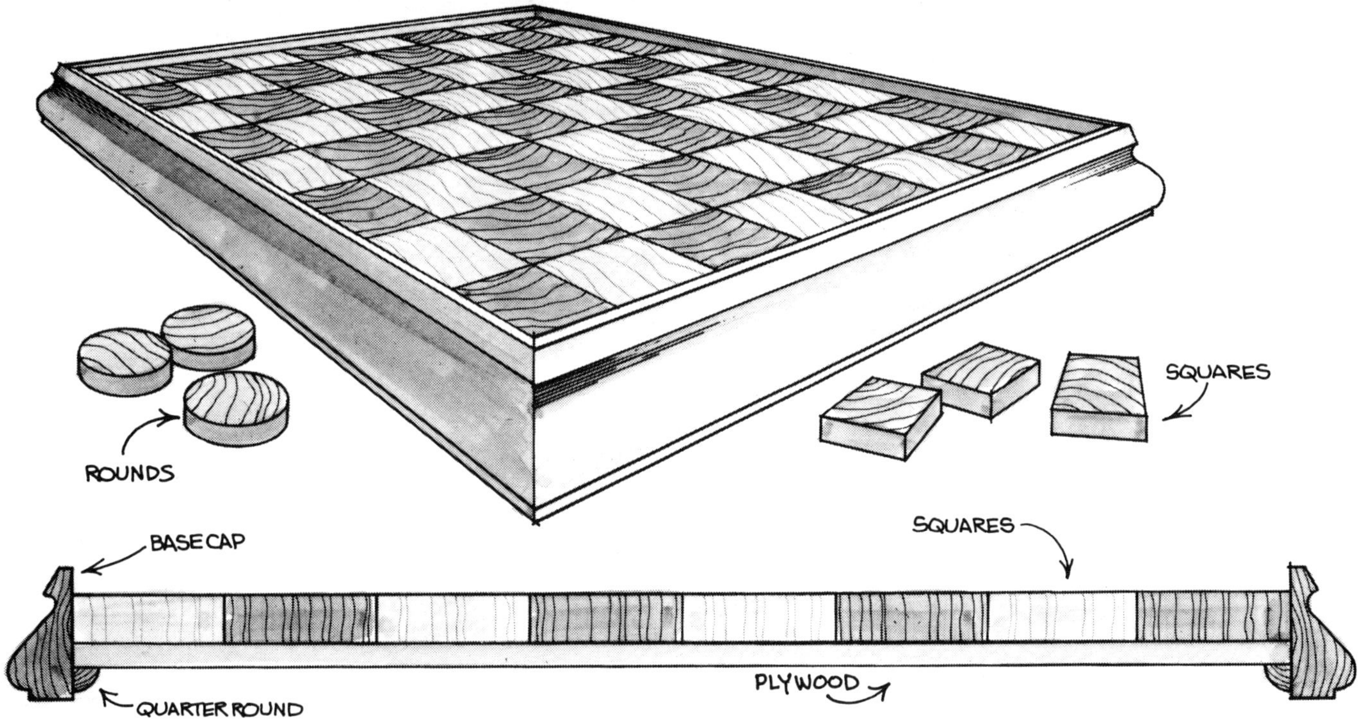

The old game of tic-tac-toe now has the more permanent look of wood mouldings. It's still played the same way, but with moveable wood "O"s and "X"s instead of paper and pencil. The playing pieces are made of moulding rounds and coves. The board is a piece of ¾″ plywood or lumber with the cross hatch lines made of screen mouldings which are glued onto the board. The set is then framed with ply cap moulding and painted or stained to suit your game playing mood.

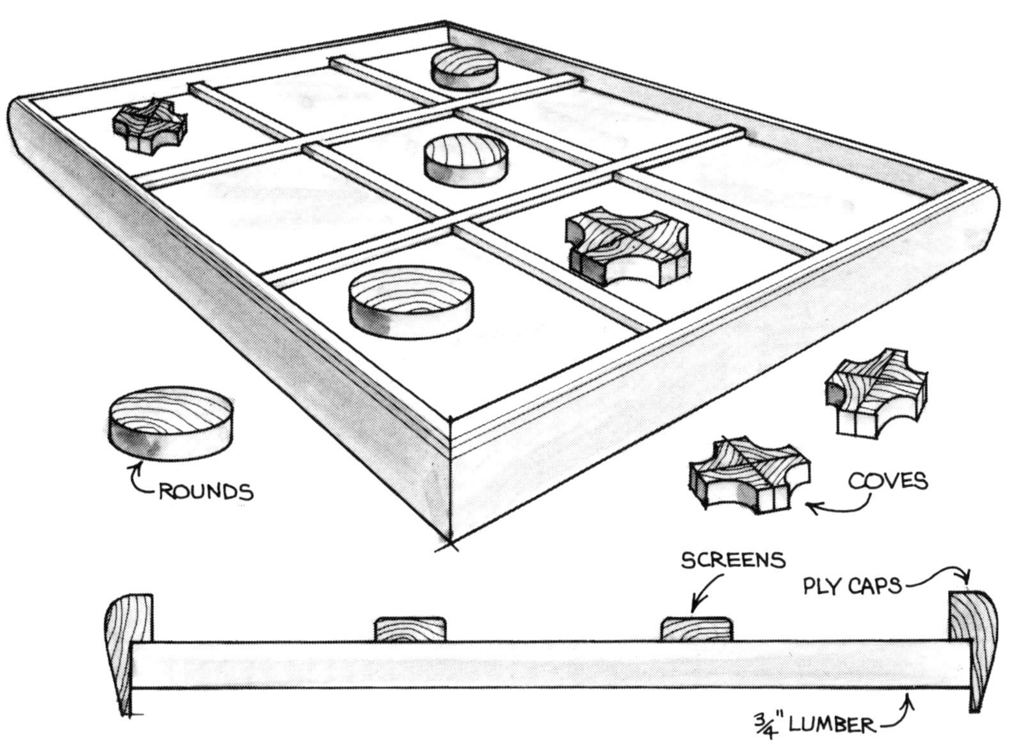

GARAGE DOORS

You can give your garage door a bright new face lift the same way you can your house door. The moulding patterns and layout designs are just as unlimited. Remember that some garage doors fold and some do not. Some are single car garages and some two car garages. Some open and close up-and-down and some do it cross ways. Be sure to keep these characteristics in mind when planning your design. Also remember that garage doors are exposed to varied weather conditions, so seal, stain or paint accordingly.

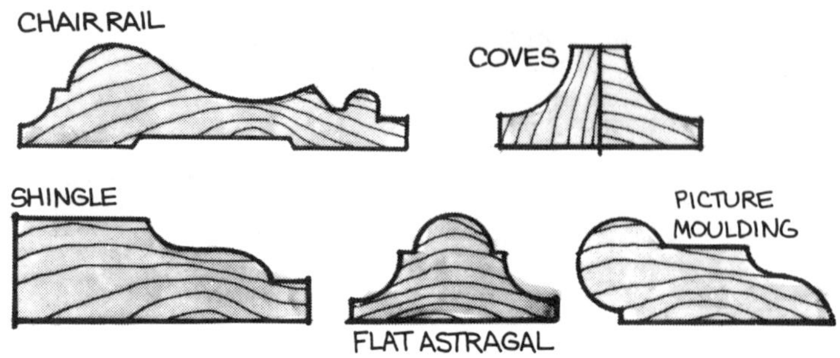

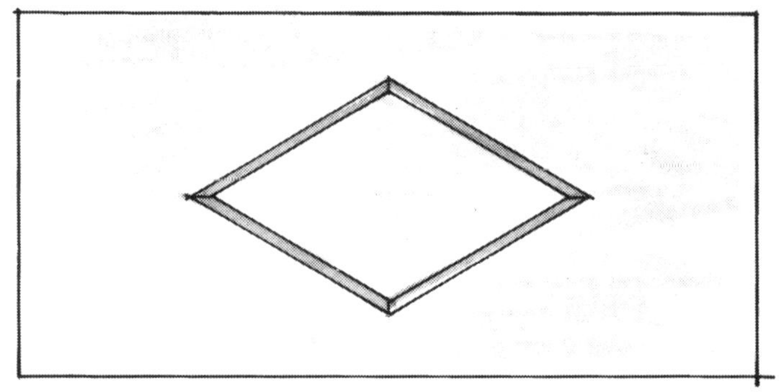

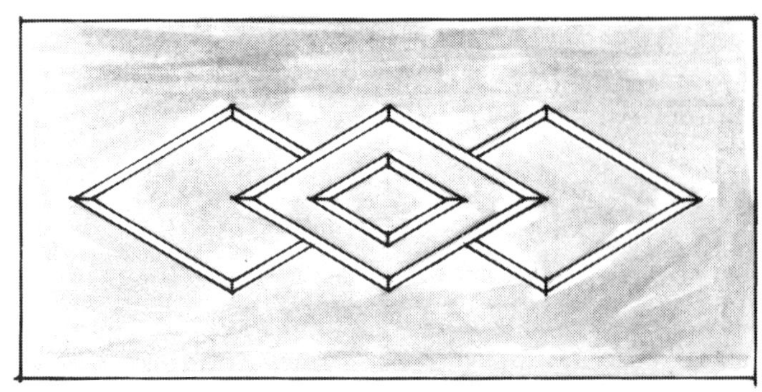

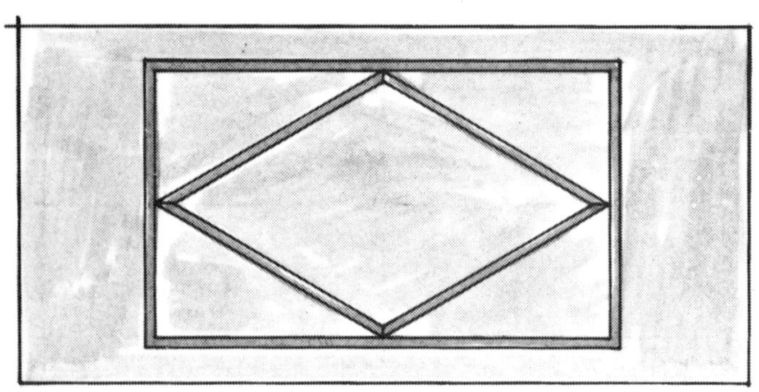

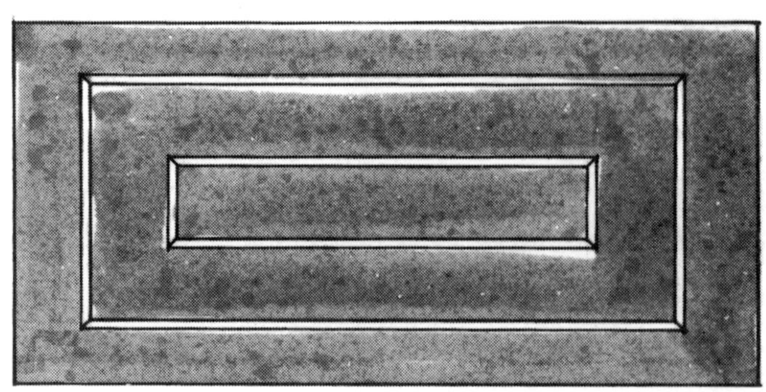

GARAGE DOORS

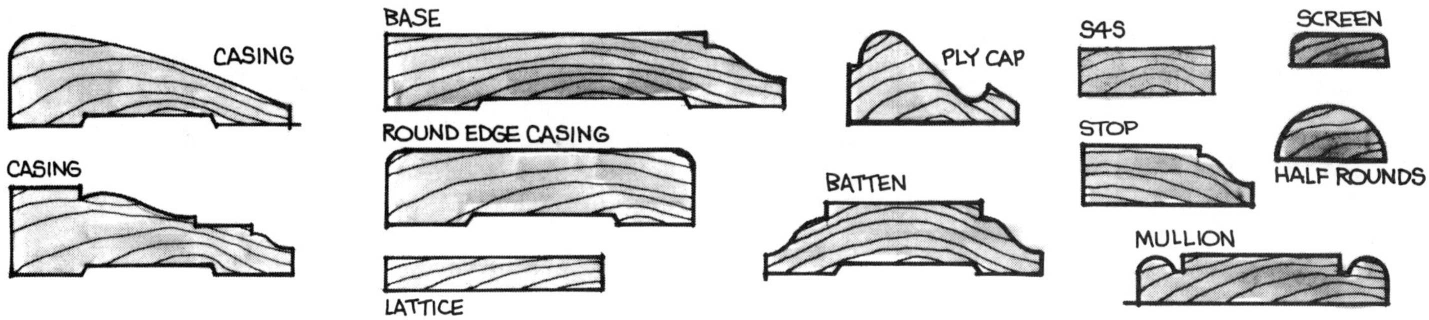

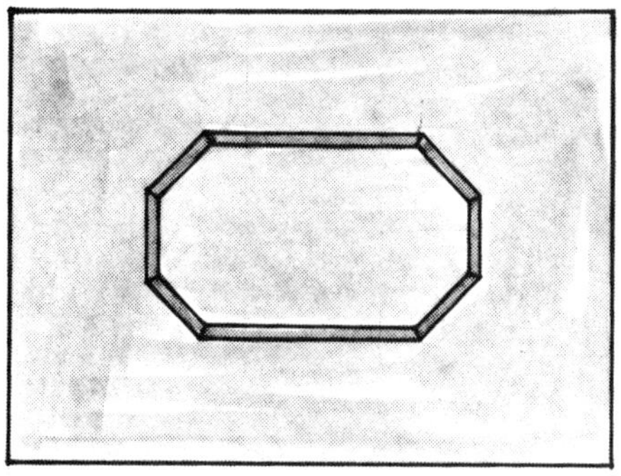

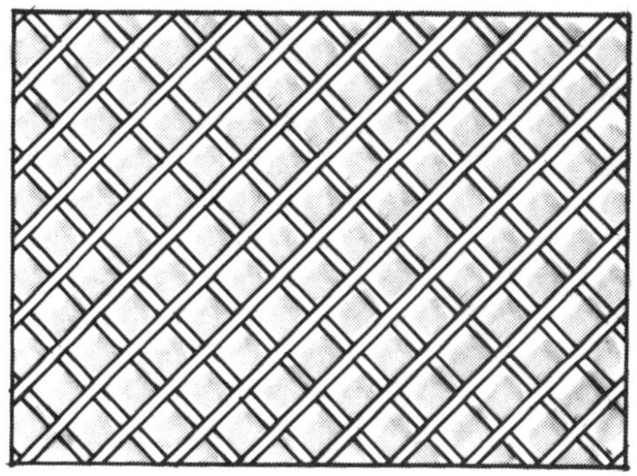

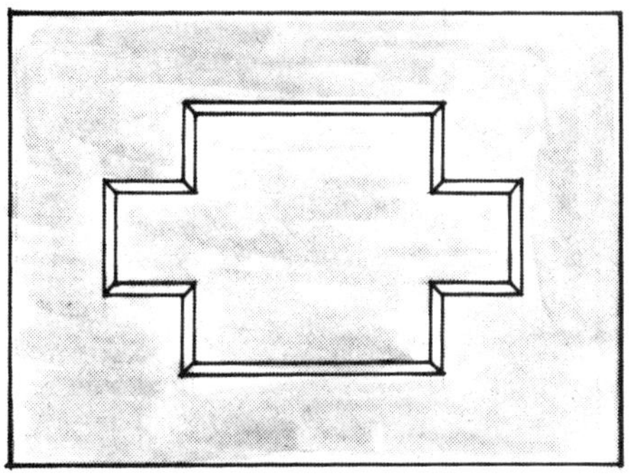

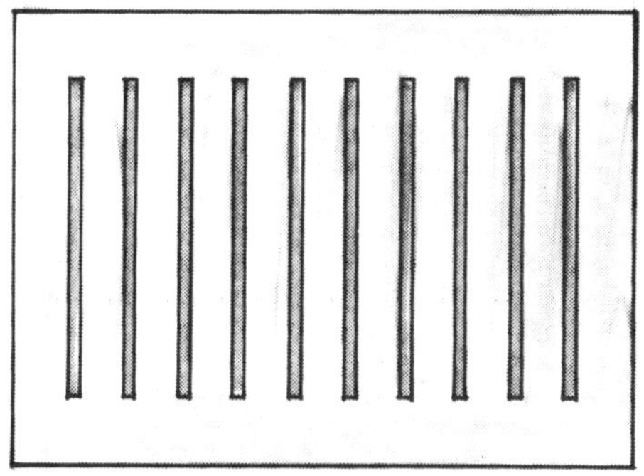

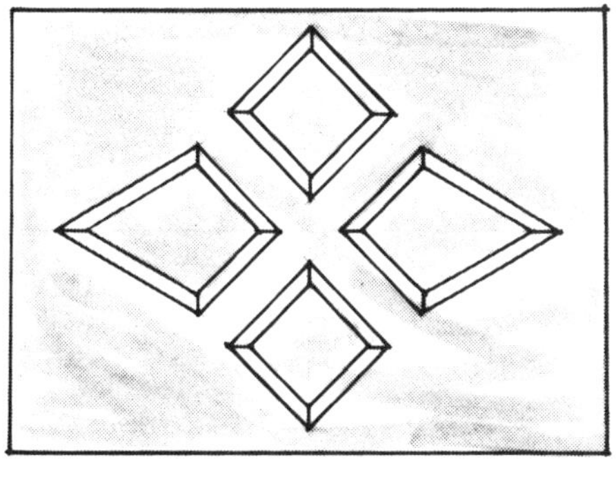

GARDEN TOOL CARRIER

Carrying seeds, pots, plants and assorted garden tools becomes a simple task when you have a carryall to do the job. The two models shown here are easily made out of plywood and wood mouldings. Good dimensions for these carryalls are about 14 to 18 inches long and 8 to 12 inches wide.

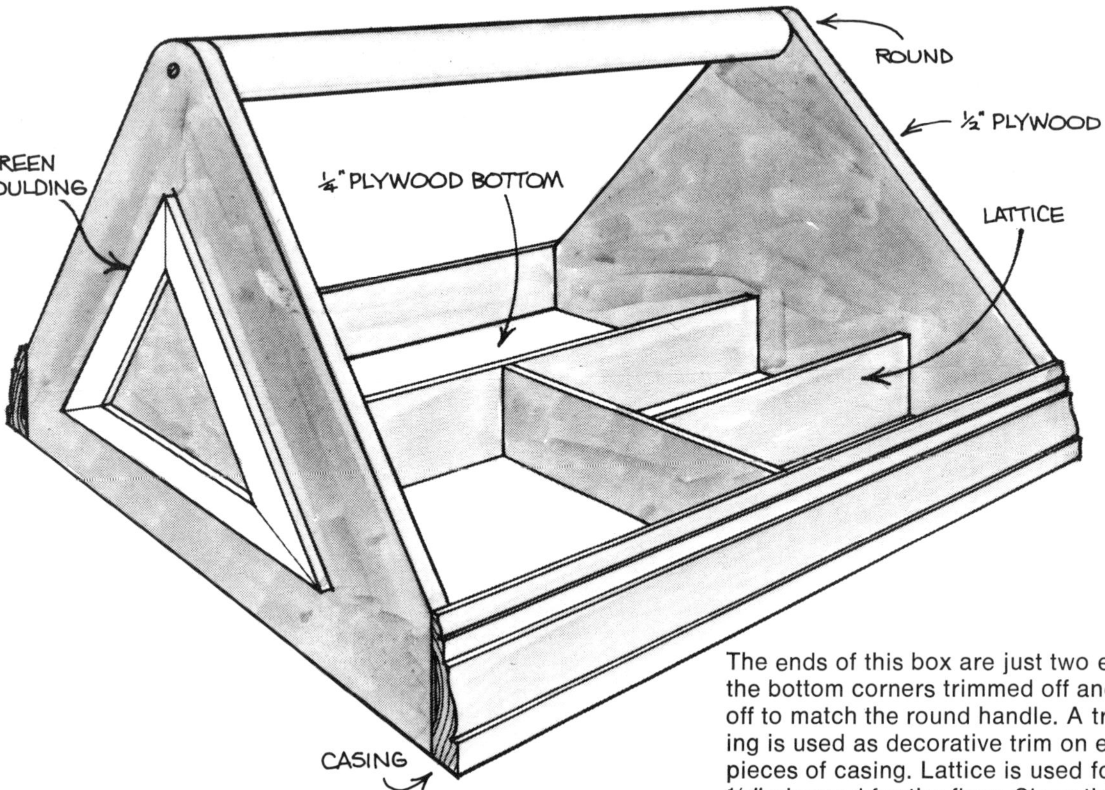

The ends of this box are just two equilateral triangles with the bottom corners trimmed off and the top corner rounded off to match the round handle. A triangle of screen moulding is used as decorative trim on each end. The sides are two pieces of casing. Lattice is used for the inside sections with ¼″ plywood for the floor. Since these boxes tend to get dirty, it's a good idea to finish them with a clear plastic or good enamel paint.

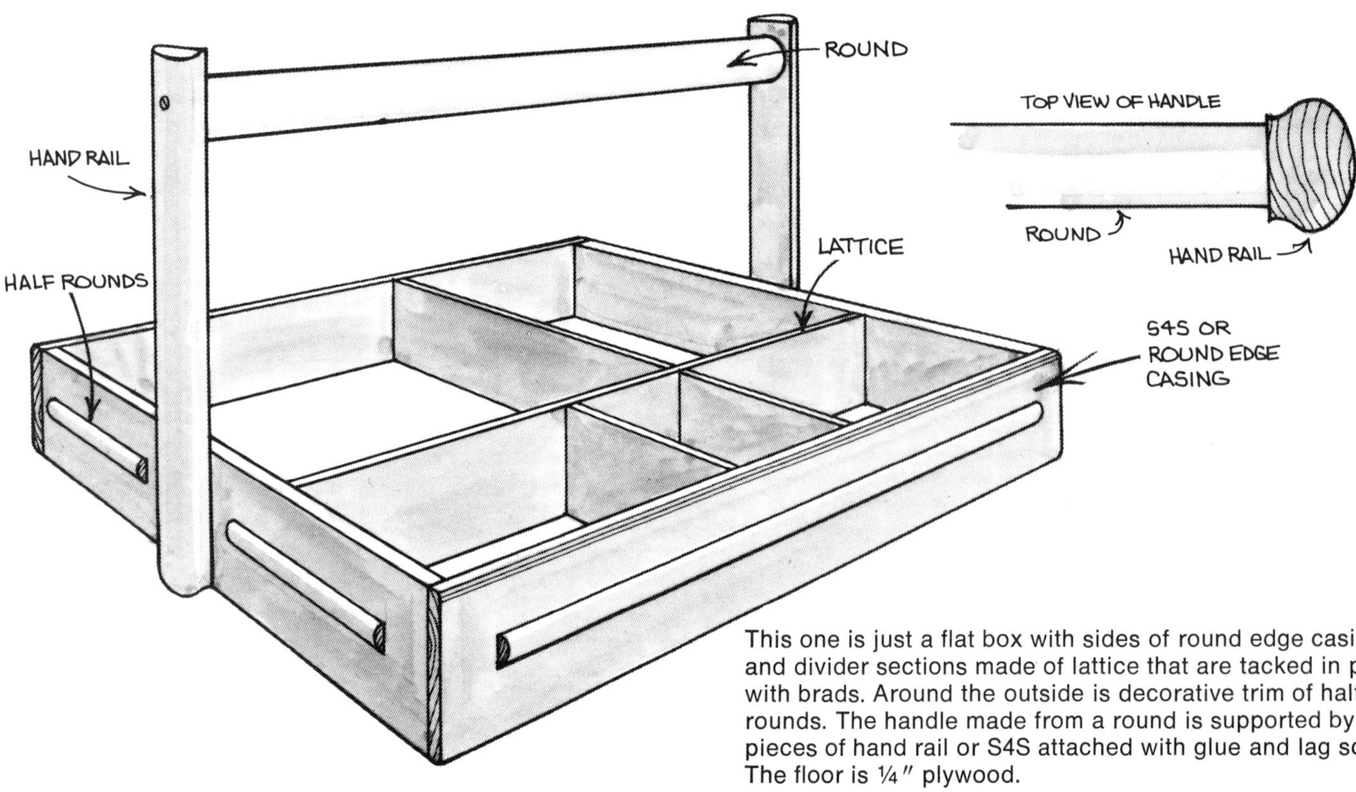

This one is just a flat box with sides of round edge casing and divider sections made of lattice that are tacked in place with brads. Around the outside is decorative trim of half rounds. The handle made from a round is supported by two pieces of hand rail or S4S attached with glue and lag screws. The floor is ¼″ plywood.

GARMENT TREE

What's more versatile than a movable coat rack? It's good for any room in your home, including your bathroom. Every office should have one, too. This simple rack can be built in less than an hour. All it uses is a 5 foot length of 2″ square, four 1 foot pieces of round edge casing, four pieces of ½″ dowel five inches long and 4 dowels three inches long, plus one decorative wooden knob to be screwed onto the top. Just attach the four legs of casing as shown with glue and nails and drill holes about ¾″ deep for the dowels. Stain or paint as desired and you have a versatile, hard working coat rack/clothes rack/towel rack, or whatever other uses you can find for it.

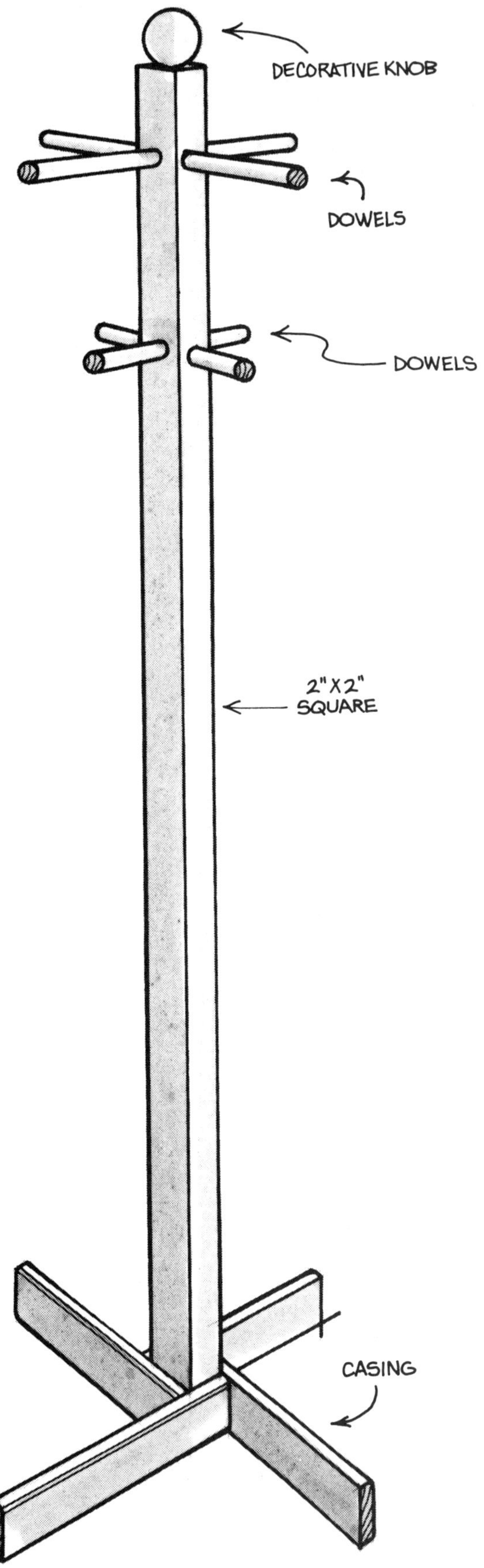

HANGERS (Banners, Macrame, Rugs, Etc.)

Here's an easy, economical and fast way to hang banners, wall rugs or large hangings of macrame . Use a wood moulding pattern that works well with whatever is to be hung. The moulding should be cut larger than the banner so some of the wood extends out each end. Stain or paint it to compliment the banner. To hang the banner, just drape it over the moulding and tack it to the back, then nail or screw the moulding to the wall, door, bannister or wherever it's going to hang. Hide the nails or screws under the banner itself.

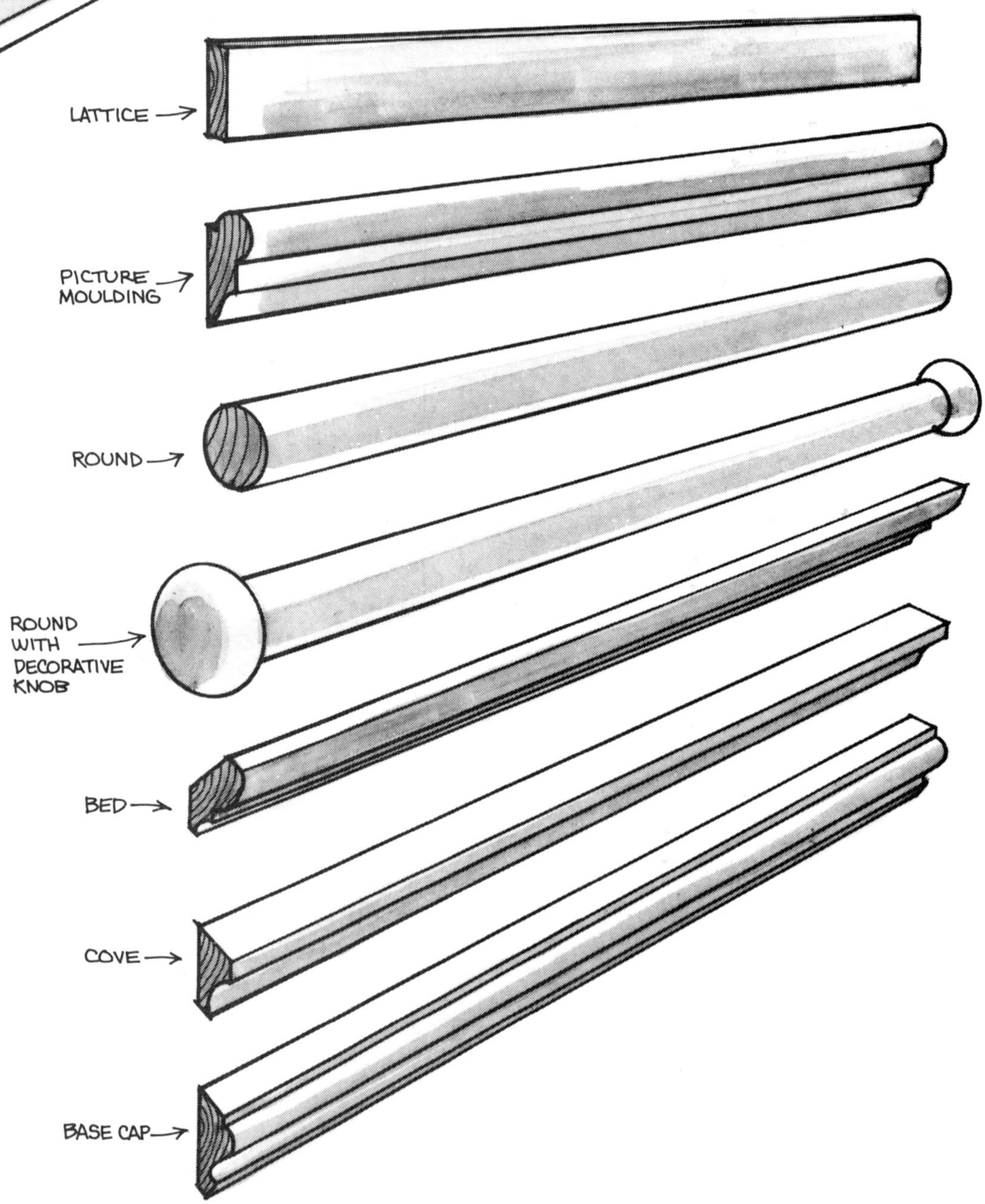

HEADBOARDS

Bedrooms should suggest warmth and snugness. Therefore, the natural warmth of wood is a perfect design element to be used when considering any change to be made in your bedroom, whether it's the walls, ceiling, closet doors, dresser or the bed. Wood moulding is an ideal solution when it comes to designing a headboard. Once again its versatility offers unlimited choices. The headboard designs shown here are all using a base of ¾″ plywood with the edges capped with S4S stock or lattice. The moulding design is then glued and tacked to the plywood surface. If the background is to be a different color than the moulding design, it's best to paint or stain that part first before attaching the moulding.

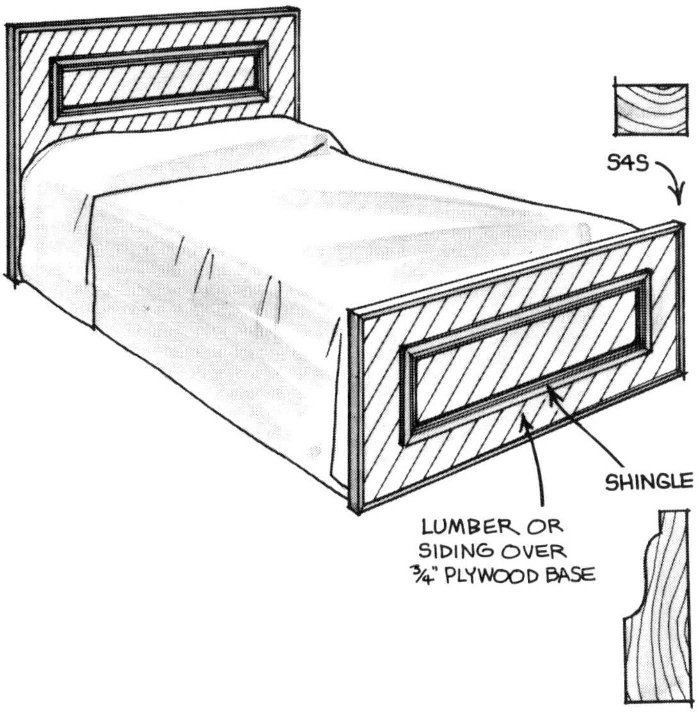

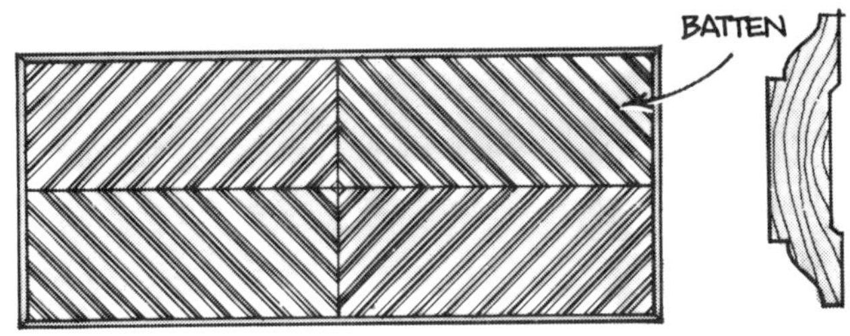

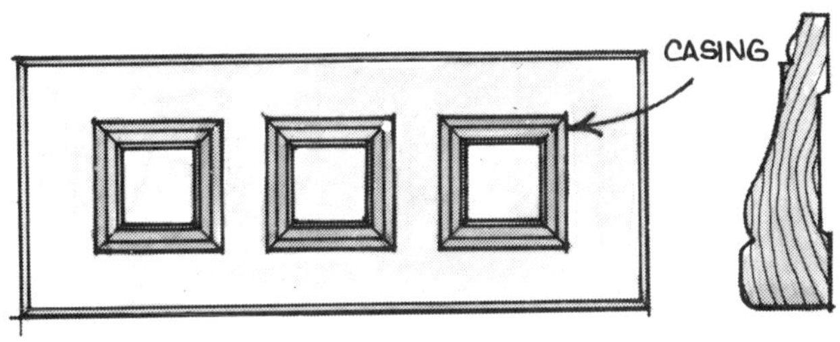

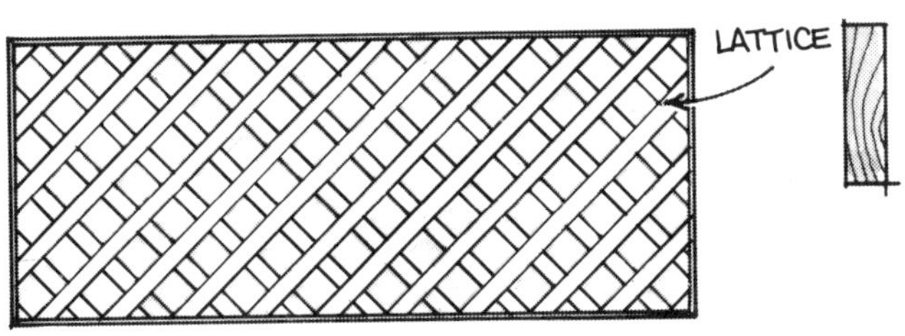

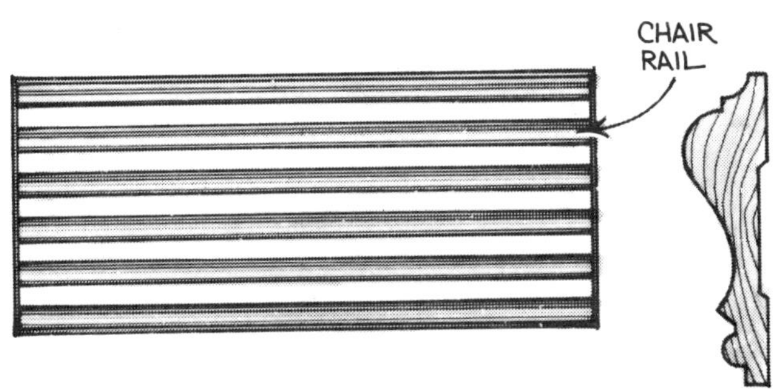

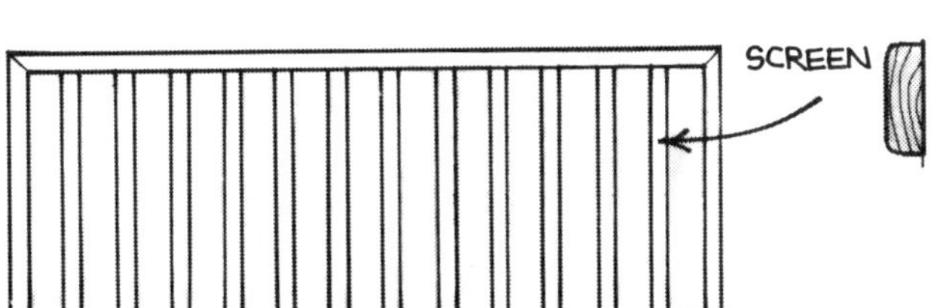

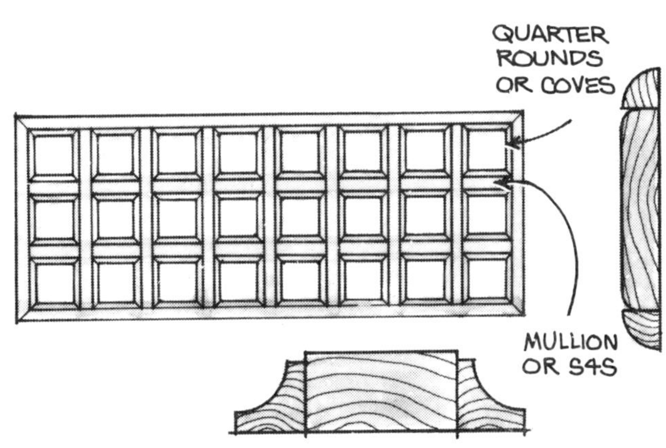

HOUSE NUMBER FRAMES

Add character and attention to your house numbers by framing them with wood moulding and displaying them against a background of siding, paneling, lumber, or whatever else your imagination might determine. How you stain or paint the frames, numbers and background will vary their appearances even further. Outdoor lighting fixtures can also be added. The numbers themselves come in all sizes and styles and can be purchased in wood, metal, plastic, plus an assortment of other materials. You'll find them at your home improvement center or lumber yard.

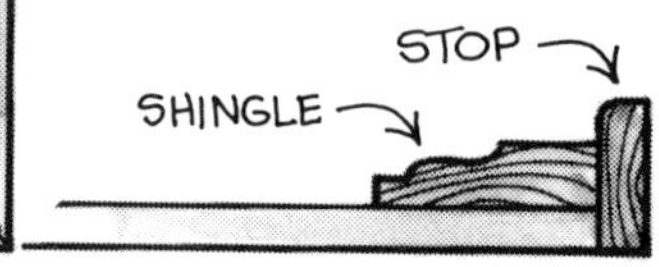

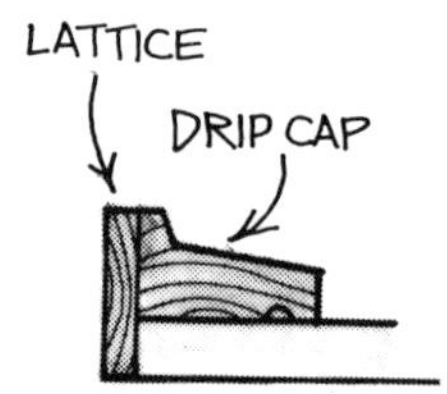

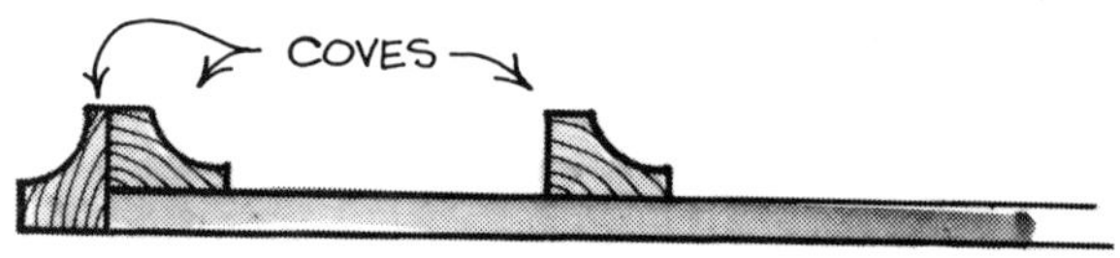

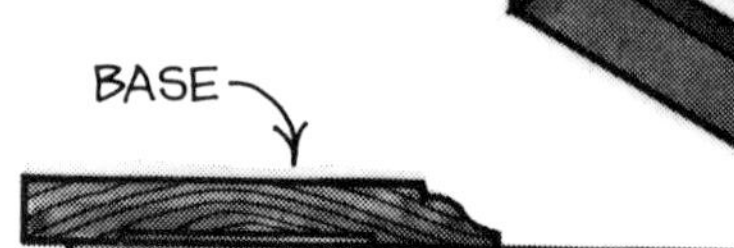

IN-&-OUT BASKETS

Here's how to clean up that paper work at home or the office. These easy to put together in-and-out baskets made with wood mouldings will simplify your paperwork. The first two models are just flat boxes with casing for the sides and plywood for the bottoms. The leg supports are made of coves or quarter rounds. Half circle notches are coped into the end panels to make access easier. The third in-and-out basket shown here is an upright box divided in the middle with the back half taller than the front half. Use hardboard or ⅜″ lumber. The lumber is easier to nail but is not as thin. After the box is put together, apply half rounds to the sides with brads and glue. Paint and stick on self-adhesive letters which can be purchased at art supply stores.

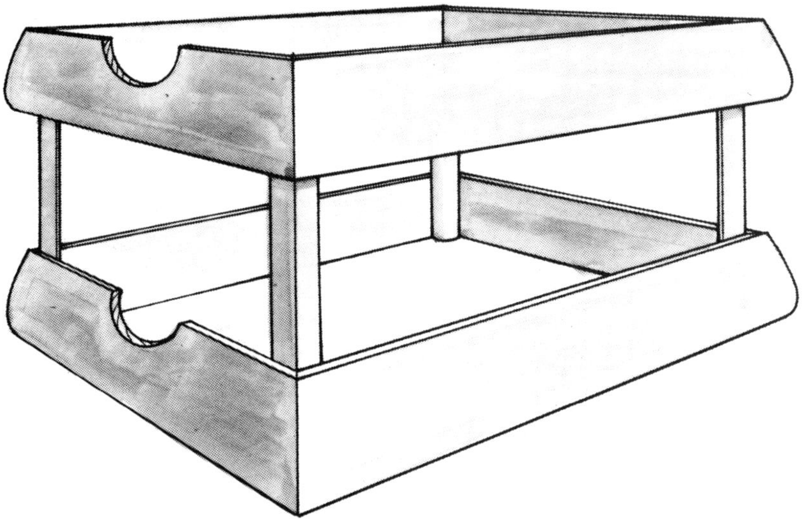

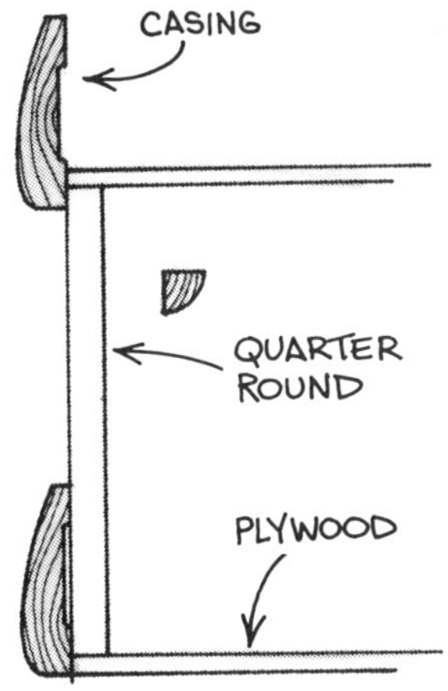

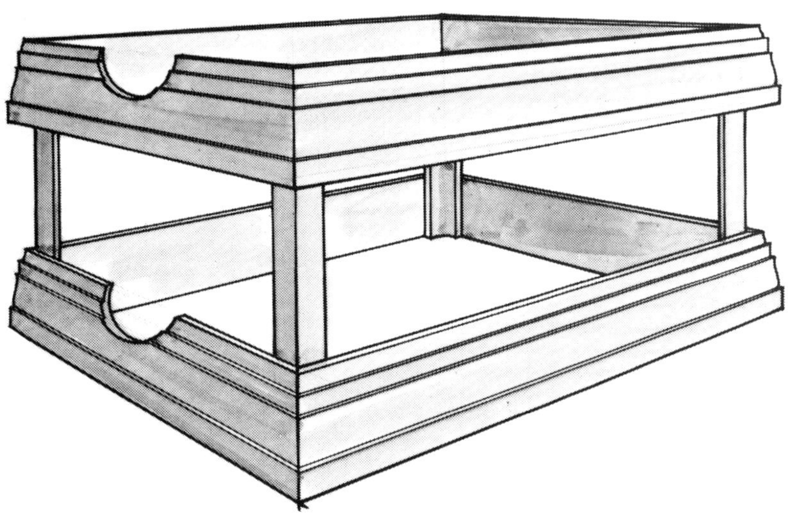

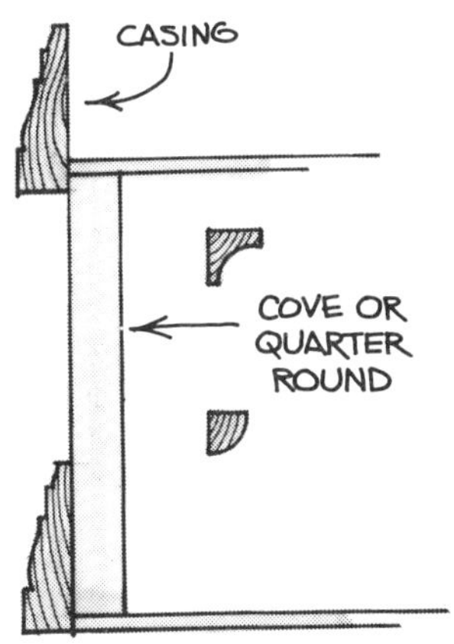

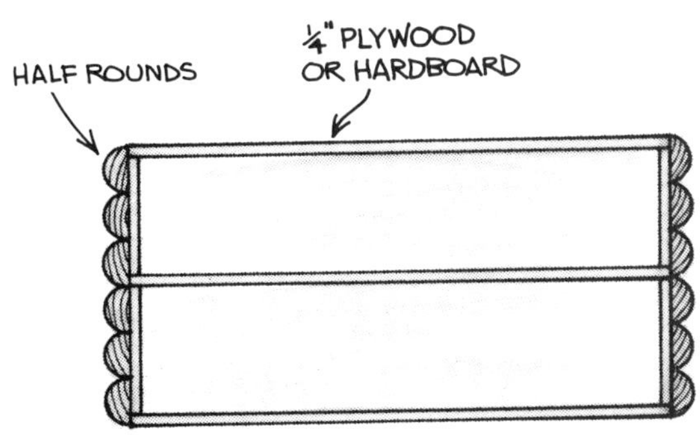

KIDDIE FENCES

Here's how to keep the toddlers where you want them. Just use a frame made of wood mouldings. The idea's been around for a long time so you've probably seen them before. It's just a matter of building a fence that slips into a doorway between the door stop and a piece of quarter round. If the doorway you want to use doesn't have a stop, add two more pieces of quarter round. The first kiddie fence illustrated here is just a frame of S4S stock or square moulding with mitered corners. Holes are drilled and dowels inserted about every three inches. The second fence is made up of a lattice grid sandwiched between two frames of lattice. Both designs are simple, but effective. Whichever one you choose for your little moppet, make sure it's painted with childproof paint.

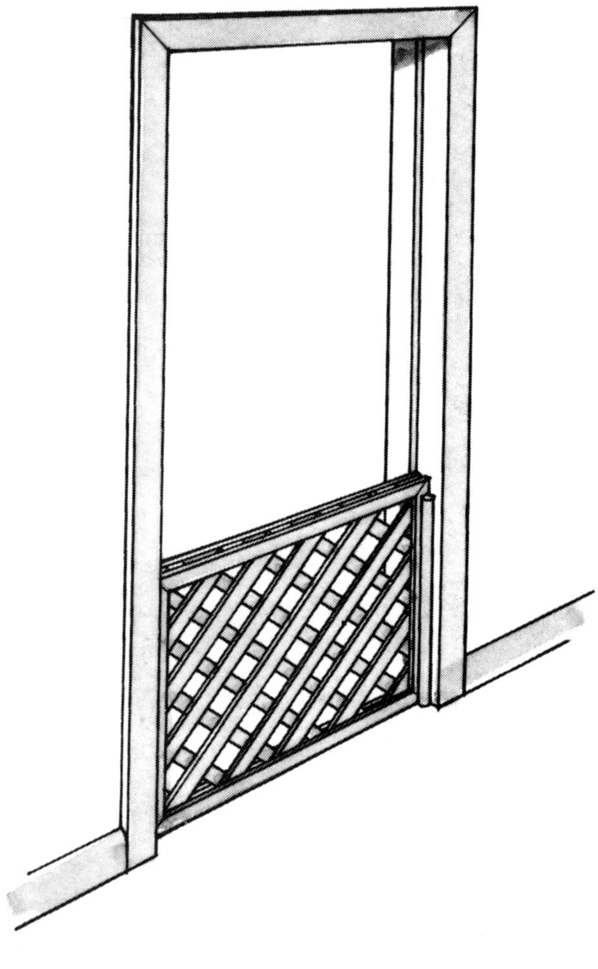

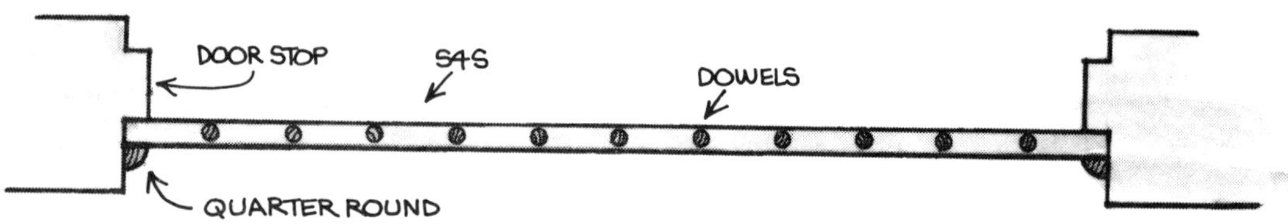

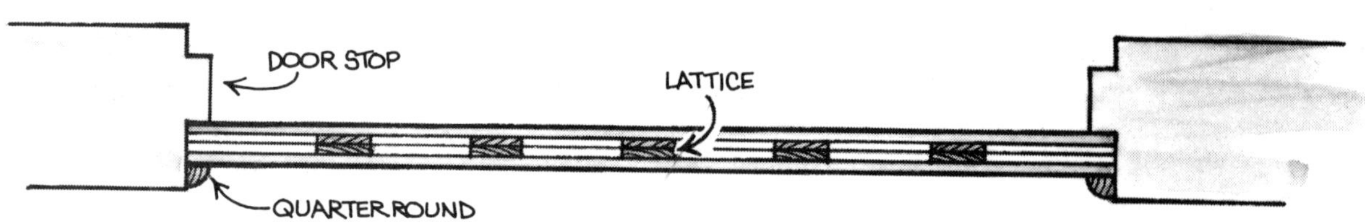

KNIFE RACKS

Here's an easy way to keep those knives out in view so you won't have to fumble around through a drawer looking for the right one. Whether you want your knife rack to hang on the wall, attach to a chopping block or sit on the counter, the right rack for your kitchen is simple to build with wood mouldings, as these illustrations show. When completed, either stain or paint the rack to complement your decore.

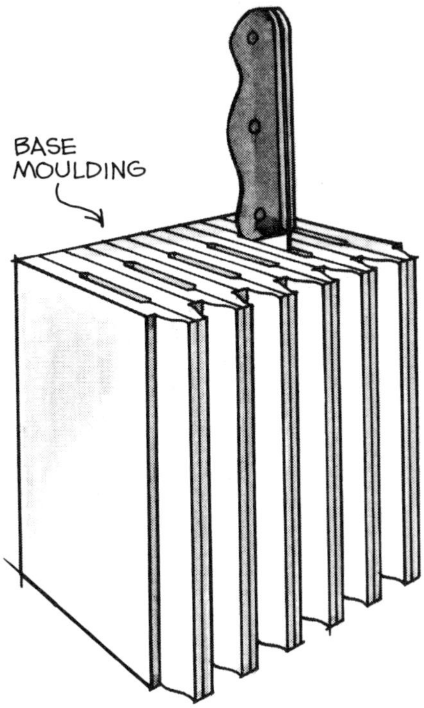

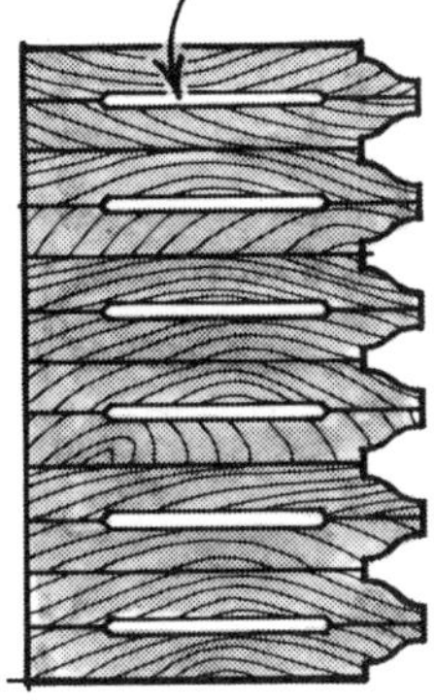

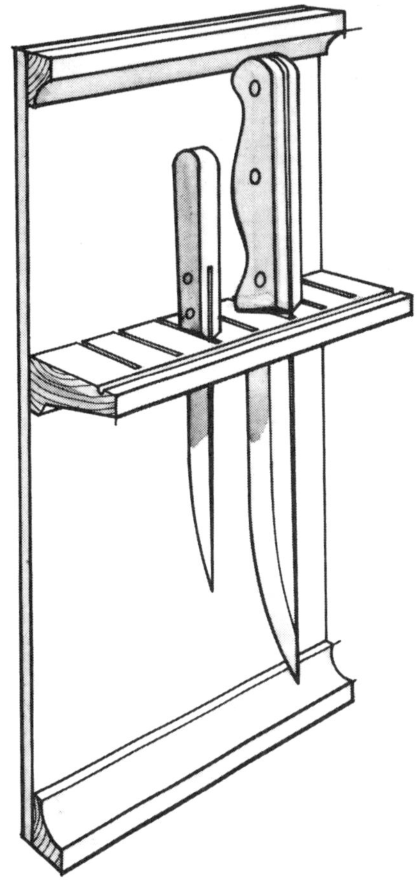

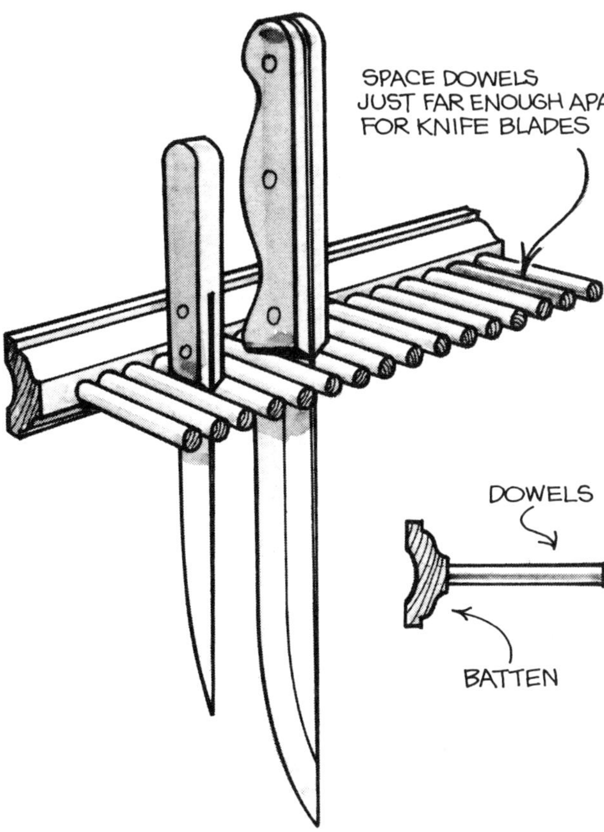

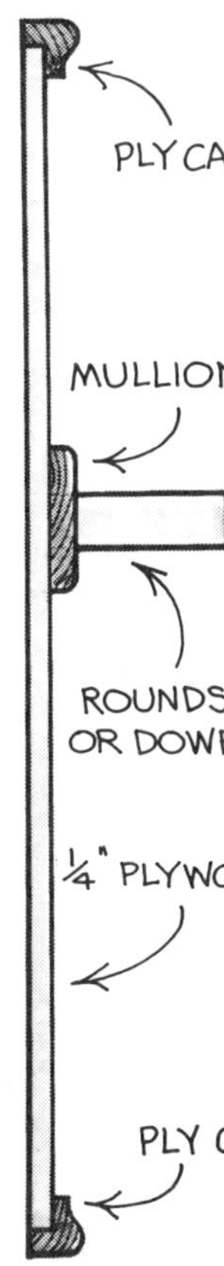

LAMP BASES

Basically lamps are all the same on the inside. They're just a socket and a switch attached to a wire that goes through a center tube and ends up with the wall plug. Usually there's also a wire harp attached to the socket which holds the lamp shade. Where lamps differ is on the outside. In most cases they're just a box. How the box is covered is what gives a lamp its uniqueness, charm, character and beauty.

The lamp shown below is a narrow box made of S4S stock with squares of casing attached. Chair rail moulding, base, flat astragals, stops, battens and shingle moulding can also be used. The top and bottom of the box is capped with a piece of S4S with a hole drilled for the center tube. Mitered stop moulding is attached to the bottom to form a wider base. A large square of lumber could also be used.

Diminishing layers of S4S stock make this a very simple lamp to build. An inside lining of plywood is used on which the layers of S4S are attached. As each row goes up, one less layer is used to give a stairstep appearance. The corners are alternated as shown. S4S is also used for the top and bottom.

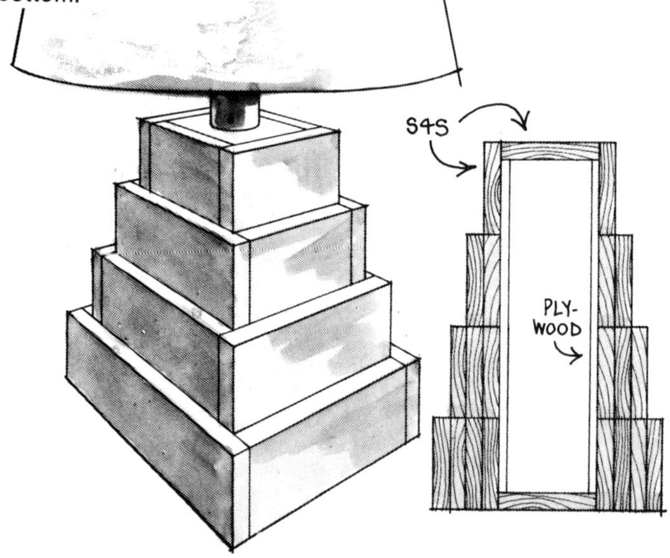

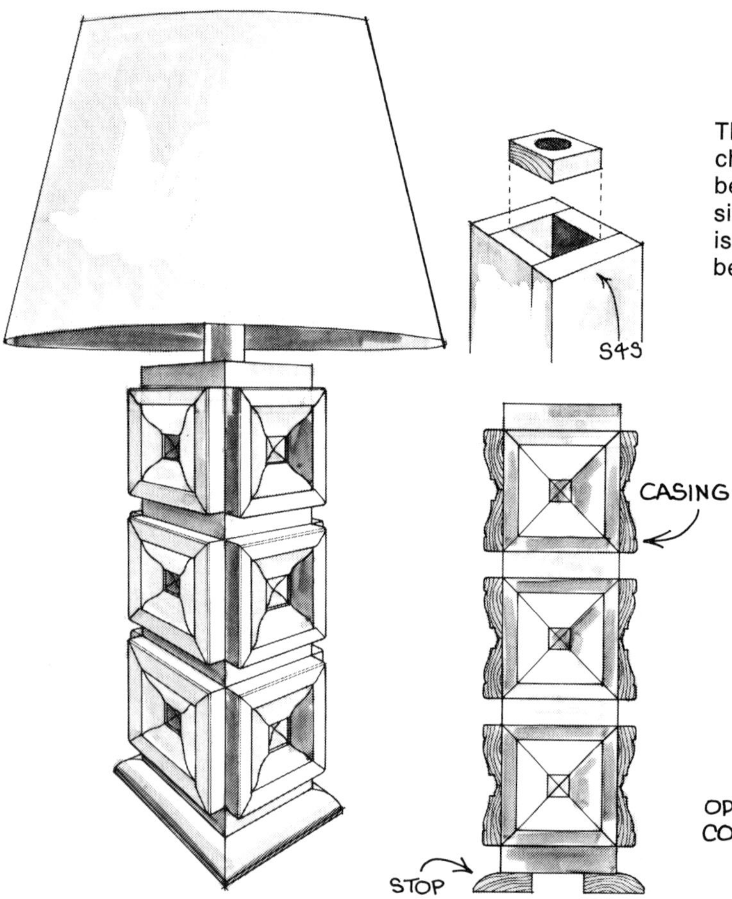

The elegant simplicity of this lamp makes it a popular design choice. Its basic structure is just a box built with ¾″ lumber and ¾″ square mouldings evenly spaced around the sides. The corners may or may not be mitered. The choice is yours. Half rounds, screens, coves and stop moulding can be used in place of squares.

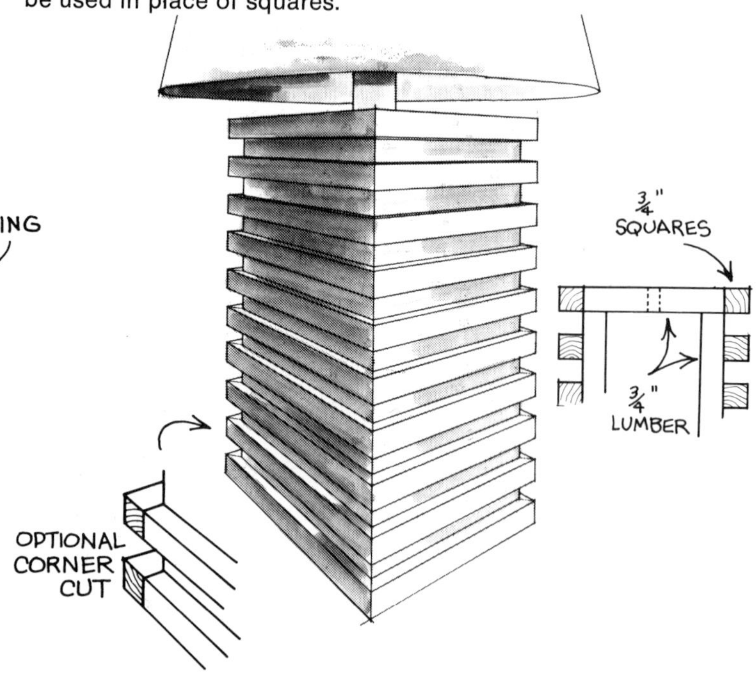

LAMP BASES

This square cube is just a plywood box with screen moulding attached to the face as shown. Small quarter rounds are used in each corner. Besides the various screen mouldings available, half round, flat astragel or chamfer strips can also be used.

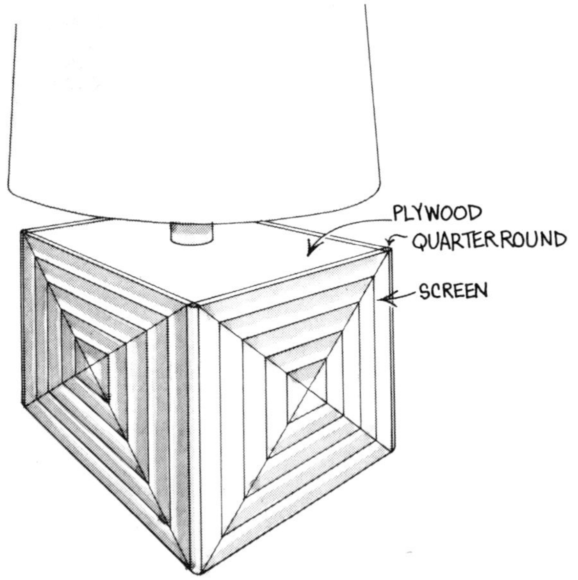

This lamp is a plywood box with ¾″ lumber for the top and bottom and is covered diagonally with screen mouldings cut at a 45° angle. Quarter rounds are used in each corner.

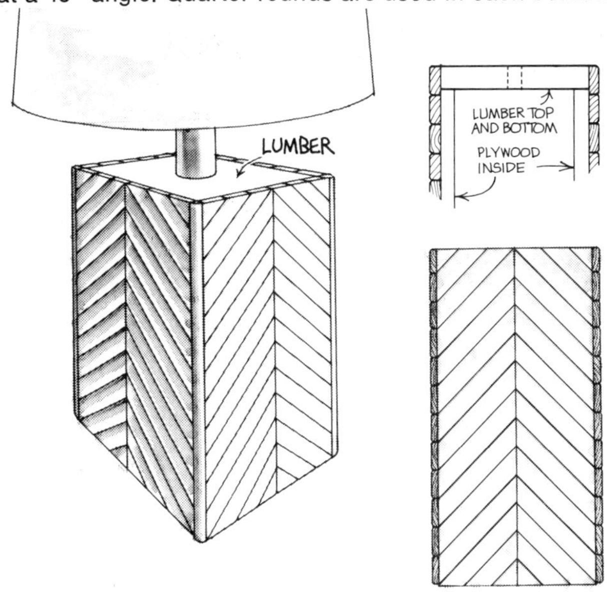

A simple box of plywood with a framework of casing is all there is to this lamp. Besides casing, base, batten, base caps, shingle moulding, coves, flat astragal, chair rails and mullions can be used.

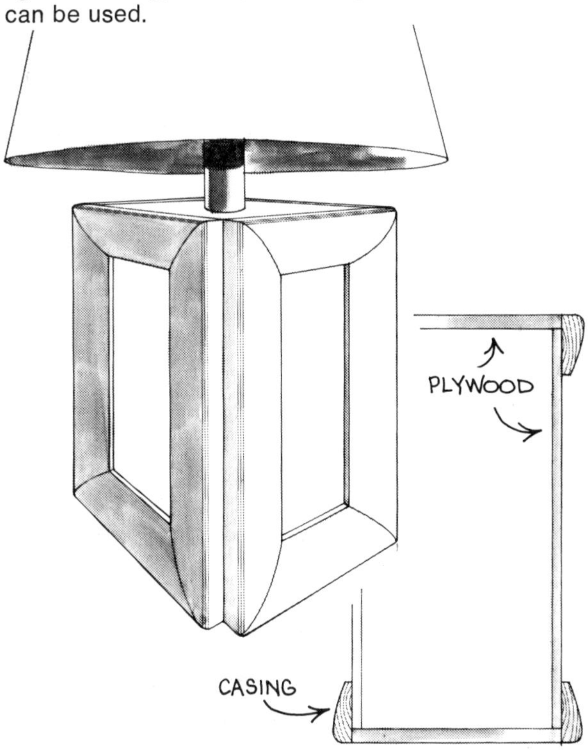

Cylindrical shapes are also popular. This lamp is just a top and bottom circle cut out of ¾″ lumber with half rounds attached. Be sure all the mouldings used here is perfectly straight, otherwise spaces will show through the lamp base. This design could also be done in a square, octagon or hexagon shape. If those shapes are desired, use plywood under the moulding to eliminate the possibility of space gaps.

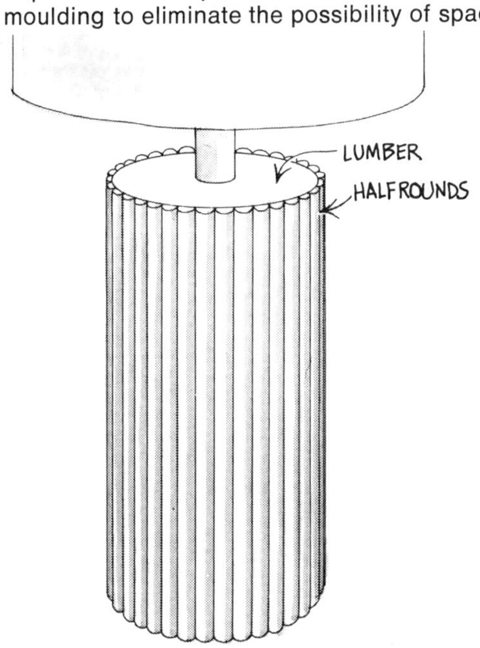

LIGHTING FIXTURES

Every room in the home, whether it's the bathroom, den, office, game room, garage, basement or living room can use lighting fixtures. Building your own is not the monumental problem it might seem. A quick look at these wood moulding designs will show you how simple they really are. The lights, fixtures and wiring can be purchased in any lighting store.

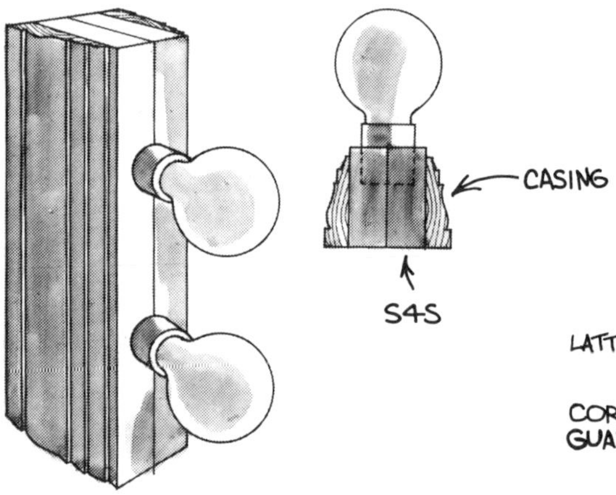

This block of lights, whether its one, two or ten lights, is easily put together. The base is just two pieces of S4S set between casing. Wide holes are bored for the decorative sockets, with small holes drilled for the wiring. Also rout out or chisel a groove in the back for the wiring.

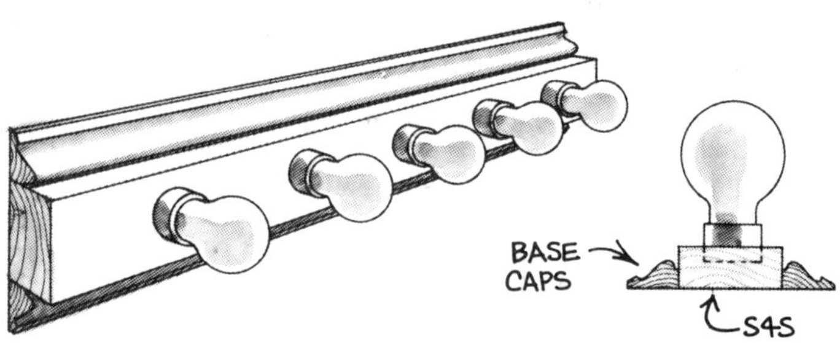

This is a variation of the same fixture, but in this one a single S4S is used with base caps on either side. It makes an excellent light for the bathroom or bedroom vanity.

This is basically the same light design except the S4S is framed with coves. Again, the quantity of lights depends on your needs and the area it's to go in.

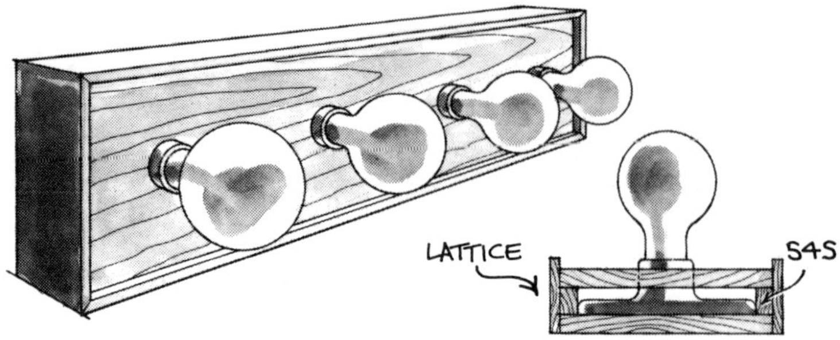

This line-up of bulbs is screwed into stock ceramic light sockets sandwiched between layers of rough sawn lumber with S4S or squares used as spacers. It is then framed with lattice, casing, stop or shingle moulding.

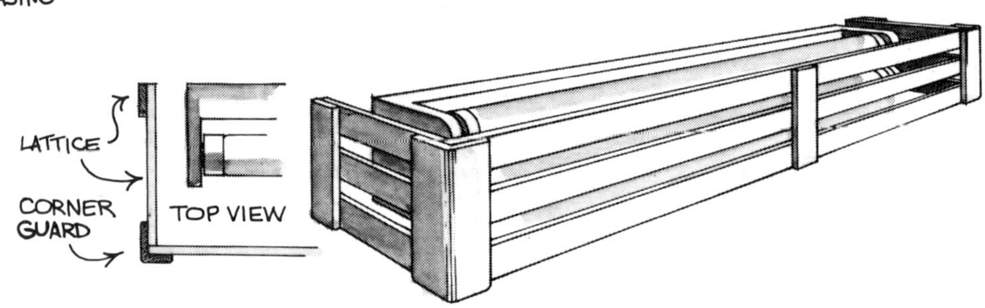

This light is just a standard fluorescent wall fixture with a facade of wood moulding extending out around it which is made from strips of lattice attached to corner guards at the two corners. Short vertical strips of lattice are also used at each end and in the center. When completed, it can be held to the wall with small corner braces.

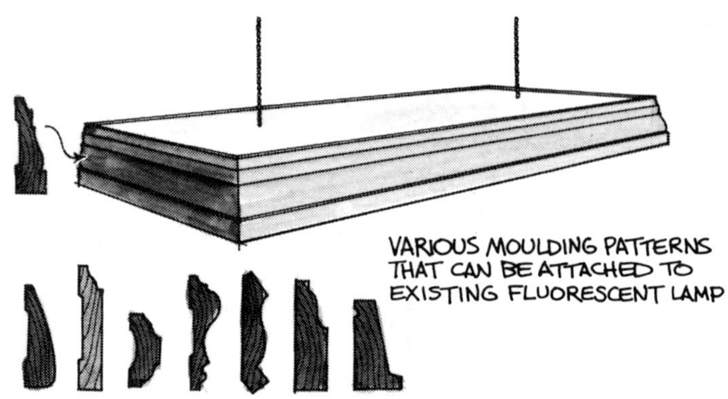

A plain, ordinary fluorescent light can easily be given new character merely by applying wood moulding around the outside. Casings, base, batten, chair rail, crown, shingle and drip caps can be used. Because the sides of the fluorescent lamp may not be perfectly straight up and down, make sure you do your measuring and mitering accurately.

LIGHTING FIXTURES

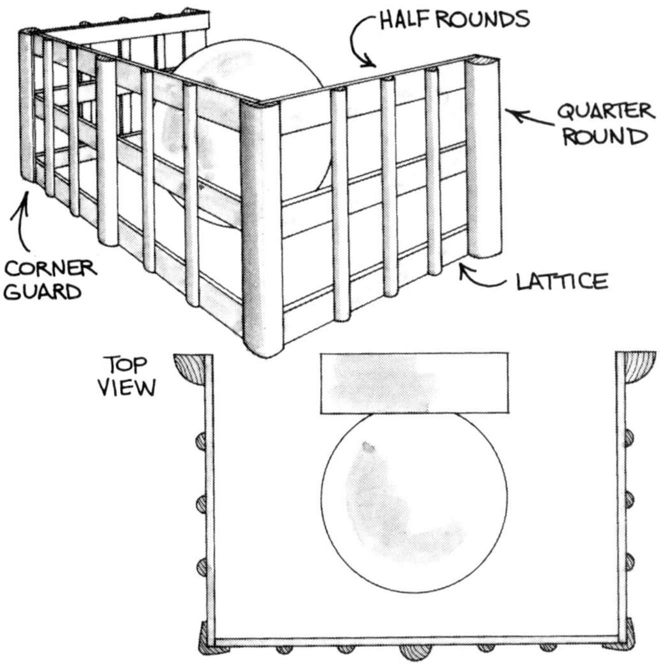

A simple round light ball attached to the wall and fenced in by wood moulding makes this a pleasant lightweight wall fixture. It's constructed of lattice with half corner guards at each corner. Quarter rounds are at each end to hold it to the wall.

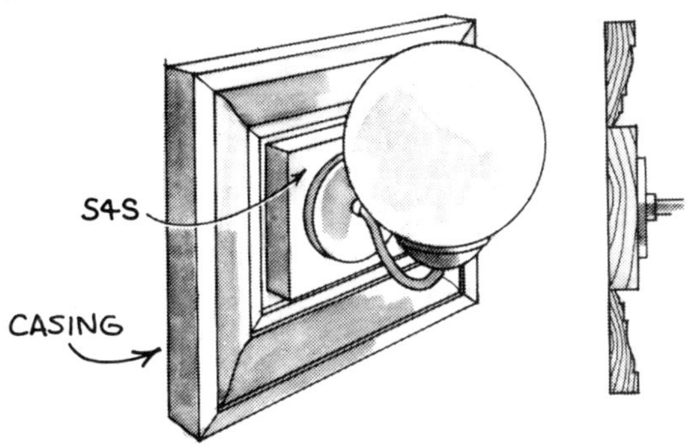

This wall light is made of a framework of casing around a square of S4S stock. The lighting fixture is one of many styles available at your lighting store.

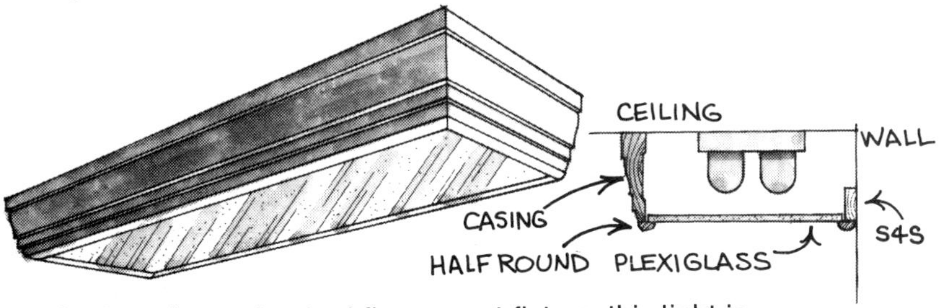

Again, using a standard fluorescent fixture, this light is attached to the ceiling with corner brackets next to the wall. It's made of either casing or base mouldings with a small half round glued and carefully tacked with brads to the lower edges of the casing. The half round must be positioned so enough lip forms a shelf for the plexiglass to sit on. The piece of half round next to the wall where there is no casing is attached as shown.

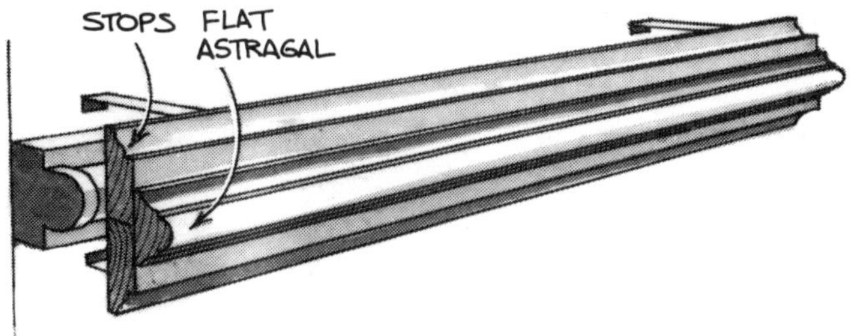

Indirect lighting is easily achieved by using standard fluorescent tubes behind facades of wood moulding. This one uses two pieces of stop moulding capped by a flat astragal and is held away from the wall with metal drapery brackets. This design allows the light to be emitted both above and below.

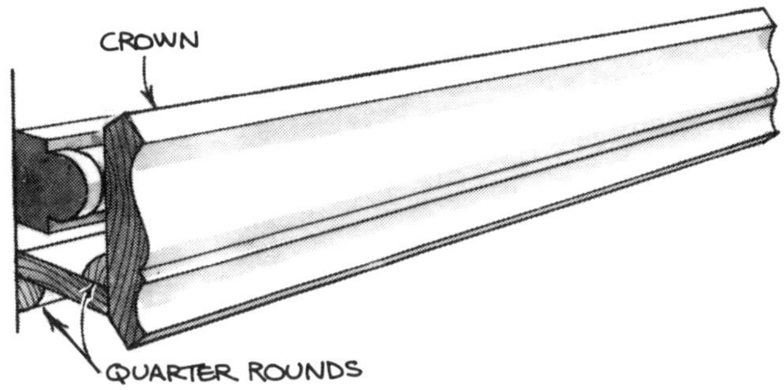

In this instance a length of crown moulding is used which is held in place with a piece of S4S stock and supported by quarter rounds. Here, light is emitted upward.

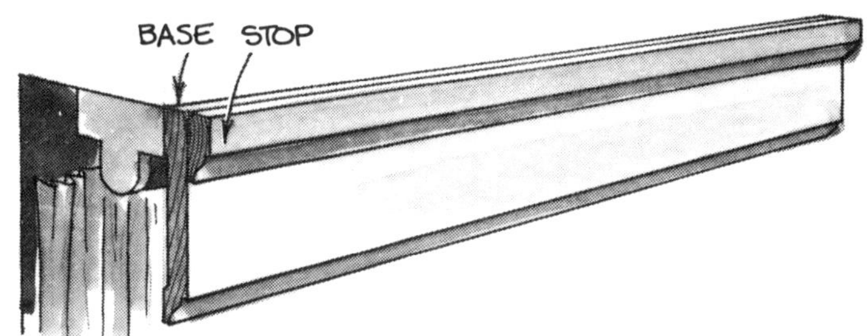

This light fixture covering is attached to the ceiling with corner brackets and is made from one piece of base moulding and one piece of matching stop moulding. It makes a good valance for both the lighting fixture and draperies.

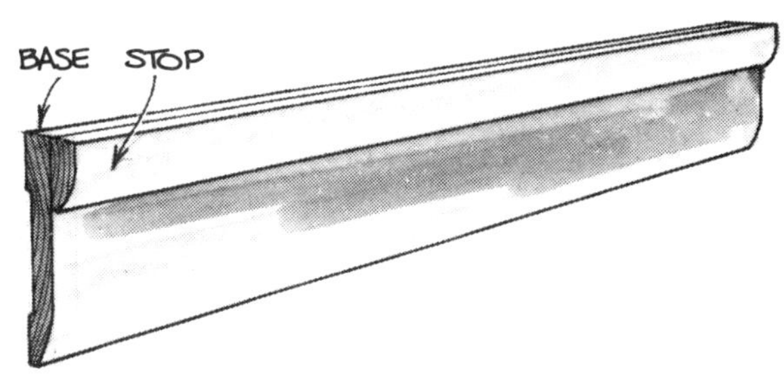

This is the same idea, but uses ranch base and ranch stop moulding.

LAMP POSTS

Now you can design your driveway or yard entrance lamp post to look any way you desire. Make it contemporary or traditional. Give it the character of your home. Modify your existing lamp post. It's all possible with wood mouldings. The lamp fixture itself can be purchased at lighting stores or home improvement centers.

The first lamp post shown here is made up of square moulding nailed to two square boards of lumber with holes drilled in the center for the pipe that holds the lamp's wiring. In all these lamp post designs, the pipe, not the wood, should touch the ground.

The center lamp post is a series of rosette squares made from shingle moulding. Casings, stops, base, base caps and battens can also be used. First build a box-like sub-structure of plywood, then make up all the moulding squares and nail them to the plywood. Add quarter rounds to the corners. Again, a hole should be drilled for the lamp pipe.

The lamp on the far right is just a plywood box with its corners capped with corner guards and a strip of screen moulding down the center.

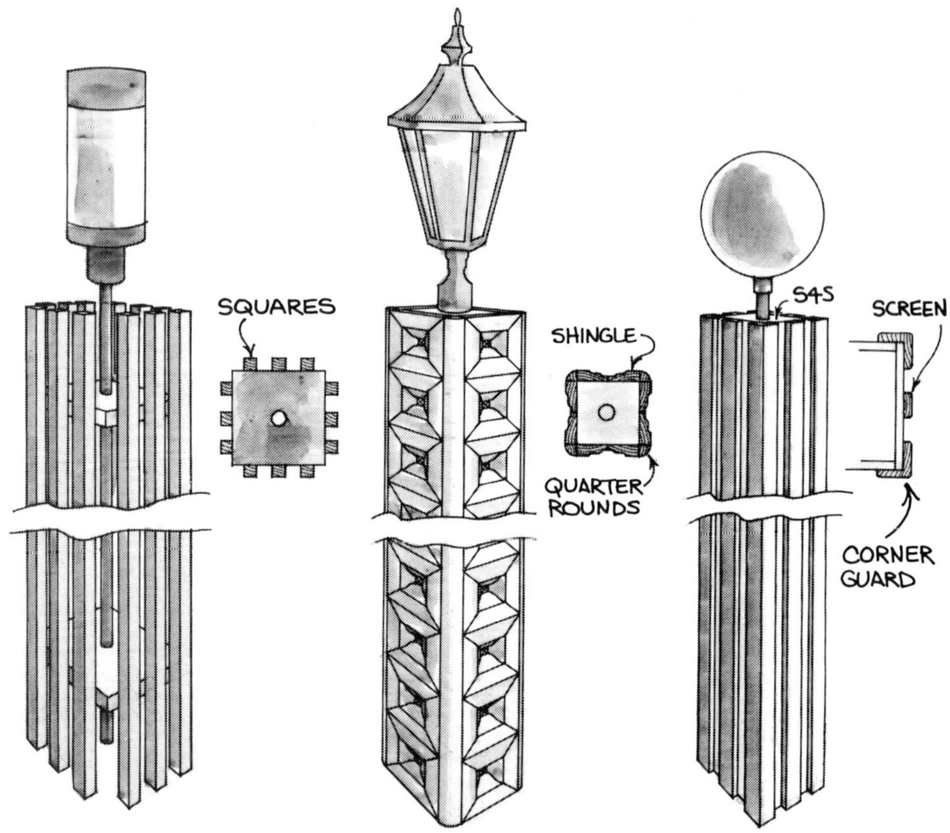

The next design is again a plywood box, but this time covered with horizontal strips of base or casing. This will require some accurate measuring because all the corners are mitered. By the time you're through with this one, you'll be a mitering master, but its clean good looks will make the extra effort worth while.

The center lamp post is merely a plywood box framed with lattice, shingle or casing depending on whether you want a contemporary or traditional look. Remember, an outside lamp post takes a beating from the weather, so use exterior plywood and the proper sealer, stain or paint.

The lamp on the right is simplicity personified. It's made up of large quarter rounds or smaller quarter rounds and squares with space for the pipe in the middle.

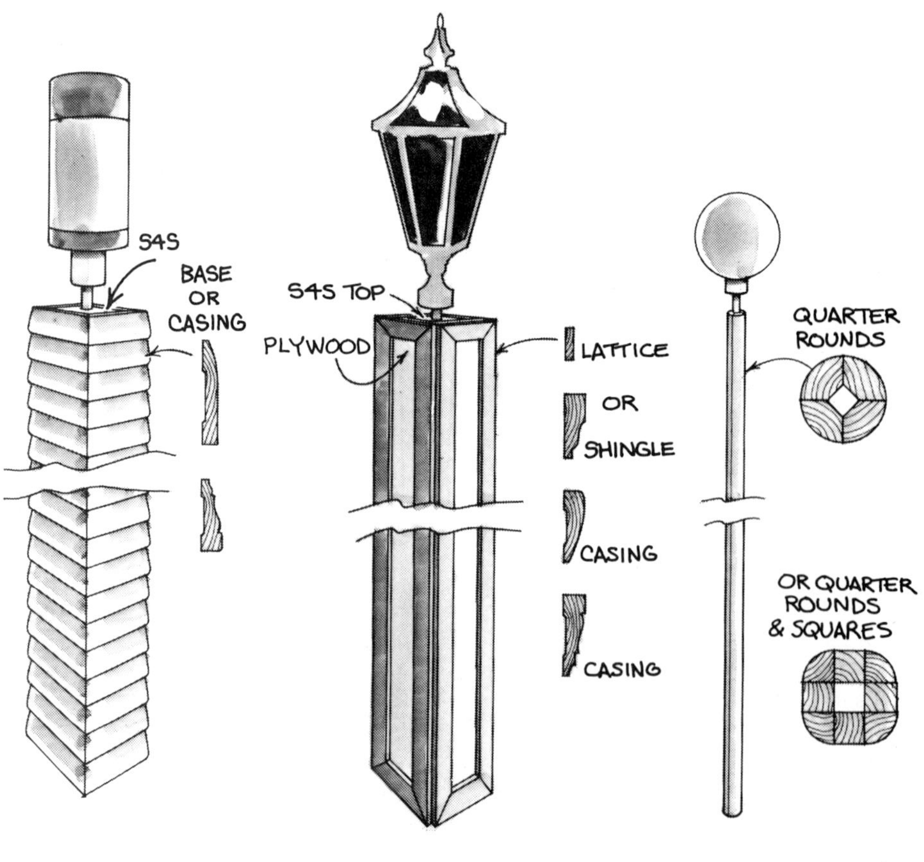

LIPSTICK HOLDER

Do all those lipsticks end up being thrown into a drawer? Here's how to keep them fingertip handy and well organized. Simply drill holes of the proper width in a piece of wood moulding. About three fourths of the depth of the wood should be sufficient. If whatever pattern of moulding you use is too thin, just add a piece of S4S stock or ⅜″ lumber of the same width under the moulding to act as a base. Cut it as long as is needed and stain or paint it to match your bathroom or vanity.

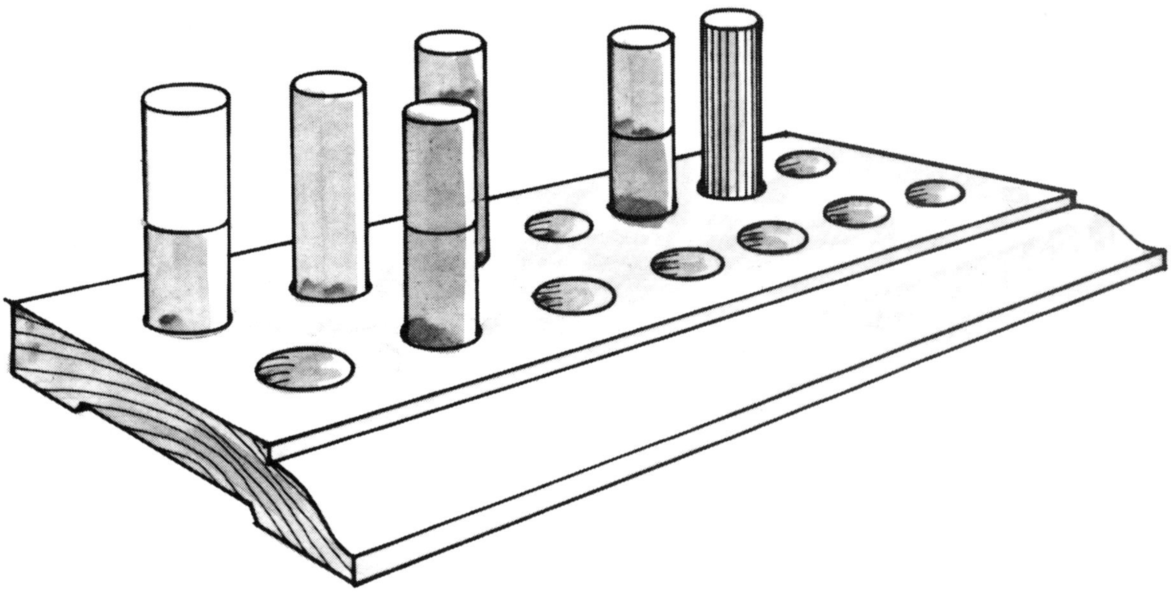

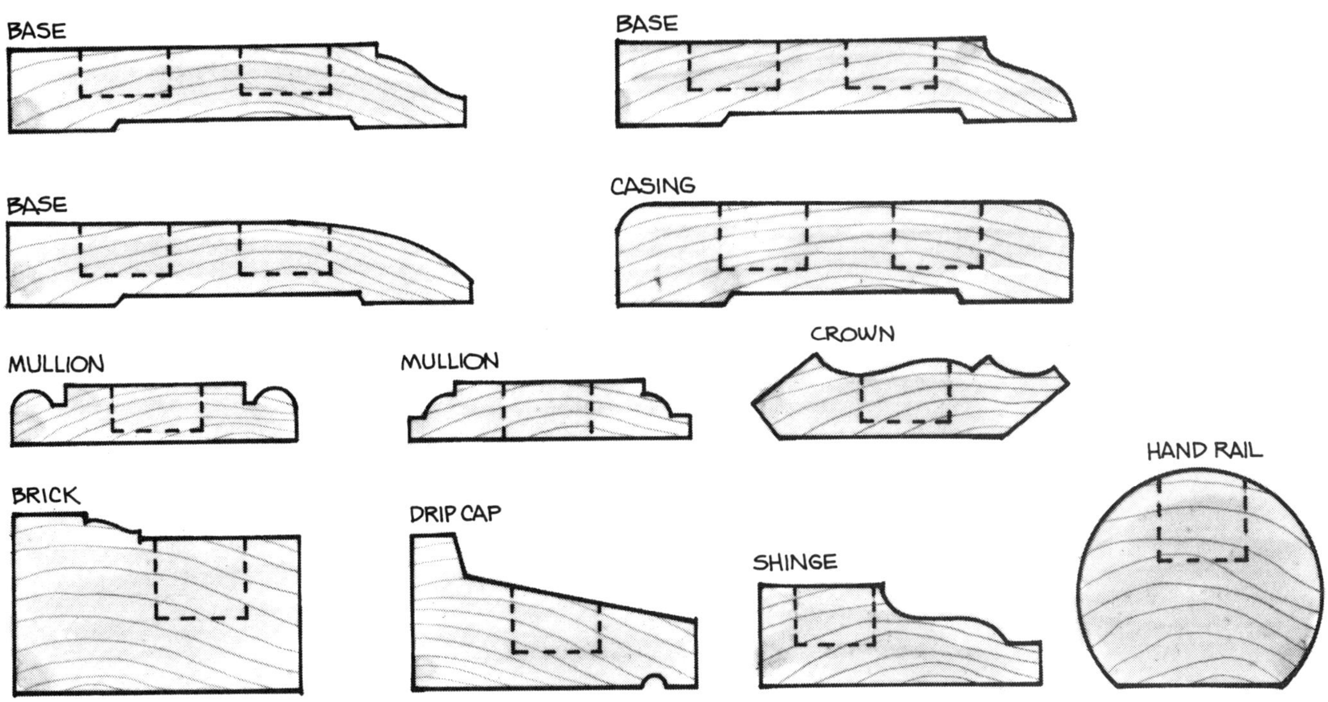

MAGAZINE RACKS

Magazines seem to have a way of cluttering up the house or office. To eliminate that problem, build a magazine rack. This one is made of two frameworks of half rounds attached to a center piece of square moulding. Use the largest size of half rounds to keep the rack from being flimsy.

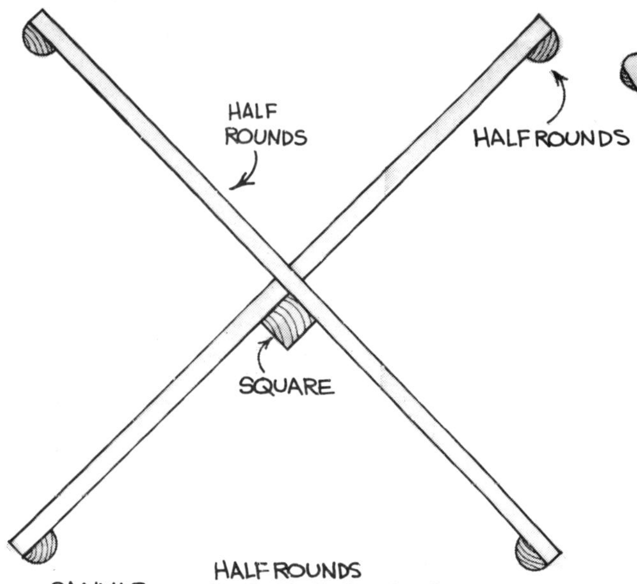

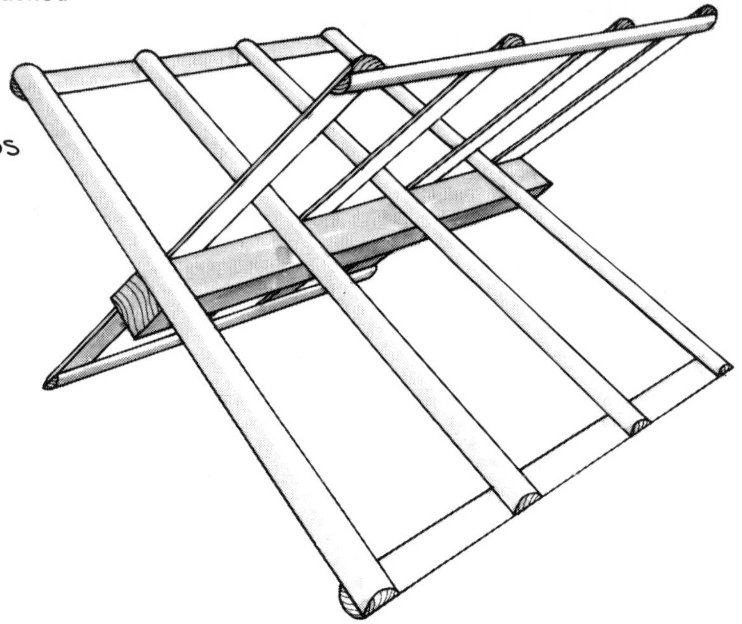

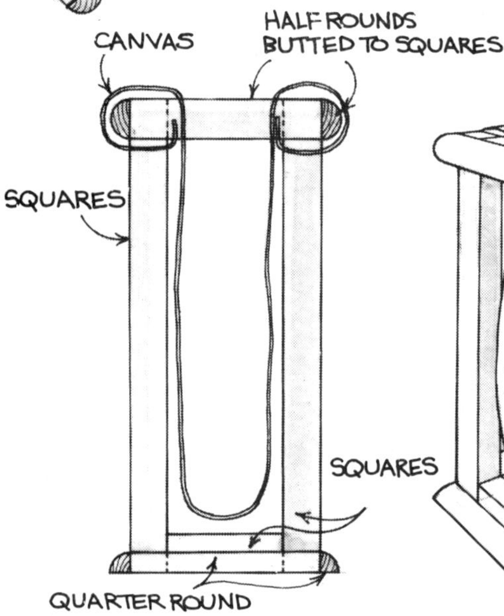

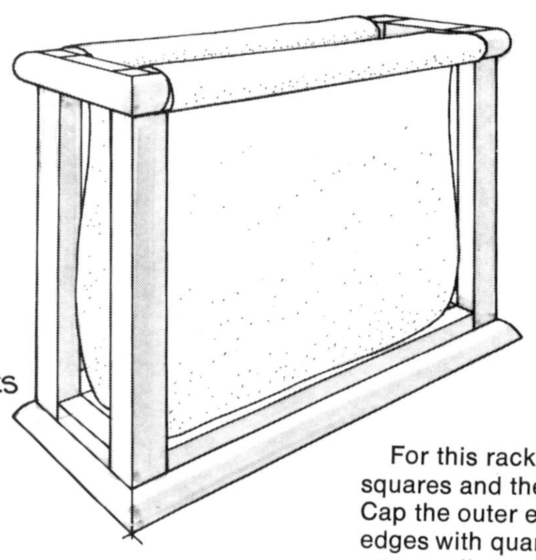

For this rack, build two rectangular frames from moulding squares and then connect them with short lengths of square. Cap the outer edges at the top with half round and the bottom edges with quarter round. To hold the magazines, tack a canvas sling in place as shown in the illustration.

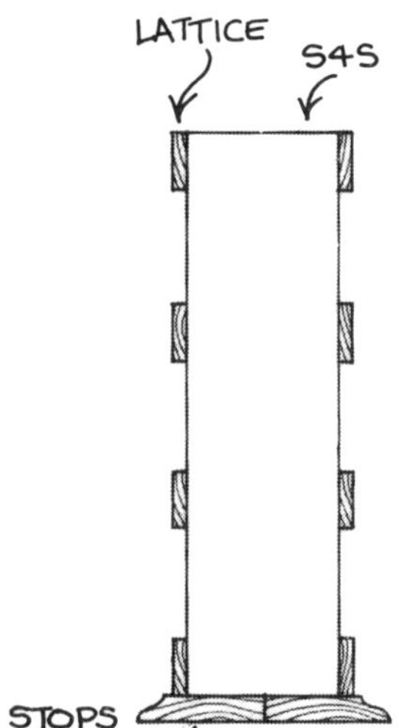

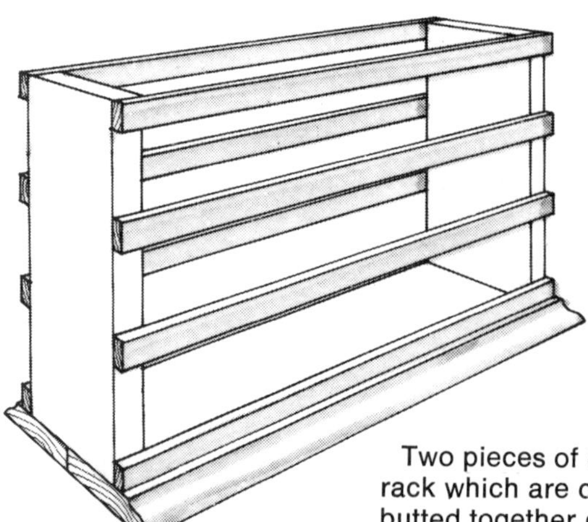

Two pieces of S4S stock form both ends of this magazine rack which are connected with strips of lattice. Two stops butted together give the rack a wide base for stability.

MAIL BOXES

Your mail box need not be a tin box that looks like everybody elses. Give it character and individuality by using wood mouldings. The idea shown here on the left is just a standard metal mailbox sitting on a piece of 1″ lumber between two fences of S4S moulding or squares. Another piece of S4S is nailed to each side for the box numbers. The mailbox flag is attached to the proper side. To keep the wooden sides of the structure from absorbing ground moisture and eventually rotting, the mailbox should be either connected to metal pipes, wooden legs soaked in creosote or should sit on cement blocks.

The center mailbox is the same basic structure, except the sides are made of exterior siding framed with corner guards. An inner framework of lattice or screen moulding is also tacked to the siding along with wooden numerals. When mounting the flag, remember it must be placed in such a way as to not interfere with the moulding or numbers.

The mailbox on the right is just a simple box made of exterior plywood with stop moulding around the top edges. Casing or base is used around the bottom. Remember, the corners must be mitered. Base is also used on top of the box for a name plate. Nail it in place from the inside of the mailbox. The letters and numbers used on the box are the self-adhesive type purchased in art supply stores. Attach a decorative hook for opening the door. A large metal pipe holds the mailbox.

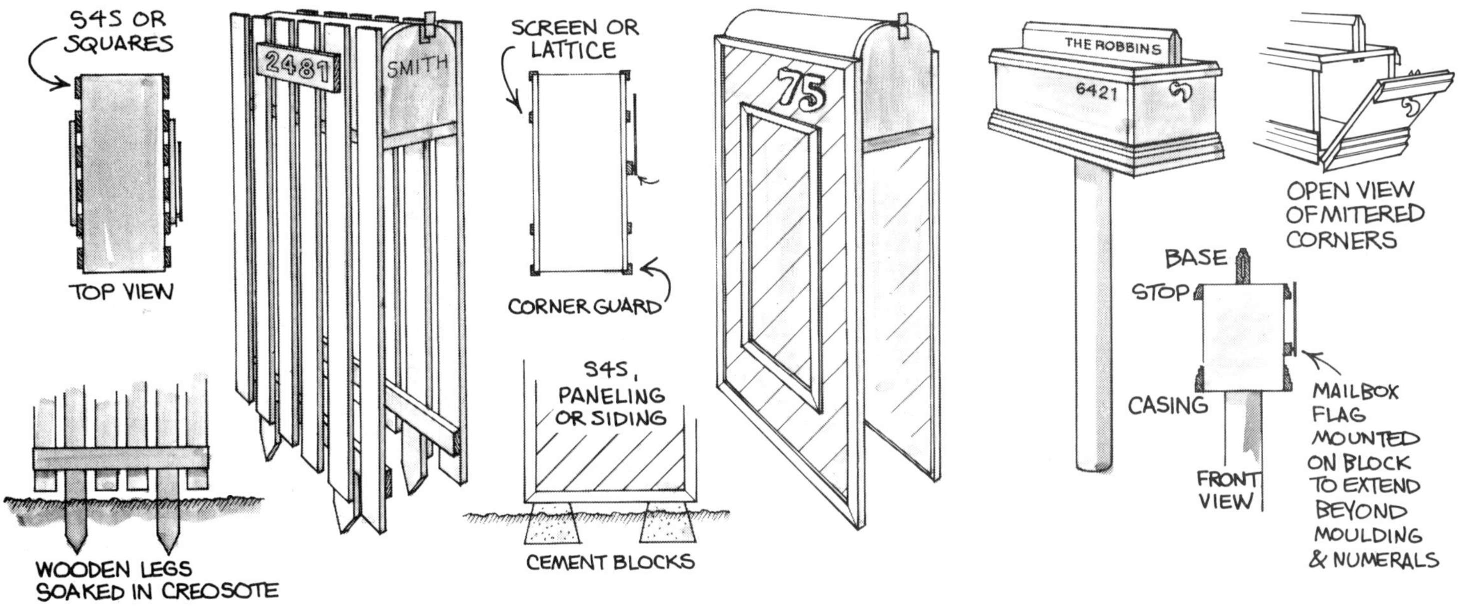

Diagonal lattice grids make up the sides of this mailbox with the edges capped by corner guards or lattice. An S4S board is attached for the numbers.

The box in the center is just a standard metal mailbox given a face lift with frames of screen moulding. The same pattern is repeated on its large wooden post. Remember to reposition the flag so it does not hit the moulding.

The last mailbox is the same kind, but with rows of half rounds attached. To make sure they stay on, it's best to glue them and nail them from the inside. The proper nails and measuring is important here. Half rounds are also used on the post.

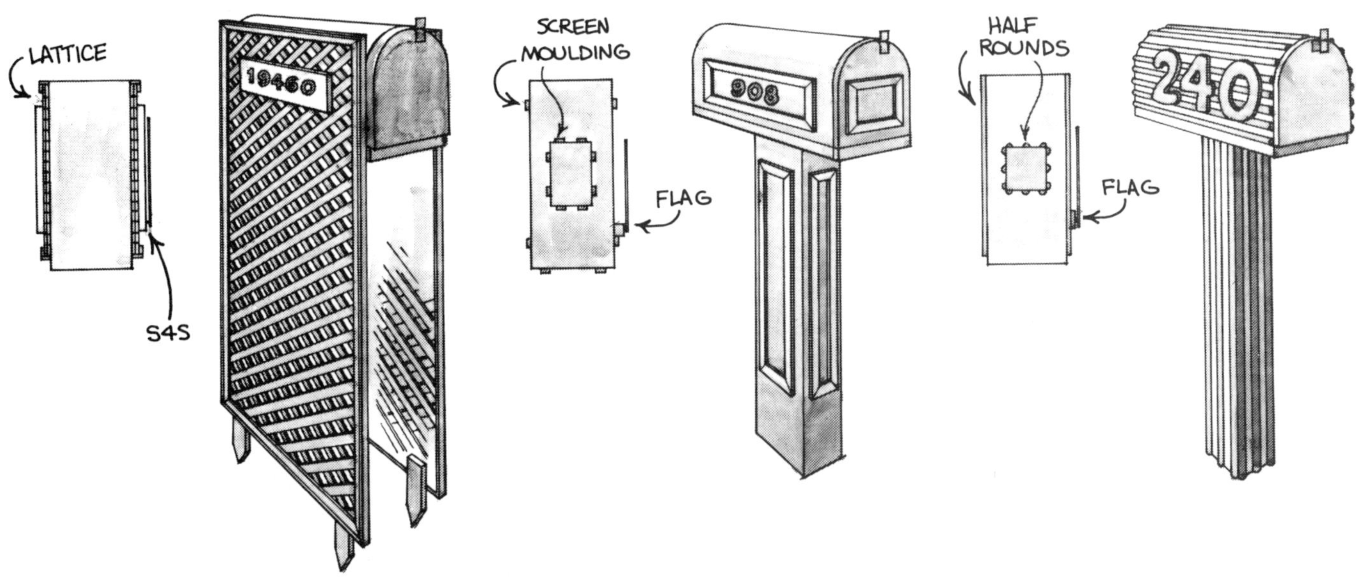

MEN'S DRESSER-TOP VALETS

A little organization can really make life easier, especially in the wee sleepy hours of the morning when you're fumbling around for all your pocket articles. Simplify that task with a man's dresser-top valet made with moulding. The first one shown here is just three large coves on a plywood base, framed with quarter round. The next is made of four drip caps, two laying down and two standing up on two strips of screen moulding. The last valet is on a plywood base with an inner framework of cove and an outer framework of stop moulding. The inside is divided into sections with back to back coves. Some coping is required here, so now's your chance to get in that coping practice you've been looking forward to.

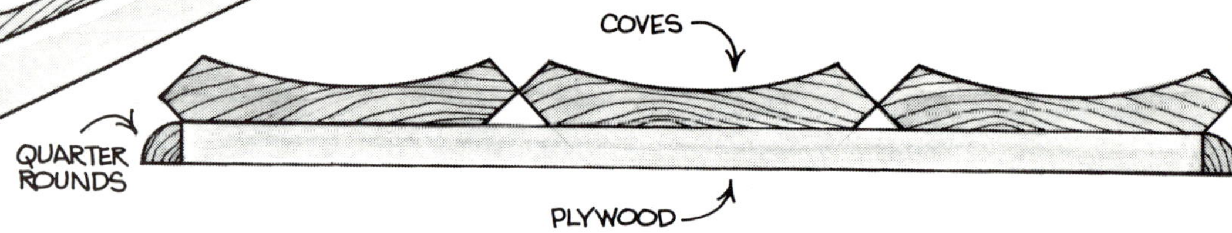

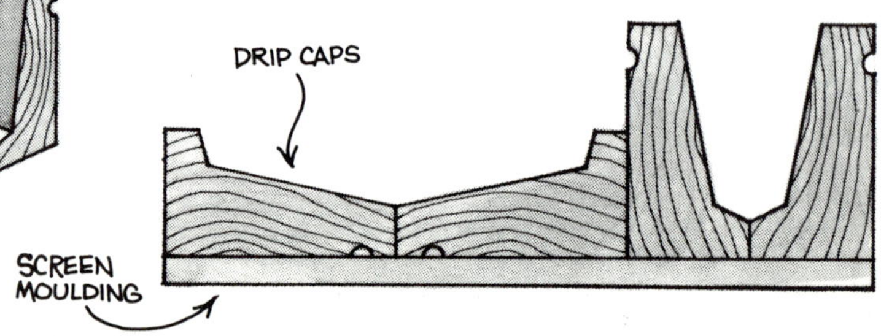

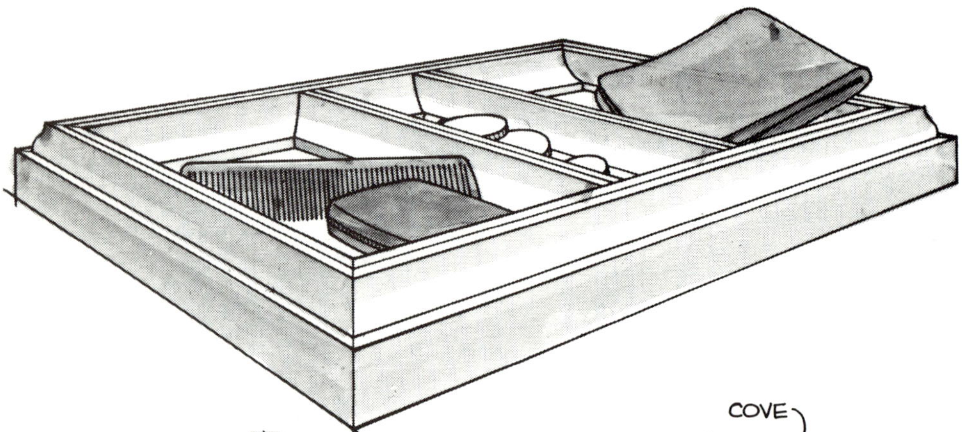

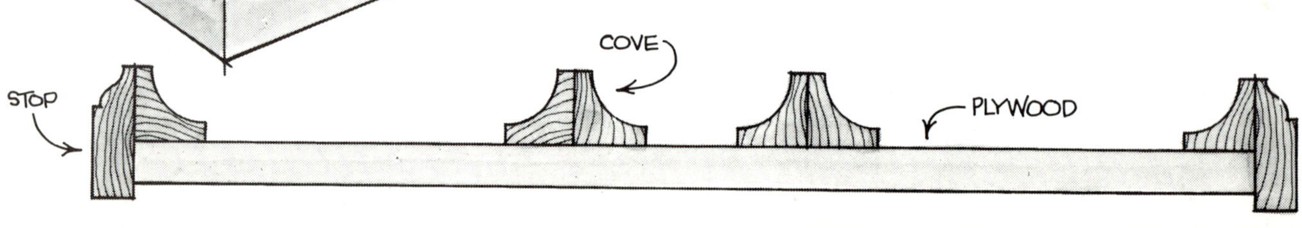

NAPKIN HOLDERS

Easy to make and practical to use are two reasons you should give this project a try. Actually, each napkin holder takes minutes to make. The first has S4S ends and base, with strips of screen on the sides and coves along the bottom. If you think that's simple, try the next one. It's just a piece of round edge with dowels. The other two styles are pieces of base moulding attached to S4S bottoms. One also has finger notches coped out. Stain or paint them an appropriate color and you'll have a cheery addition to your tabletop.

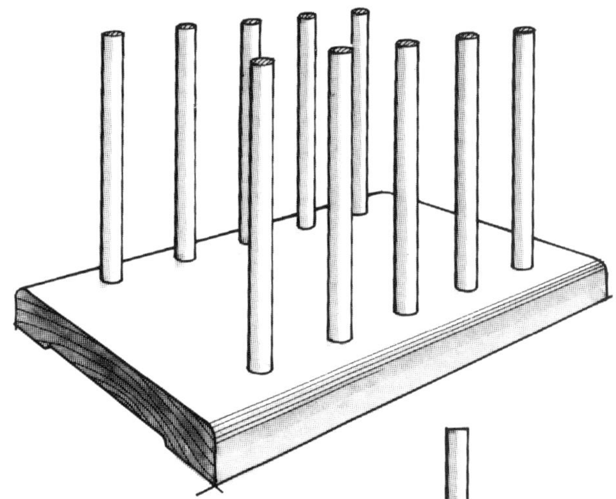

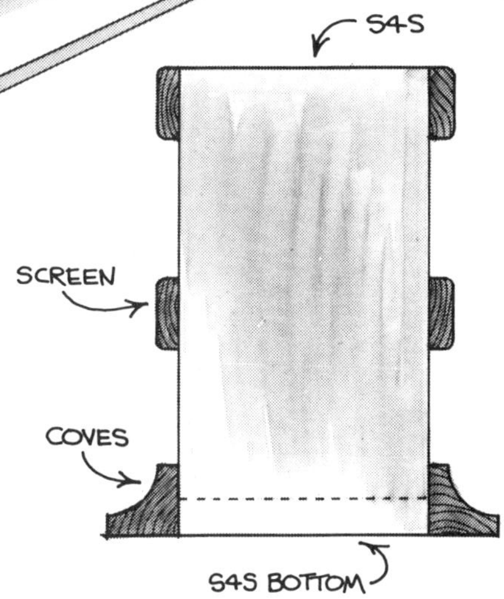

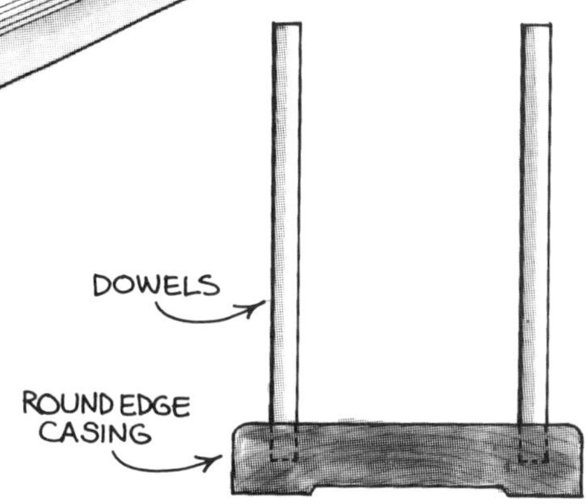

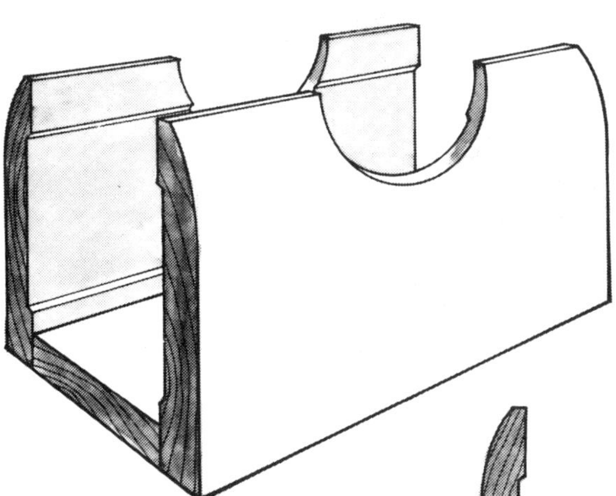

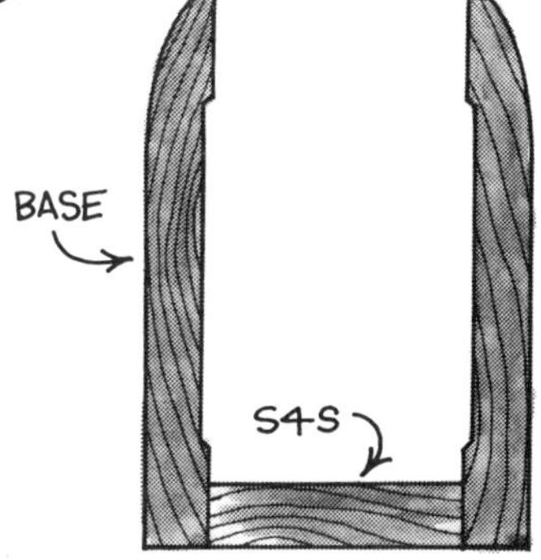

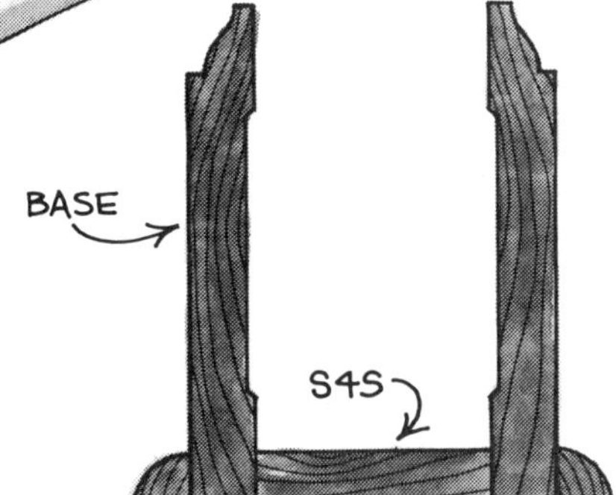

NAPKIN RINGS

Hand made napkin rings may not end up as the main topic of conversation at your dinner table, but they will set your guests to talking. The uniqueness of wood moulding is easy enough to add to your table. Choose the design shown here that you prefer, or all of them if you wish, and start cutting. A napkin ring need not be very long. Anywhere from ½″ to 2″ is sufficient. After the pieces have been cut to size you can glue them together. It's best to set them on end when gluing. Either clamp them in place or gently stretch a rubber band around them for drying. Keep them in place overnight. Then just sand off the rough spots and they're ready for staining or painting and impressing your dinner guests.

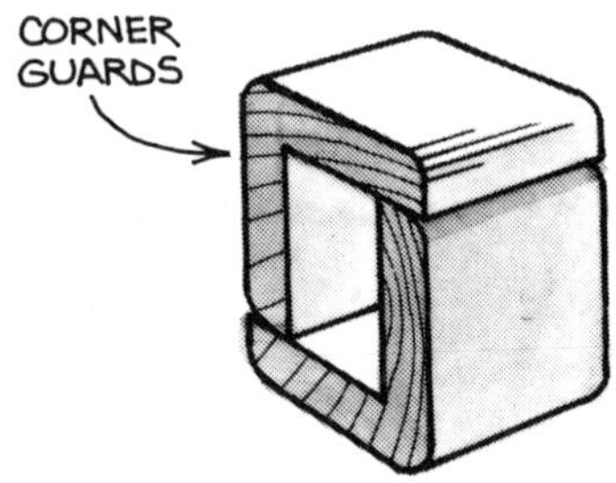

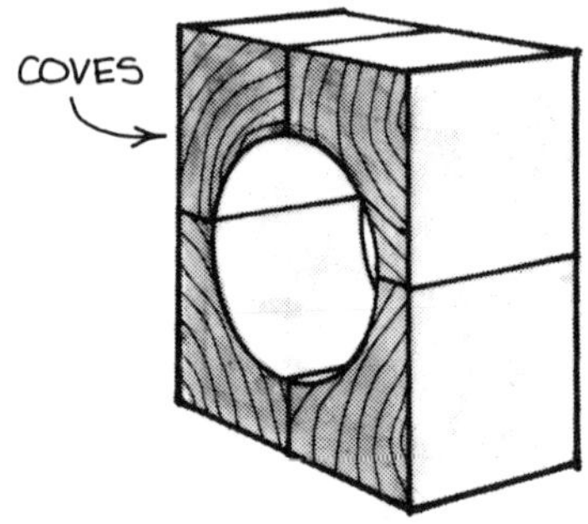

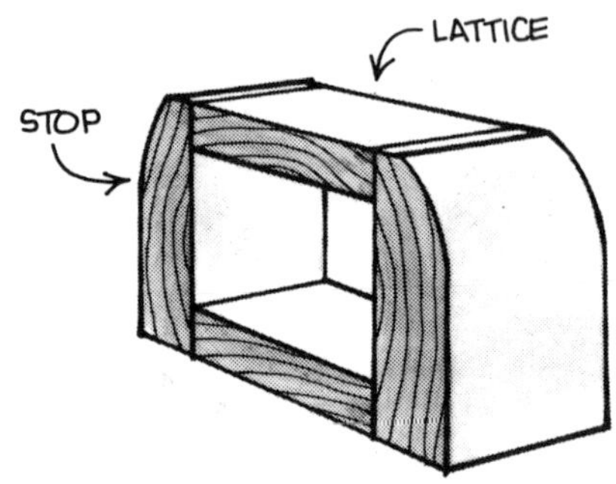

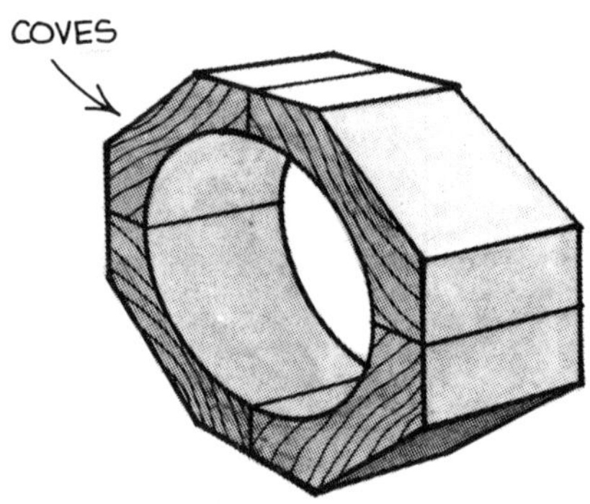

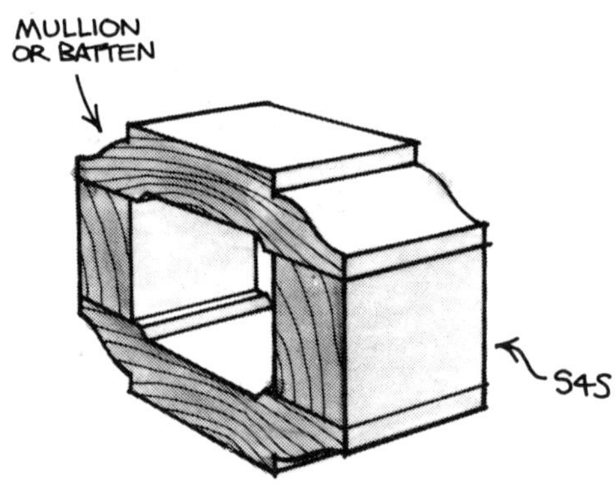

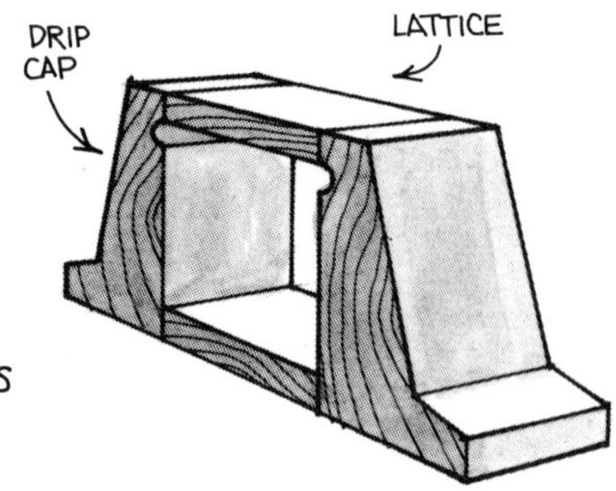

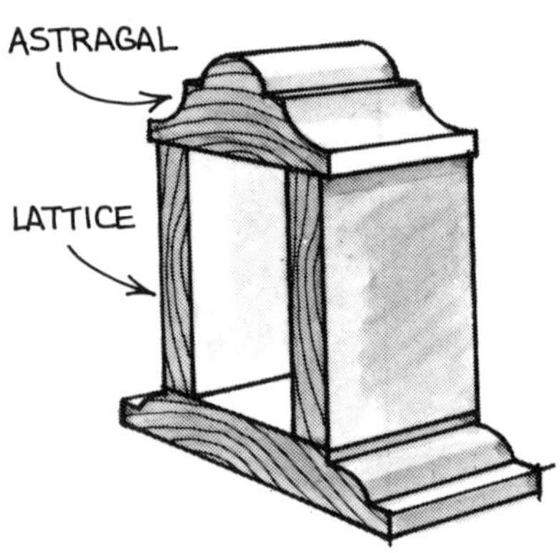

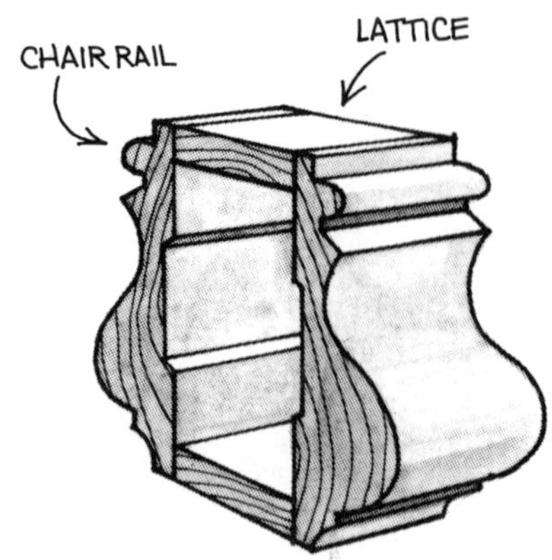

NECKLACES/NECKLACE RACK

Necklaces of wood? Why not? They've been made out of everything else. Wood that has been stained or oiled to a high polish can make a handsome necklace, especially considering the large variety of moulding patterns available. The styles shown here are just a random sampling. Choose the type you want, cut to size, carefully drill the holes, sand and stain. A metal neck band can be economically purchased at your neighborhood hobby or craft shop. And there you have it, a one-of-a-kind necklace of wood moulding.

The handy necklace rack illustrated below is simply a sheet of 3⁄8″ plywood cut to the size you need, depending on how extensive the jewelry collection is. Drill holes for 1⁄2″ dowels spaced to hold whatever type of necklace necessary. Frame the whole board with either cove, stop, wainscot or ply cap moulding. Tack picture brackets to the back for hanging.

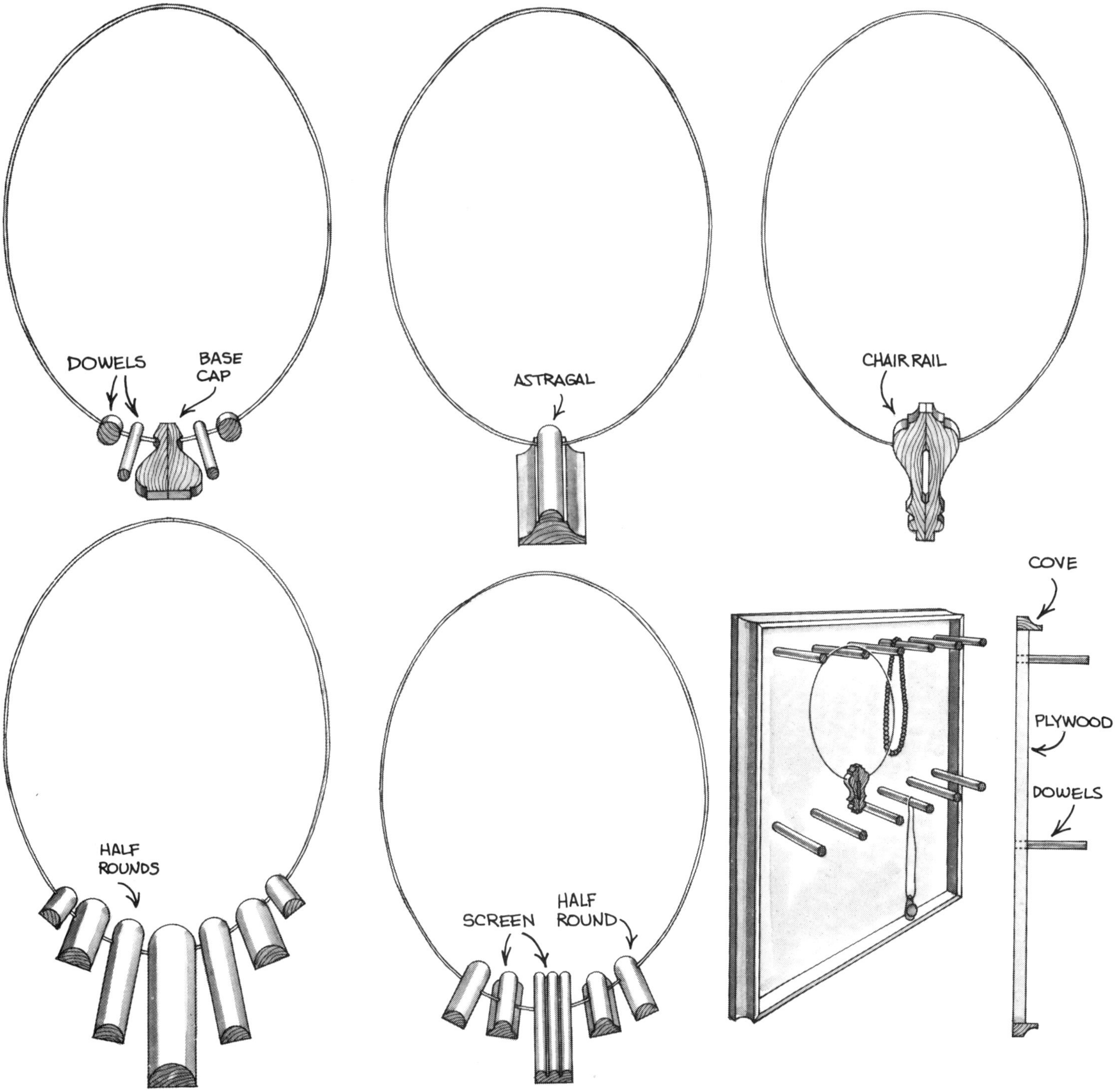

OPTIONAL PATTERNS :

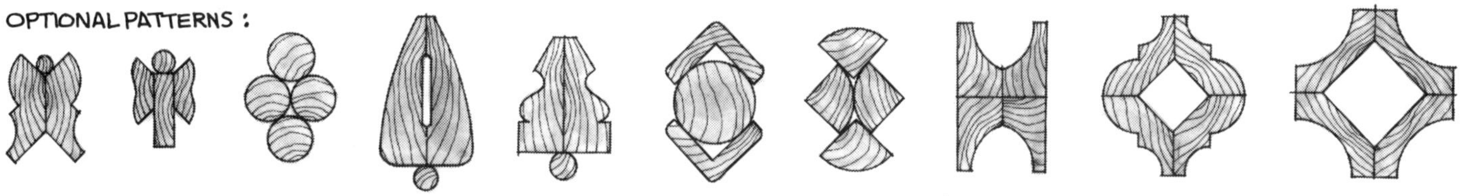

NOTE PAD HOLDERS

Looking for something to write on can be a nuisance at times. Eliminate that annoyance by having one of these handy note pad holders by your desk or telephone. These first two containers can both be built in less than an hour. The first is just a box with sides of casing and a front and back of 3/8″ lumber with coped half circles. The second holder has S4S or 3/8″ lumber for the sides. The backboard is also lumber. Two casings butted against each other with a screen moulding cap make up the front. Round off the corners of the backboard and cope the sides to curve upward. If these two note pad holders are to be hung on the wall, drill holes in the back piece.

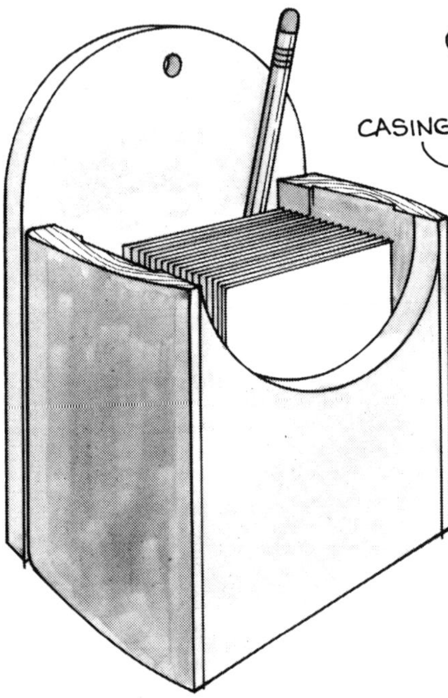

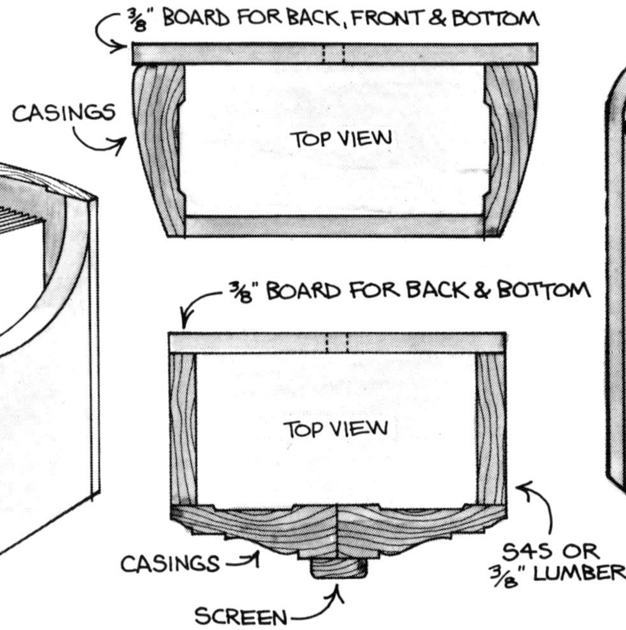

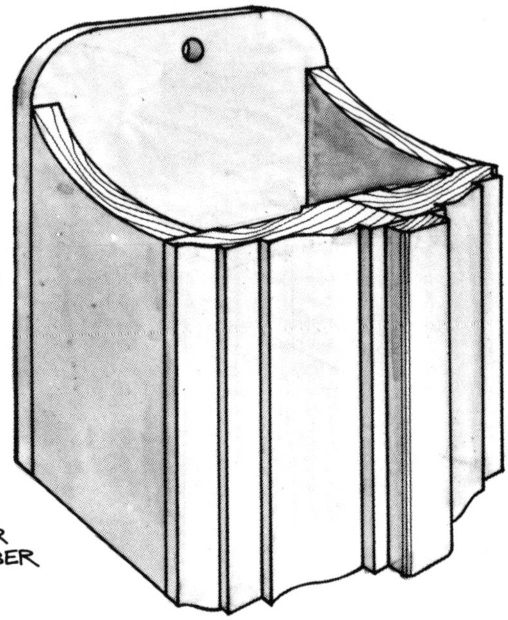

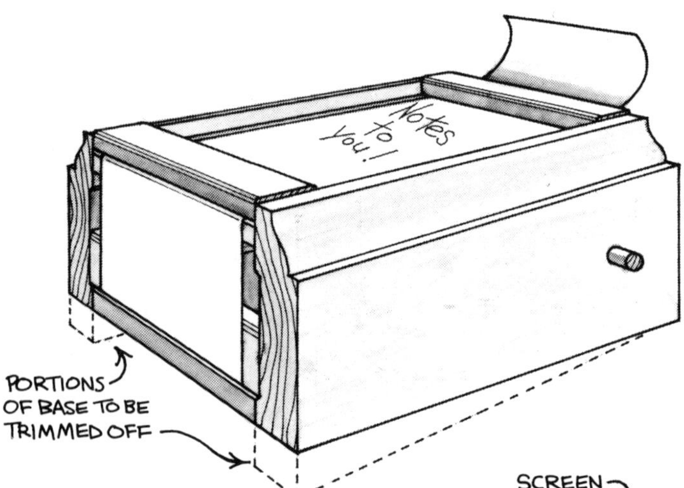

The third model shown is a little more complex. Instead of using a note pad or cards, this holder uses a roll of adding machine paper. The sides of the box are pieces of base moulding with a bottom section diagonally trimmed off so the paper holder will slope. The two sides are connected by a top and bottom made of lattice or S4S stock. Two pieces of screen moulding are connected slightly above the top piece to act as paper guides. Another guide is placed at the lower end slightly above the floor of the box. Drill holes in the sides to insert the dowel that holds the paper roll. The dowel should be slightly smaller than the holes so it will slip out easily in order to insert the roll. The paper is then fed through the space between the floor and the lower end guide, up the outer end and through both guides on top. The far upper guide also acts as a tear off tab. When all the pieces are cut to size, glue and nail together. Stain to match your desk or paint to match your telephone or kitchen.

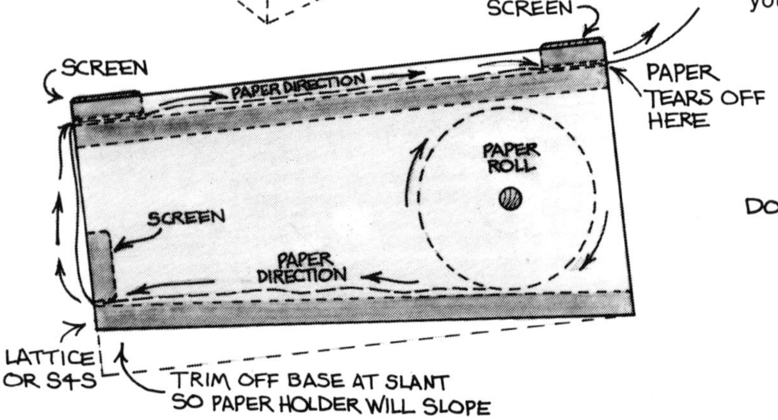

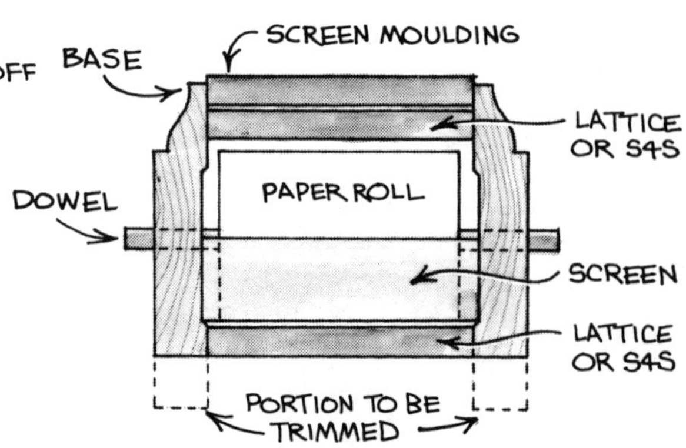

PAPER TOWEL HOLDER

Unfortunately most paper towel holders have that stamped-out plastic look. That's because they're usually stamped out of plastic. If you prefer a towel holder with a rich traditional look, build one out of wood mouldings. The first model illustrated below utilizes 3/8″ lumber for the sides and back. The front is made of two traditional stops capped by a flat astragal. One stop is also set below the towel holder. Drill holes for the round that holds the paper roll. The round should slip out easily for loading. If you prefer a more contemporary paper towel holder, use ranch stops in place of the traditional stops and screen moulding in place of the astragal. Before cutting, measure a full roll of paper towels so your dimensions are correct.

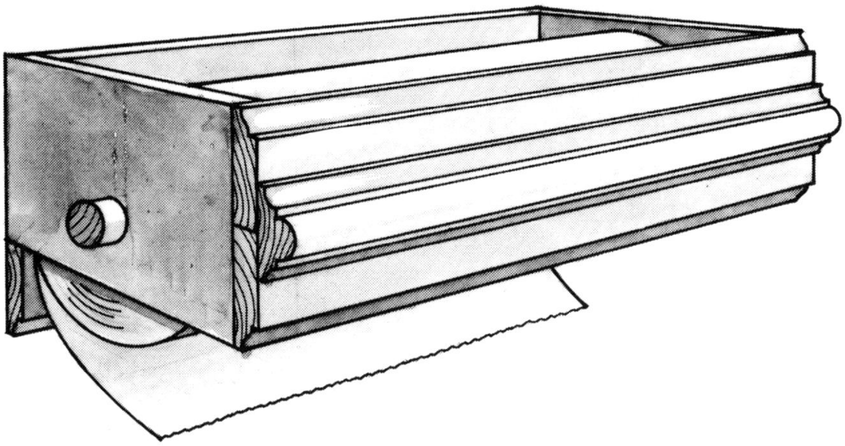

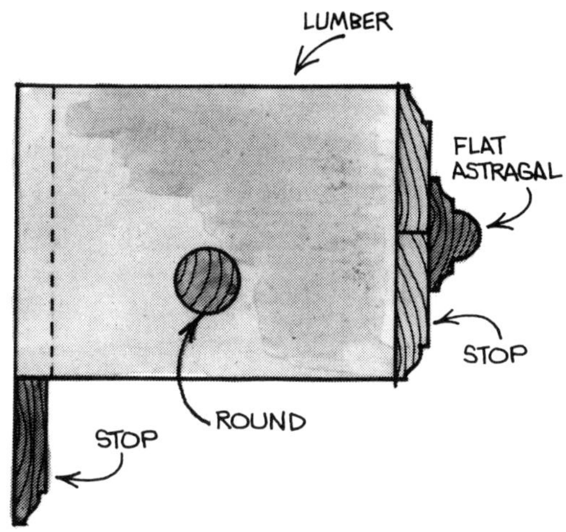

The sides of this paper towel holder are made of base which have been cut at 45° angles upward to form the ends of the shelf. The lower corner has been coped round to match the paper towels. Shelf and backboard are both 3/8″ lumber. Drill the holes in the same manner as previously mentioned.

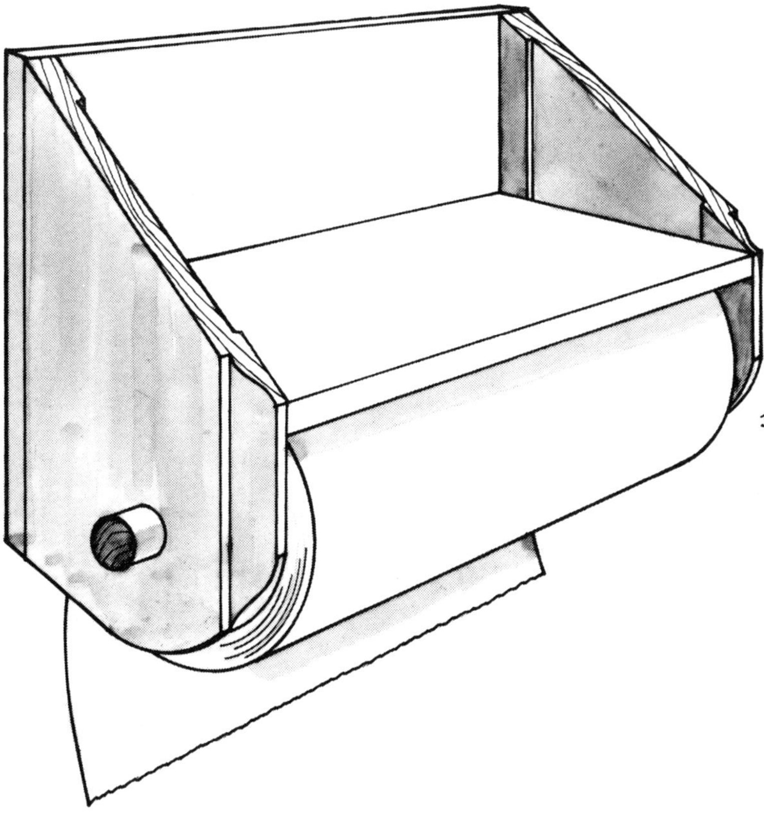

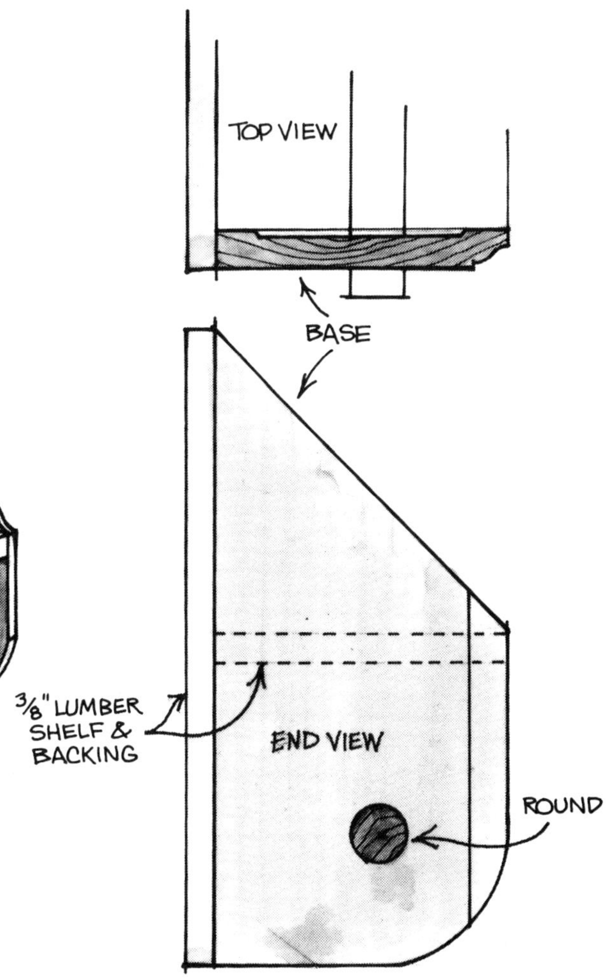

PEDESTALS

A pedestal for your favorite sculpture, plant or vase is easy to accomplish, as are most of the projects in this book. The pedestal on the left is merely a tall box made of ⅜″ lumber sides and bottom. At the lower end, the pedestal is framed with S4S which has quarter round on top and cove moulding below. The corners are mitered, so measuring must be accurate. S4S and cove are also used at the top of the pedestal, with the addition of crown moulding. The top platform itself can be made from a square of lumber, plastic laminate, tile or even a square of marble. The size of the pedestal will, of course, depend on what you want to display and where it's to be used.

The second pedestal shown is a variation of the first, except the bottom area uses the S4S and coves twice. On top, the S4S is used three times and the cove twice. The top platform in this case is inset. Again, it can be made of lumber, laminate, marble or tile. If tile is used, it would set on a base of lumber or plywood. The pedestals shown here are but two suggestions. For other variations, try other moulding combinations. The ideas on page 15 might also be useful.

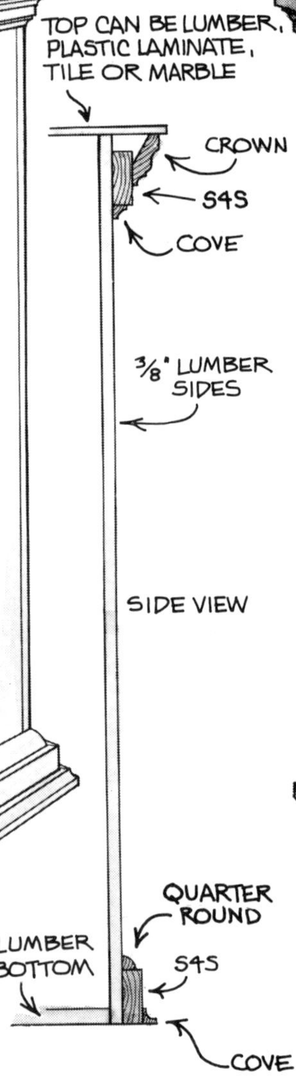

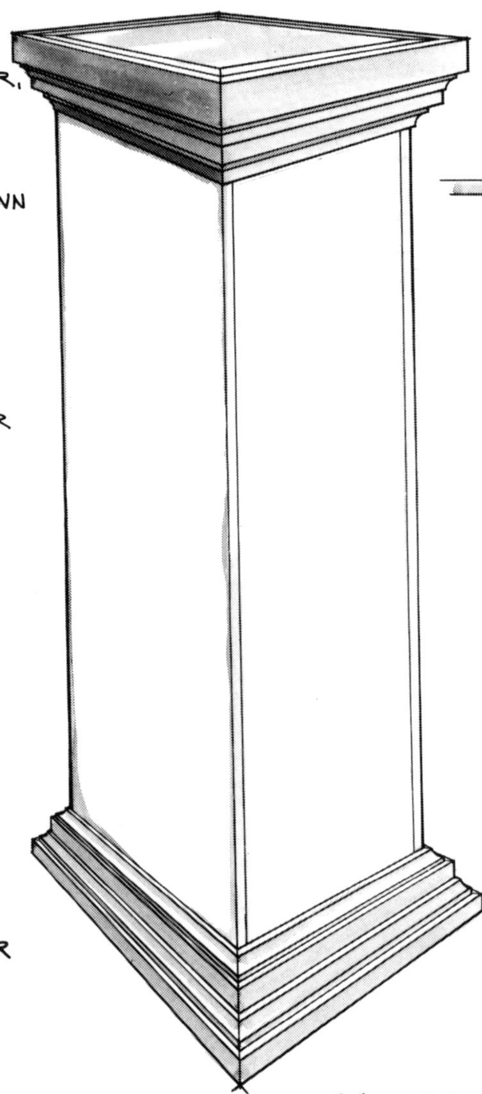

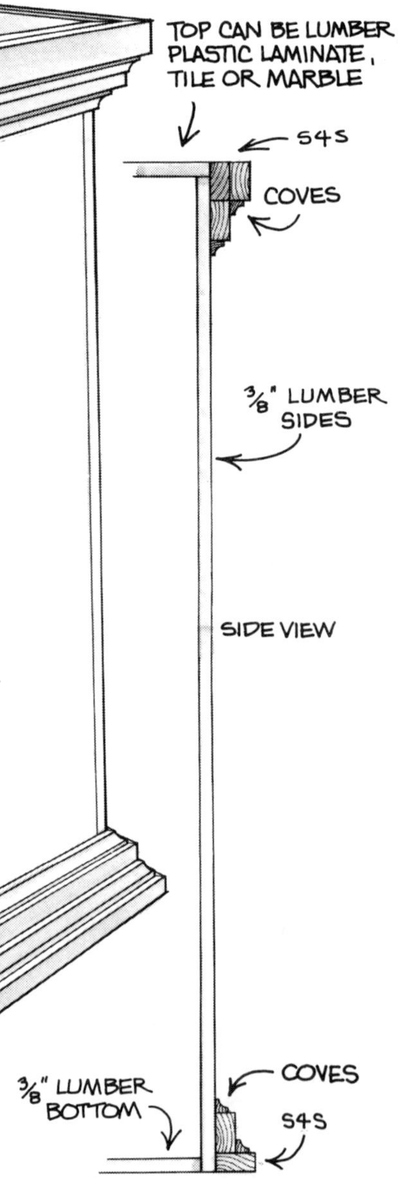

PEN & PENCIL HOLDERS

Pen and pencils are useless if you can't find one when you need it, but keeping them in an organized location is simple when you have a holder made out of wood moulding. Each of the holders shown on this page are so simple they can be put together in a matter of minutes.

The first is merely a drip cap attached to a piece of base moulding with holes drilled to the appropriate size and distance apart. Its length depends on your needs. The next holder is even simpler, a single piece of hand rail cut to the size you want. The third idea is a group of random length rounds with drilled center holes. When gluing these together it's a good idea to either tape them or put a large rubber band around them 'til the glue has dried overnight.

The two styles of upright pencil holders shown are just as simple to build. One is merely four pieces of batten glued and nailed to each other with a bottom of S4S or lumber added. Be careful when nailing this one. There's not much area to work with, so use small brads. The last pencil holder is made up of two coves attached to two S4S sides and an S4S bottom. If coves are unavailable in your area, crown will work as well. You could also use casing, chair rails, base, batten or mullion.

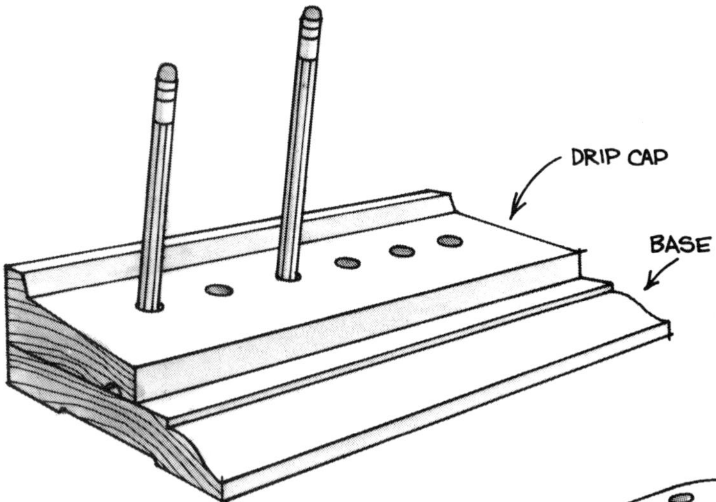

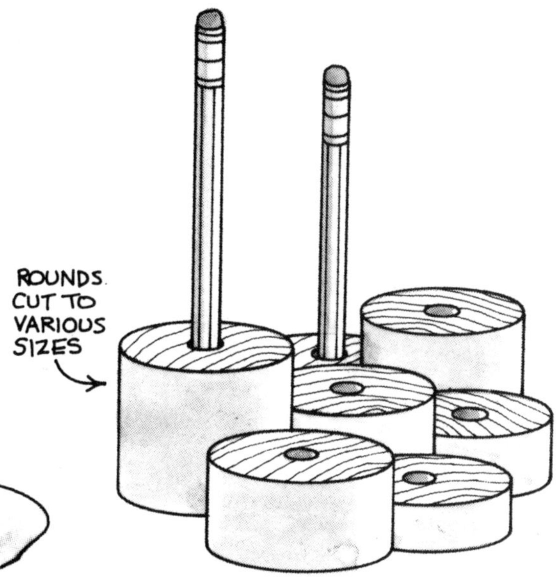

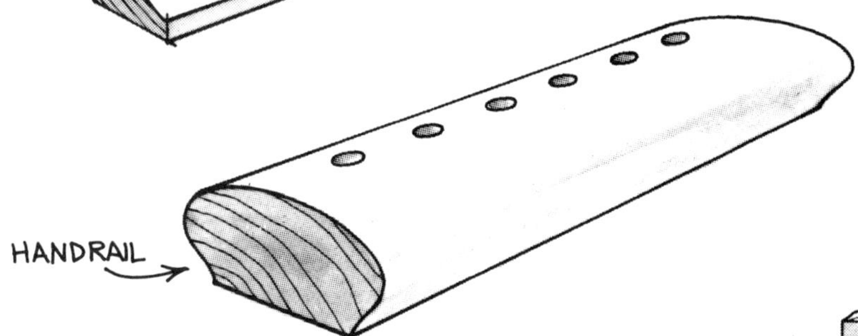

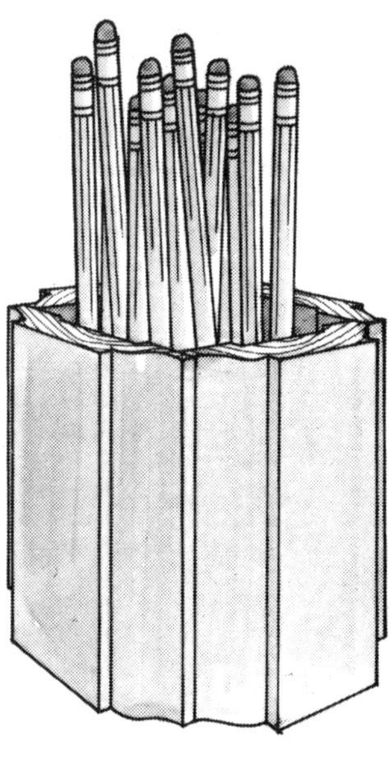

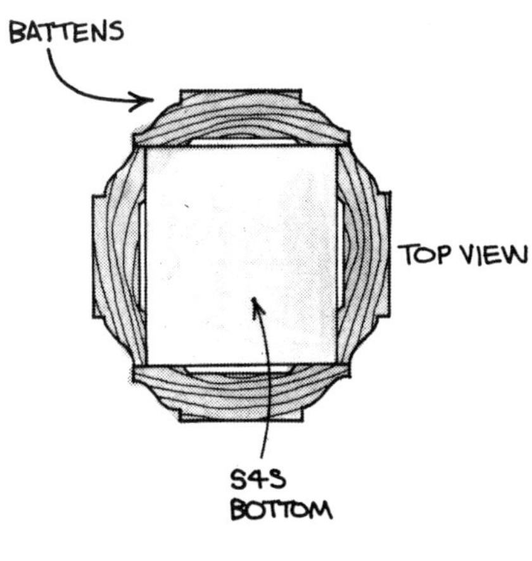

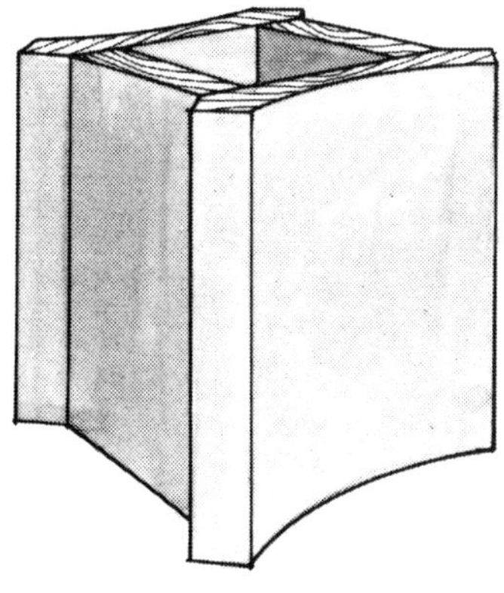

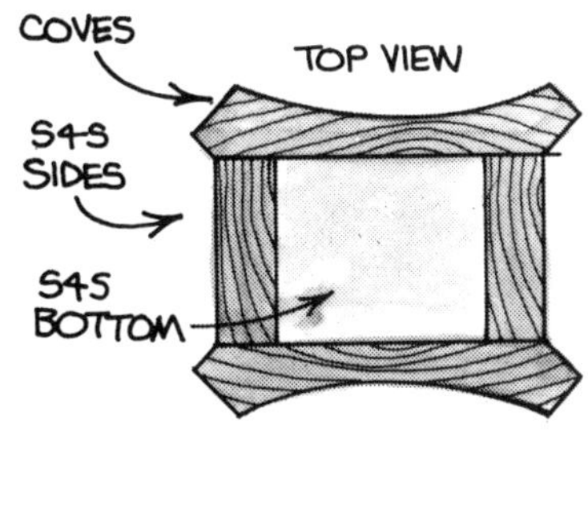

PENCIL BOX

Remember when all the kids used to take a pencil box to school? Well, they still can, or if you feel this box is a little too classy for school classes, maybe you can use it at home to hold a variety of treasures. Pencils and crayons are, of course, a natural for the kids, but for the grownups, it's an ideal little holder for stamps, buttons, sewing spools, bills or maybe even old love letters. To make it, cut the top and bottom from ¼″ hardboard, plywood or lumber. All four walls are made of lattice. Base cap moulding is used on the sides over the lattice walls. Corner guards are then attached above the base caps with enough space to allow the lid to slide back and forth. If desired, you can drill a finger hole in the lid. In case base caps are not available at your local lumber yard, you could also use cove, picture moulding, screen, stops, mullion and shingle moulding. If this project ends up being a pencil box for the kids, paint it a bright happy color or maybe make each piece of moulding a different color. If you prefer it a little more subdued, stain it to show off the beauty of the wood.

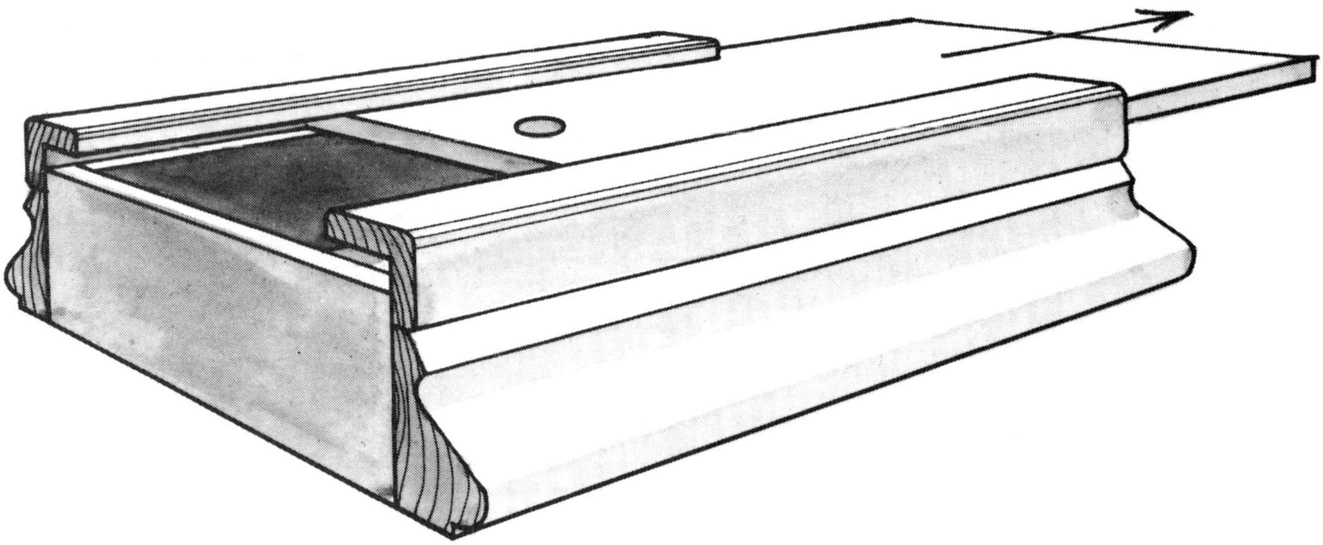

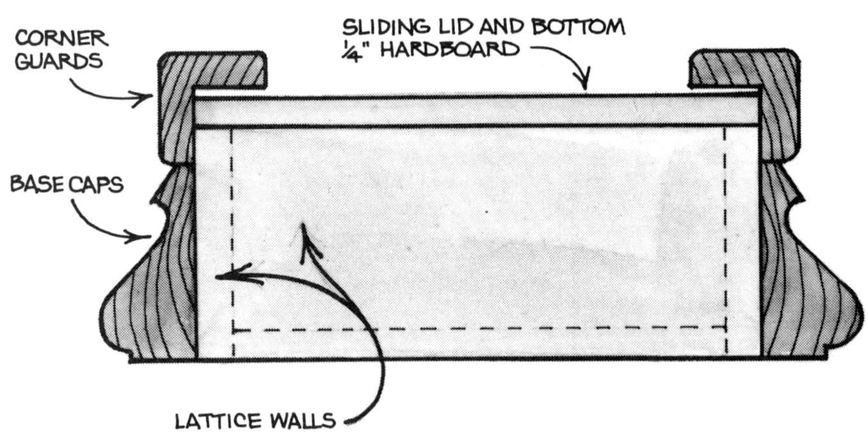

PHOTO DISPLAY WALLS

Here are two ways to show off your favorite photos and add drama to your wall at the same time. The first display is a group of vertical strips of large cove moulding running from floor to ceiling. Between the pieces of cove, quarter round is nailed to the wall to form a shelf on which the photos sit. Each end of the quarter round is mitered to fit behind the lip of the cove. The mats of the photos should be slightly larger than the spaces between the coves so they won't fall out when in place. When inserting the photos, just bend them slightly and they'll pop into position. The distance between the coves will be determined by the size of your matted photos.

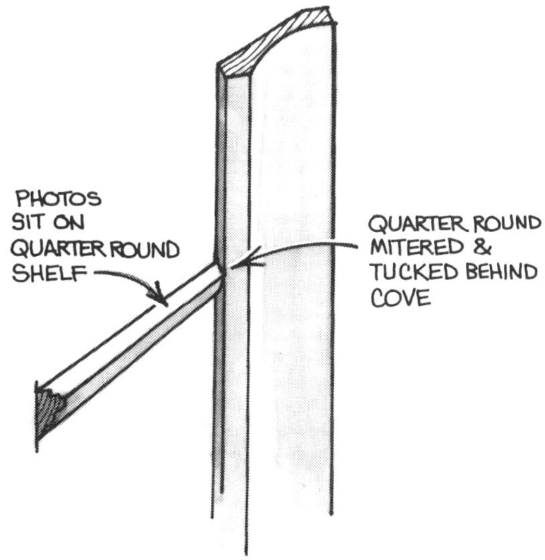

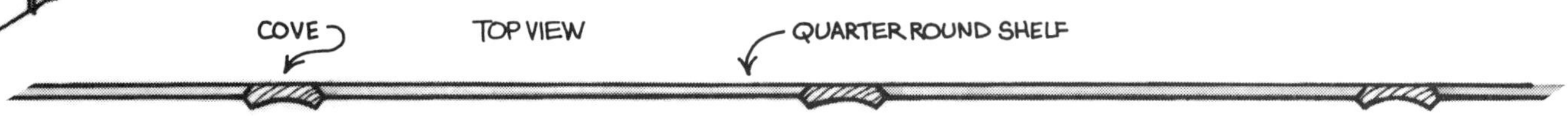

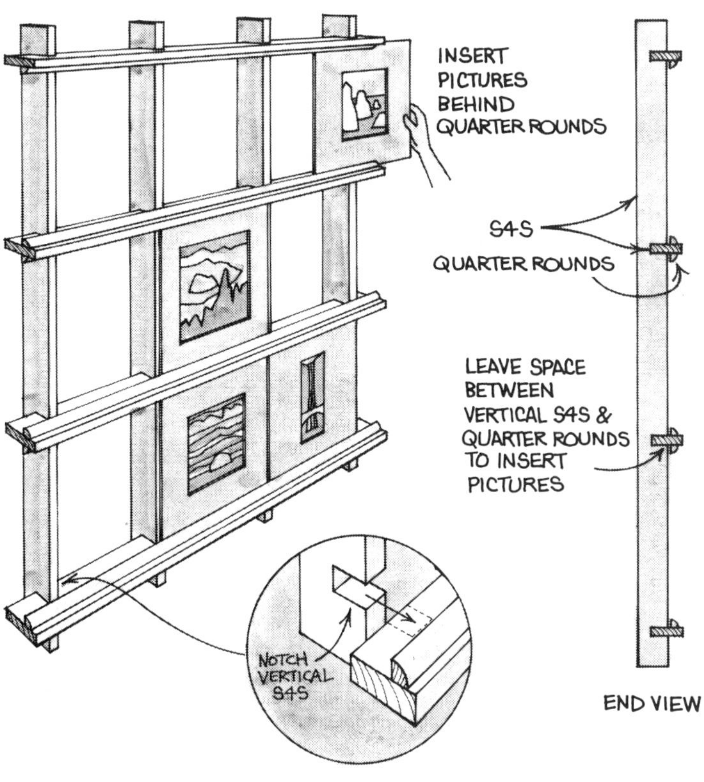

The second photo display rack hangs on the wall, takes up a smaller space and is portable. The uprights and the shelves are all made of S4S stock. The size and spacing will depend on how big of a display you want. The shelves fit into notches cut in the vertical pieces of S4S. Attach them with either nails or screws from the back. To hold the matted photos in place, pieces of quarter round are attached to the shelves, top and bottom, as shown. Make sure enough space is left between the verticals and the quarter round to insert the pictures. Use picture brackets on the back of the verticals for hanging on the wall. For other dramatic ideas to dress up your walls, see WALLS, starting on page 118.

PICTURE FRAMES (Mirrors, Etc.)

Picture frames have two purposes: to present and to protect the picture. The presentation is achieved by the size, shape and color of the frame. The protection is given by the cover and security the frames give the picture, along with the mat and glass, if any. When making your own frames, remember the design of the moulding you choose should relate to the picture, whether it's traditional or contemporary. Now turn your creative talents loose and enjoy the excitement of designing your own frames. The variety of moulding pattern combinations shown below are but a few of the many possibilities. One of these or one of your own design will be just right for your picture.

The first step in making the frame is to combine the different mouldings so the finished profile is completely put together. Do this before any cutting is done. Start by adding the smaller pieces to the larger ones. Use glue and brads. Careful nailing is required here to prevent cracking. Clamp or tape together overnight. Next, saw off the end of the moulding at a 45° angle. Carefully measure the longest side of the picture, add ⅛″ to this measurement to allow your picture to fit into the frame easily. The long side of the frame should be cut first. By doing this, if a mistake is made on the long piece, it could be used for a short piece. Transfer this measurement to the moulding starting from the edge of the miter at a point where the picture inserts into the frame, (the rabbet), mark and cut a 45° miter at the opposite end of the piece just cut. You will need another piece exactly like this one for the opposite side. Measure the second piece from the first so that the length will be identical. When cutting the 45° miter on the ends of the pieces, it is wise to try to clamp the pieces to the miter box so

PICTURE FRAMES (Mirrors, Etc.)

a nice cut can be made. Now measure the short sides of the picture and cut two pieces in the same way as before. You are now ready to assemble your frame. Take one side piece and the bottom piece and coat the ends to be joined with glue, place carefully in your corner clamp. When the corner is aligned just right, tighten the clamp so as to exert sufficient pressure to hold the pieces rigidly enough so that you càn nail the corner together. Now drive two or more brads through the corner from each side allowing the heads of the brads to protrude slightly. At this time, make certain that the frame fits the picture. (Here is where four corner clamps come in handy.) Follow this same procedure on the other three corners and allow the glue to dry thoroughly. Check for squareness by measuring diagonally across the frame each way. The two measurements should be equal. If not, a slight push on the long diagonal corner will square up your frame. Tack a strip of wood on the back diagonally from corner to corner to hold the frame square while the glue is drying. Carefully finish driving the brads with a small nail set so that the heads are about 1/16″ below the surface. Fill the nail holes with a little wood filler and let dry. After the glued-up frame is completely dry, sand it lightly with fine sandpaper. You are now ready for staining or painting in whatever manner you choose. See page 5 for hints on finishing. The designs shown here, by the way, also make excellent frames for mirrors.

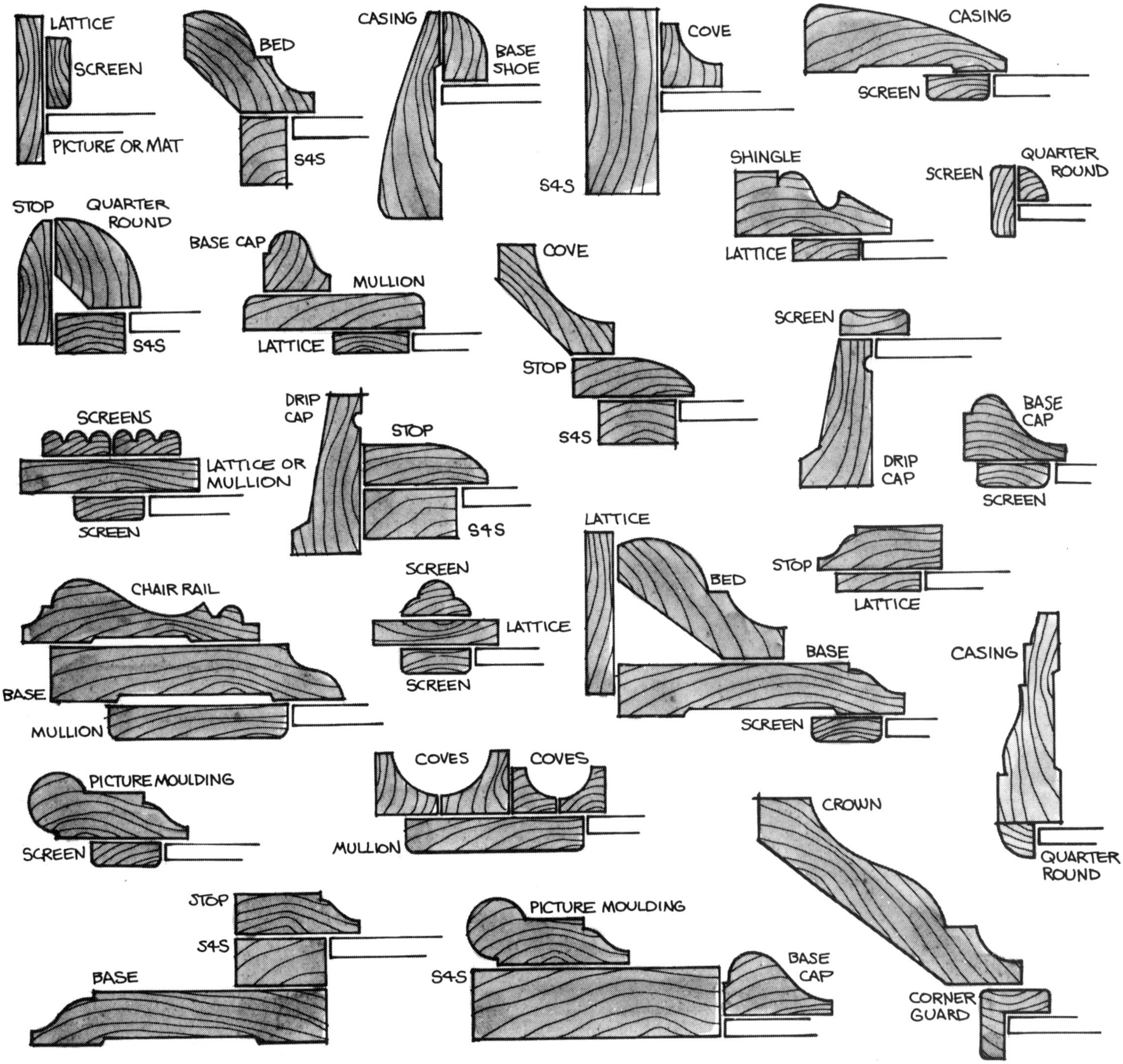

PICTURE WINDOW

Liven up that bare wall space with a window of beautiful pictures. Use either photos clipped from a scenic calendar or use family photos. Simply glue them in place on a sheet of 1/4″ plywood or hardboard, then frame it with base caps. If that particular pattern isn't available in your area you can use coves, beds, crowns, picture moulding, ply caps, shingle or drip caps. Miter the corners at 45° angles. Use screen moulding as dividers to separate the pictures. Stain the moulding before putting it together to keep from getting stain on the photos. Use picture fasteners for hanging.

PIPE STANDS

The rugged good looks of a pipe deserves to be displayed on a rack that shows the warmth of wood moulding. That pipe smoker in the family will be pleased to show off his collection on one of these handsome racks.
The first one illustrated is made with two pieces of base, back to back, attached to a bottom board of 3/8″ lumber.

Holes are drilled through the base moulding to insert 3/8″ dowels to hold the pipes in place. The dowels extend to both sides of the rack so two rows of pipes may be held. The number of dowels and distance apart depends on the pipe collection. The leading edge of the bottom board is capped with base shoe to keep the pipes from slipping.

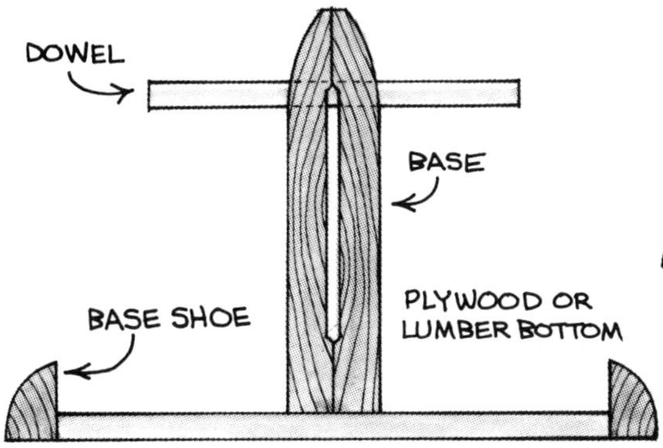

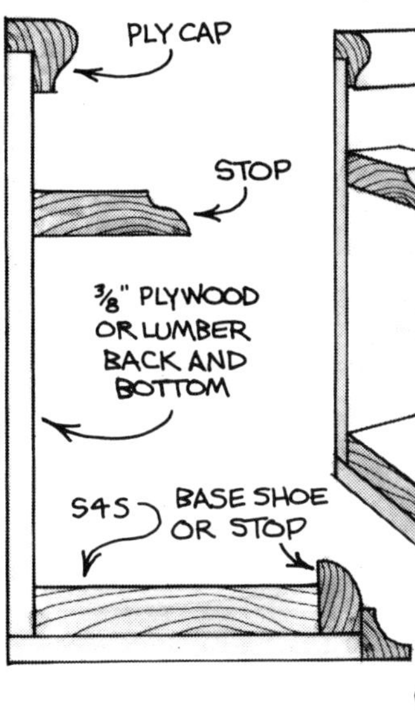

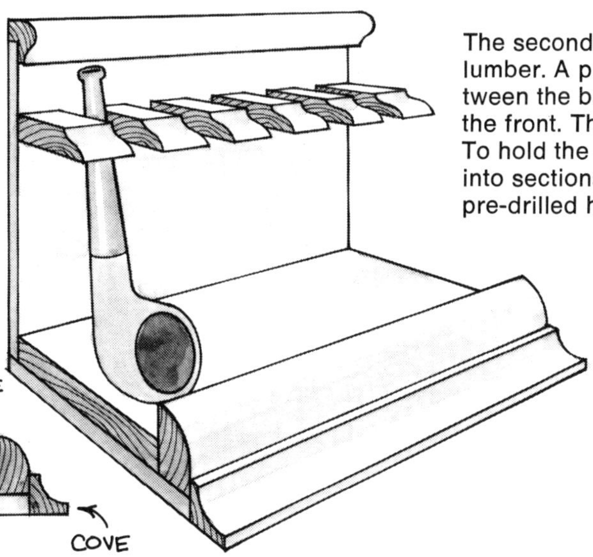

The second pipe stand has a backboard and bottom of 3/8″ lumber. A piece of S4S stock sits on the bottom piece between the back and base shoe or stop and a piece of cove in the front. The top of the backboard is capped with ply cap. To hold the pipes in place, a piece of stop moulding is cut into sections and attached from the back and screwed into pre-drilled holes.

The third design has a backboard of base moulding with a piece of base cap in front. The pipe holding section is just a piece of corner guard notched to hold the pipes. Care must be taken when cutting the holes to keep from splitting the corner guard. One of the easiest ways to do it is to first temporarily tack a piece larger then needed onto another piece of wood, then gently drill 3/4″ holes in the corner guard and cut away the excess wood with a sharp saw and trim to the size you need.

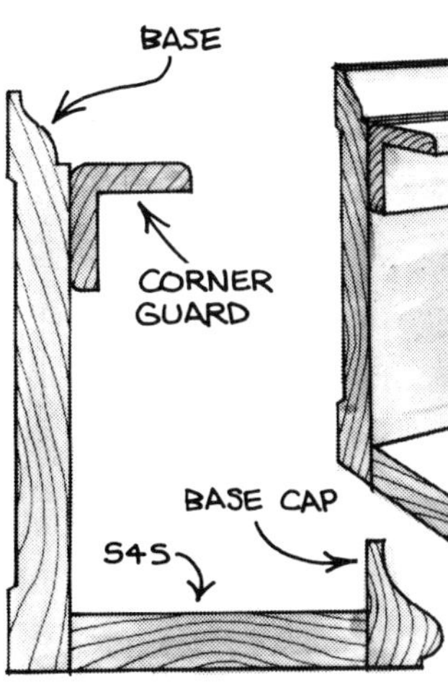

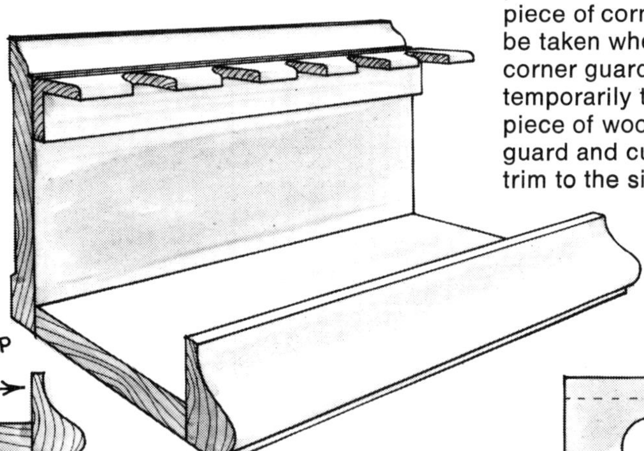

TOP VIEW OF CORNER GUARD MOLDING NOTCHED TO HOLD PIPE STEMS

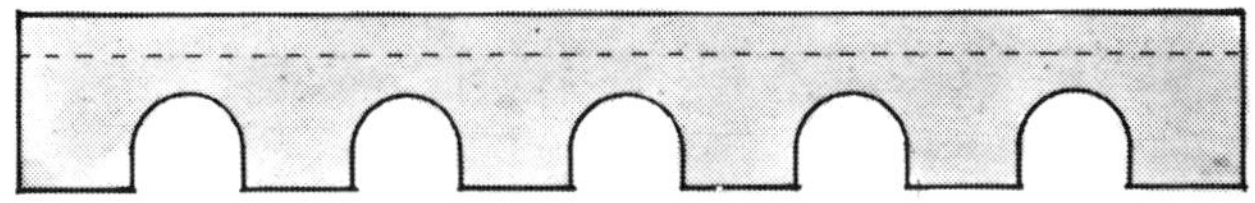

PLAQUES & AWARDS

Have you ever had to make a plaque or an award trophy for your school, company, club or organization? With wood moulding the job becomes a simple task. The first step is purchasing a walnut board for a backdrop. It can be purchased at a trophy shop and sometimes at a hobby shop or lumber yard. If you want it beveled, buy it already pre-beveled or do it yourself if you have the correct type of saw, router, planer or sander. If the plaque you choose is framed with moulding, then beveling is not needed. Most of the plaque designs shown are various moulding patterns simply attached to the background, either framing or outlining in someway, the metal plate that contains the engraved message. The moulding patterns mentioned here are just suggestions. Many other faces may be substituted. Use what you wish or what you find available in your area. Before applying the moulding to the backboard, stain or oil it, then hand rub it to a high gloss to complement the walnut background.

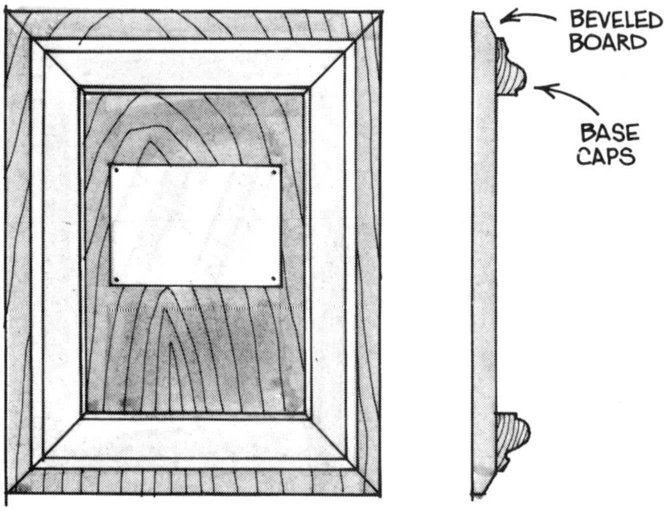

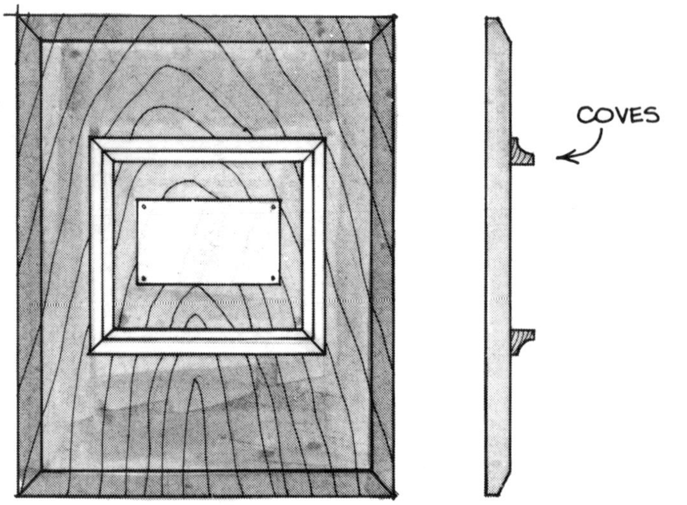

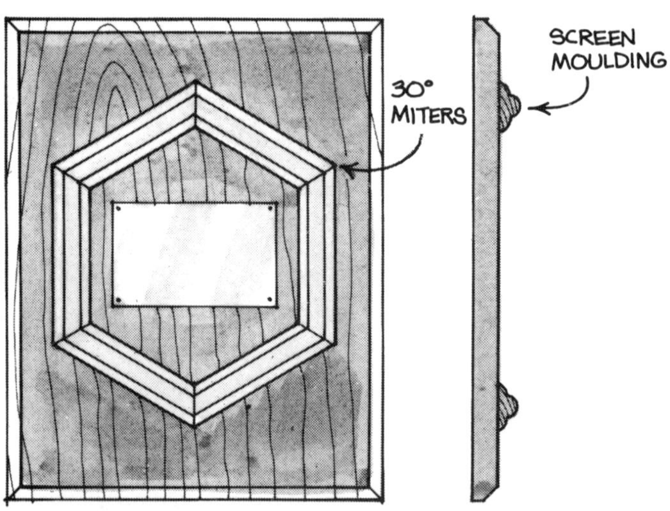

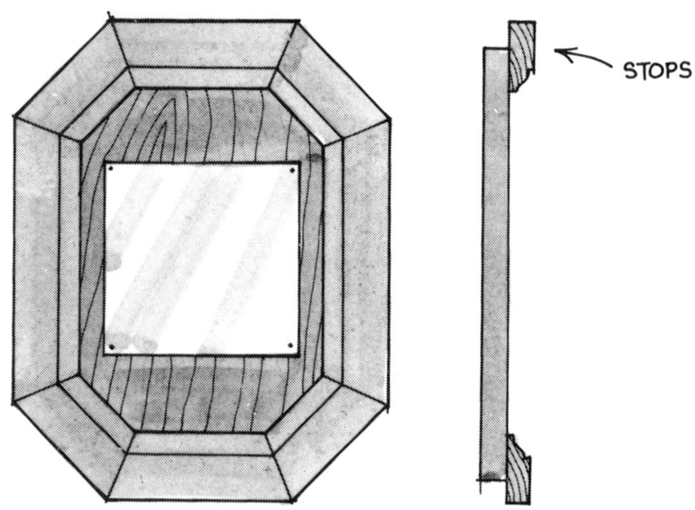

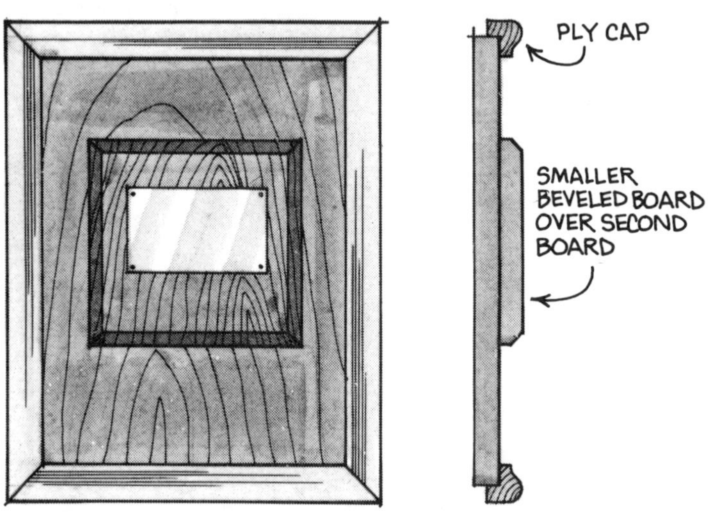

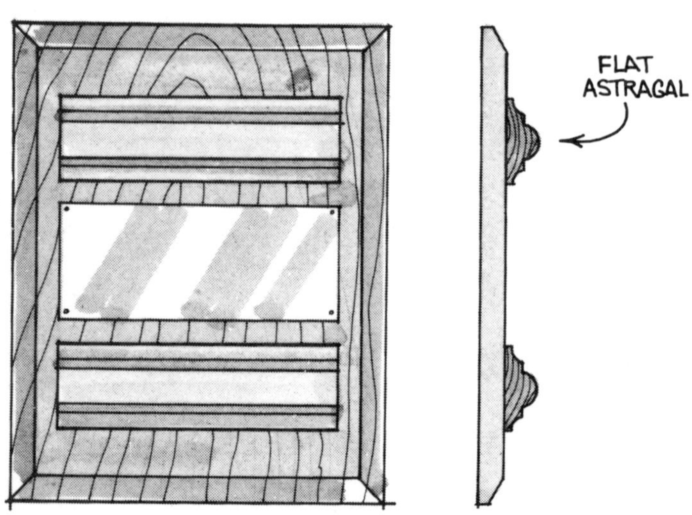

PLATE RAILS

A plate rail is merely a shelf with a lip attached to its edge to keep the plates from slipping off. They're easy to build, yet add a rich look of tradition to your kitchen or dining area. To construct the first one pictured here, start with a piece of ¾″ lumber and cut it to the length you wish. The width of the board need not be much, since it's just meant to hold plates, perhaps 4″ at the most. A piece of quarter round is attached to the top to hold the plates. The shelf itself sits on a piece of crown moulding with a 45° return miter, which merely means the moulding turns the corner and goes to the wall. This closes the moulding so the open end of the crown isn't exposed. The moulding ends of fireplace mantels are usually handled in this way. See FIREPLACES on pages 42 and 43. The shelf is hooked to the wall by wall brackets before the crown ends are attached.

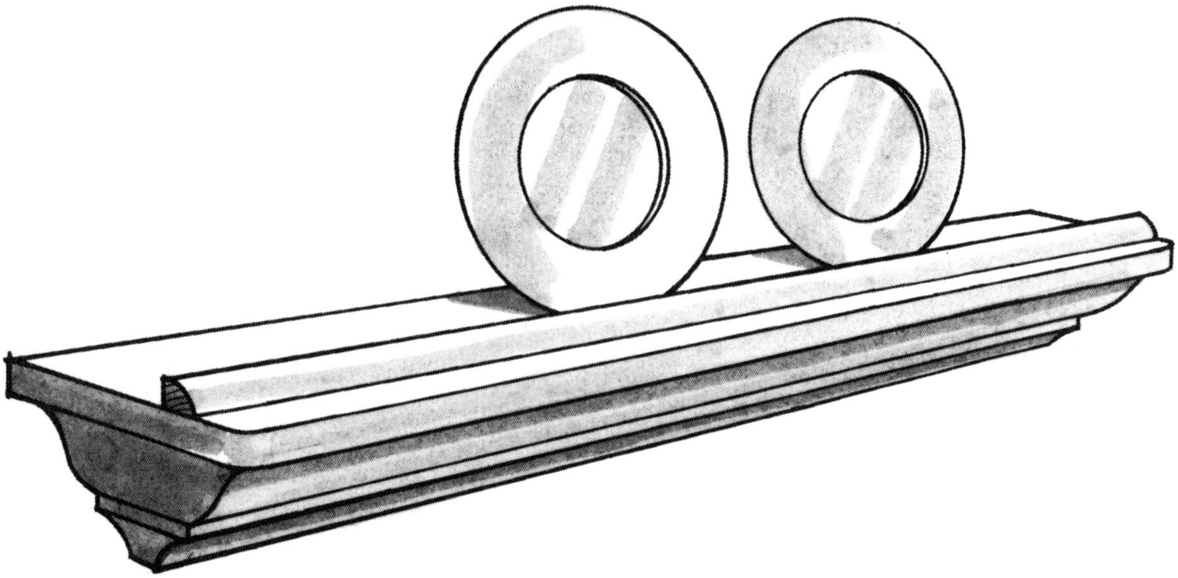

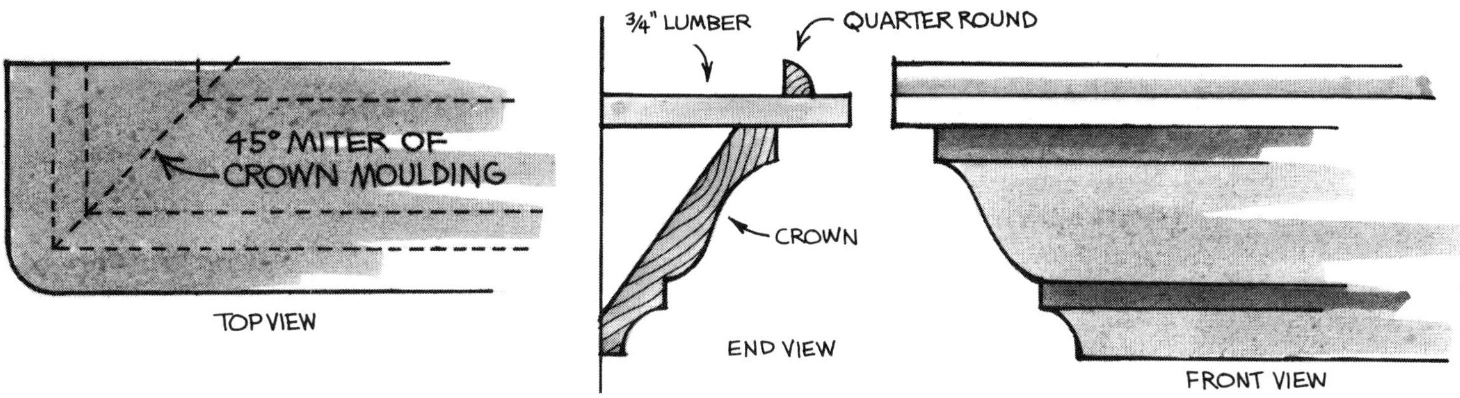

The second plate rail is designed to have its ends exposed, thereby making it an easier shelf to build. Both the backboard and the shelf are pieces of base with the top piece screwed into pre-drilled holes onto the lower piece. A base cap is then tucked in under the shelf for further support and for decoration. A strip of cove on top holds the plates in place. This shelf is attached to the wall with screws before the base cap is nailed into place. Matching stop moulding could also be used in place of the base cap.

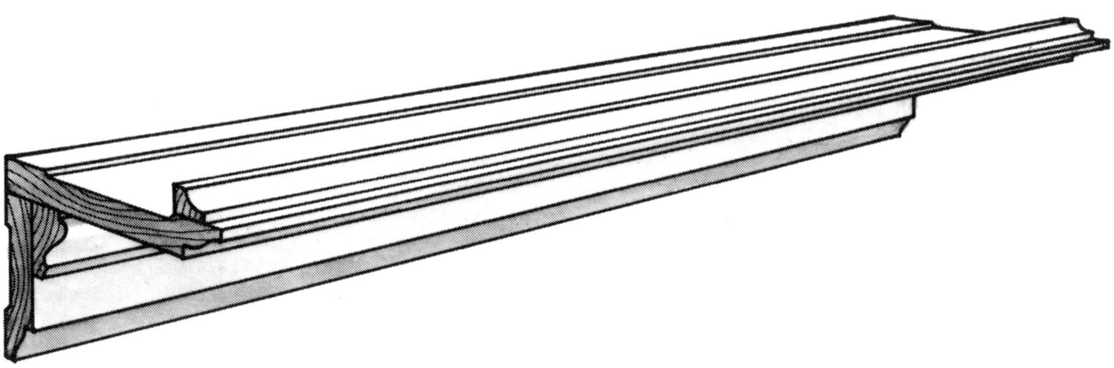

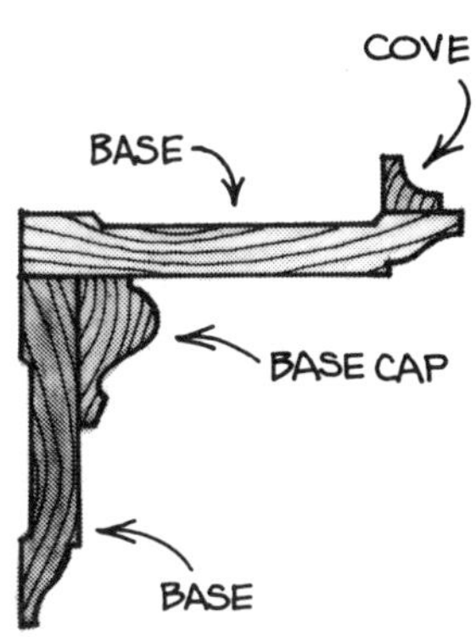

PLANTER BOXES

Just like the plants themselves, planters can come in hundreds of shapes and sizes. Although the planters illustrated here are merely a start, one of them might be just right for your favorite bit of greenery. Make them any size you wish, just keep in mind what's going to go in them.

The large container below is a box made from ¾″ exterior plywood with strips of lattice or screen moulding decorating the sides. S4S spacers are attached to the bottom, or if you want the planter to be more portable, connect casters to it. If it's to be used outdoors drill drainage holes in the bottom. Whether used indoors or out, this type of container makes a perfect home for a large plant or small tree.

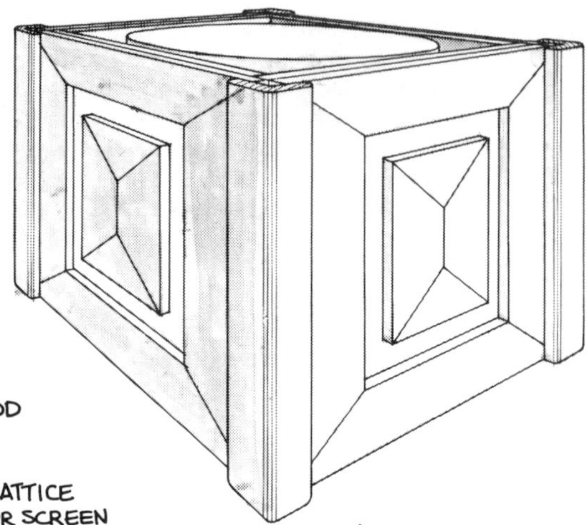

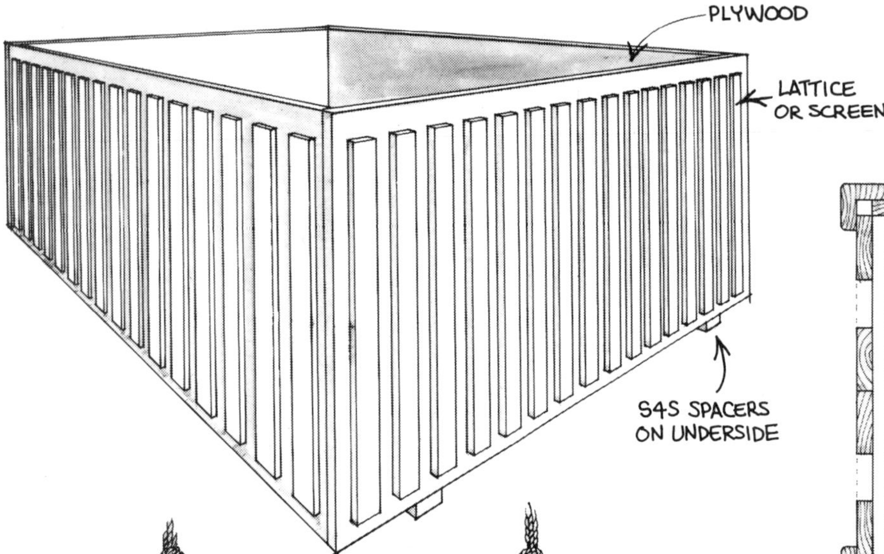

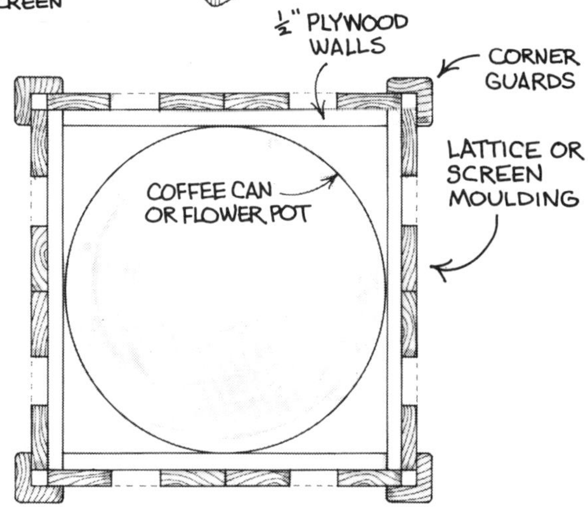

The planter above is just another plywood box, this time framed with lattice, plus a rectangle of mitered lattice placed within the frame. Corner guards are then fastened to each corner.

The two plant hangers on the left are both extremely simple. One is made of two base mouldings connected by dowels and the double-decker is merely two platforms made of ¾″ squares. Hang with rope, yarn or chain.

The box below is simply decorated with screen moulding on the sides.

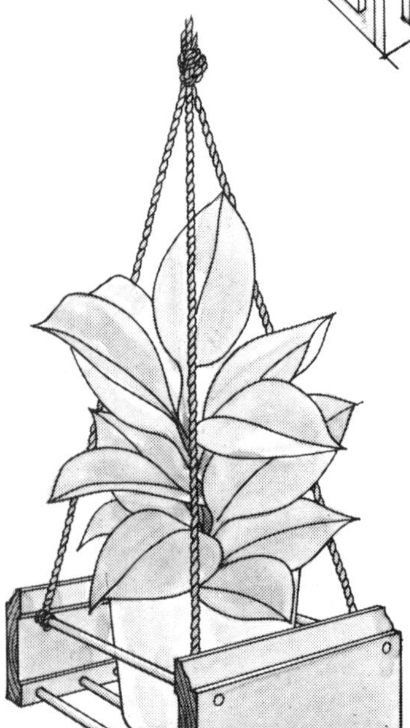

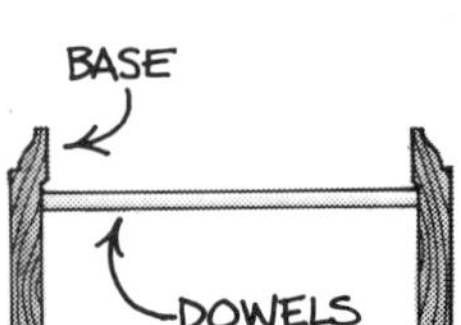

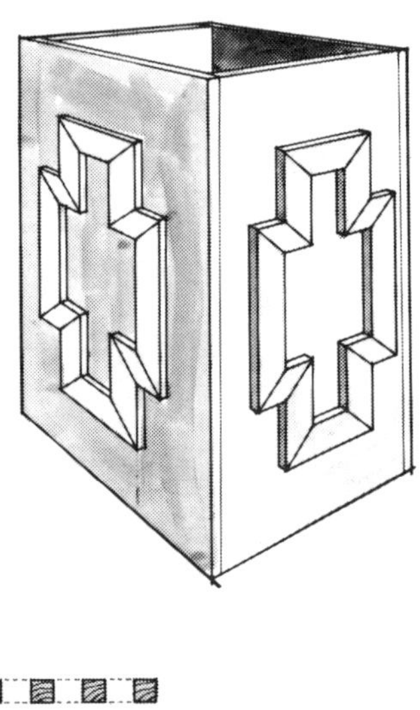

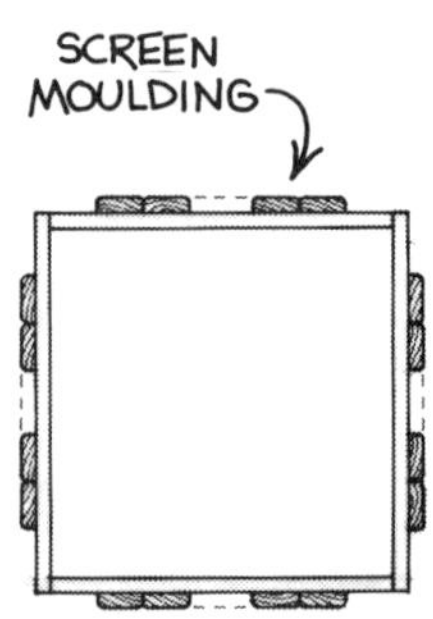

PLANTER BOXES

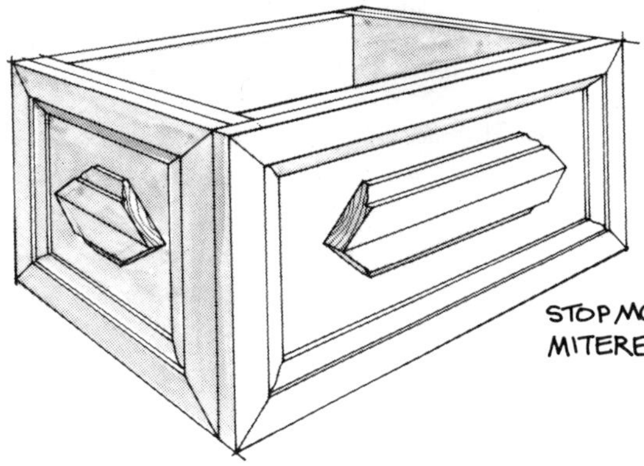

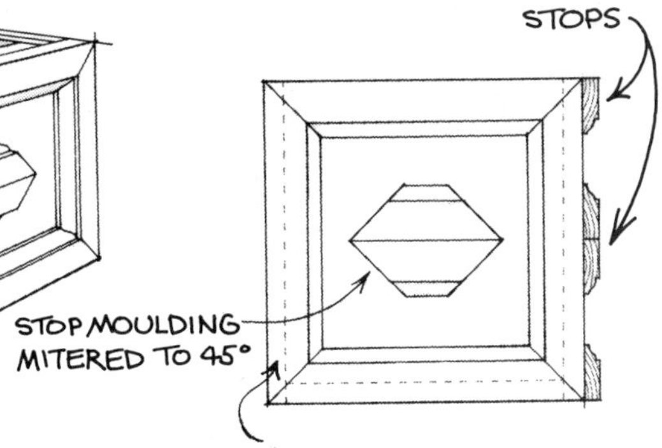

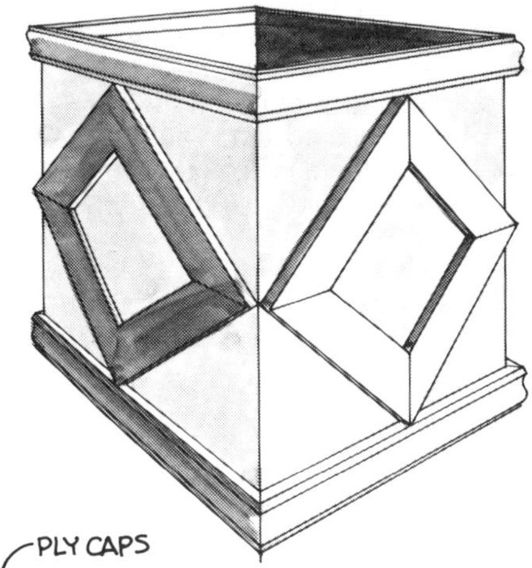

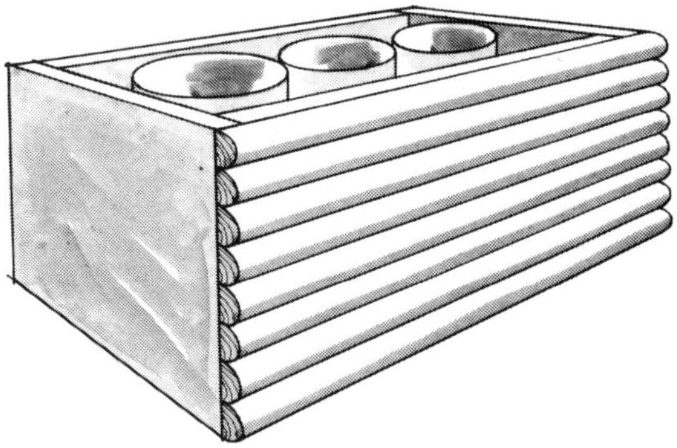

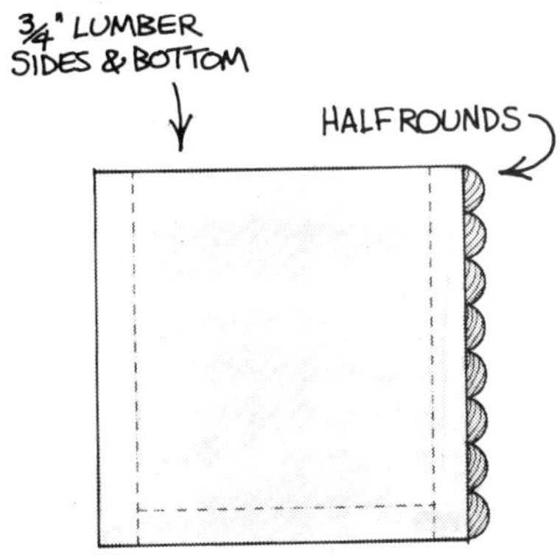

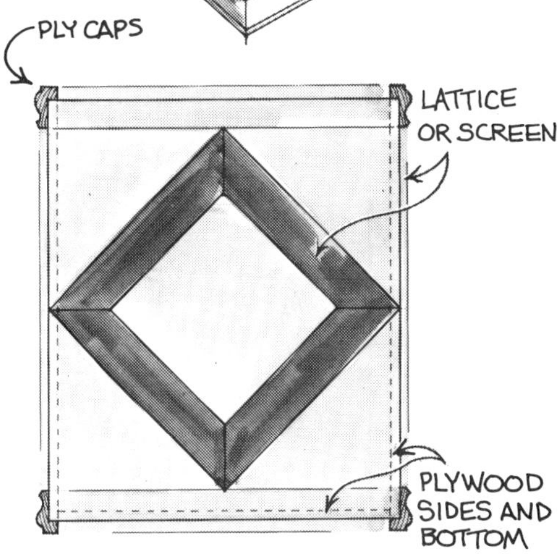

The two containers above make excellent window planters. Both are boxes built the same way, but decorated differently with moulding. The top one uses stop moulding to frame the ends and then the front. Two pieces of mitered stops are butted within the frames. The second window box merely has a series of half round attached to the front.

The square box above uses ply cap or wainscot to cap the top and bottom edges. A diamond shaped square of lattice or screen is applied to each side.

The planter to the left is again, just a box, but with an exception. This one has a mirror attached to double the effectiveness of the plant. The sideboards of the planter are screwed to the shingle moulding mirror frame from the back. The mirror itself sits between the framework and a plywood backing in a space created by the screen moulding spacers. A strip of shingle moulding runs across the front of the box, or if you wish it to relate to the mirror more closely, frame the front of the box the same way the mirror is framed.

Another quick and easy planter may be made just by attaching half rounds to a coffee or shortening can, then spray paint it some delightful color.

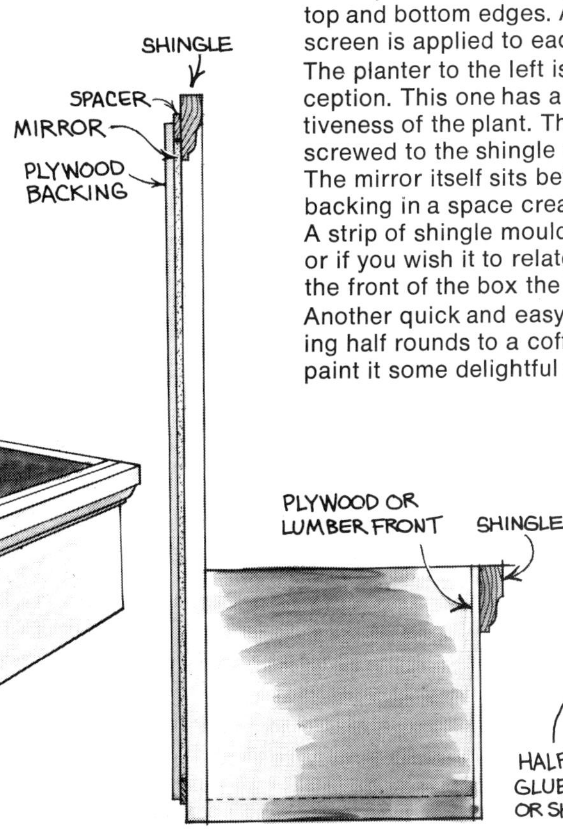

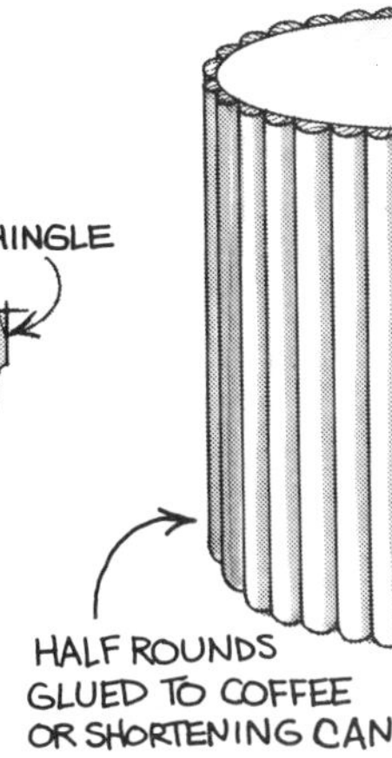

PORCH PILLARS

Plain porch pillars are probably one of the easiest things to change. Most pillars are 4″x4″ or 6″x6″ posts, and as these samples indicate, a little moulding can make a big difference. The first pillar pictured here merely has a corner guard attached to each corner. The next has a series of small frames applied to all sides. The third pillar has one long frame per side. The next one just has a strip of moulding centered on each of the four surfaces. The last pillar has two strips of moulding evenly spaced between each corner guard. Again, a variety of patterns can be used. Besides the corner guards, you can use lattice, screen, stops, mullion, panel, flat astragal, batten or casing. The character of your house, of course, will determine how the pillars are finished.

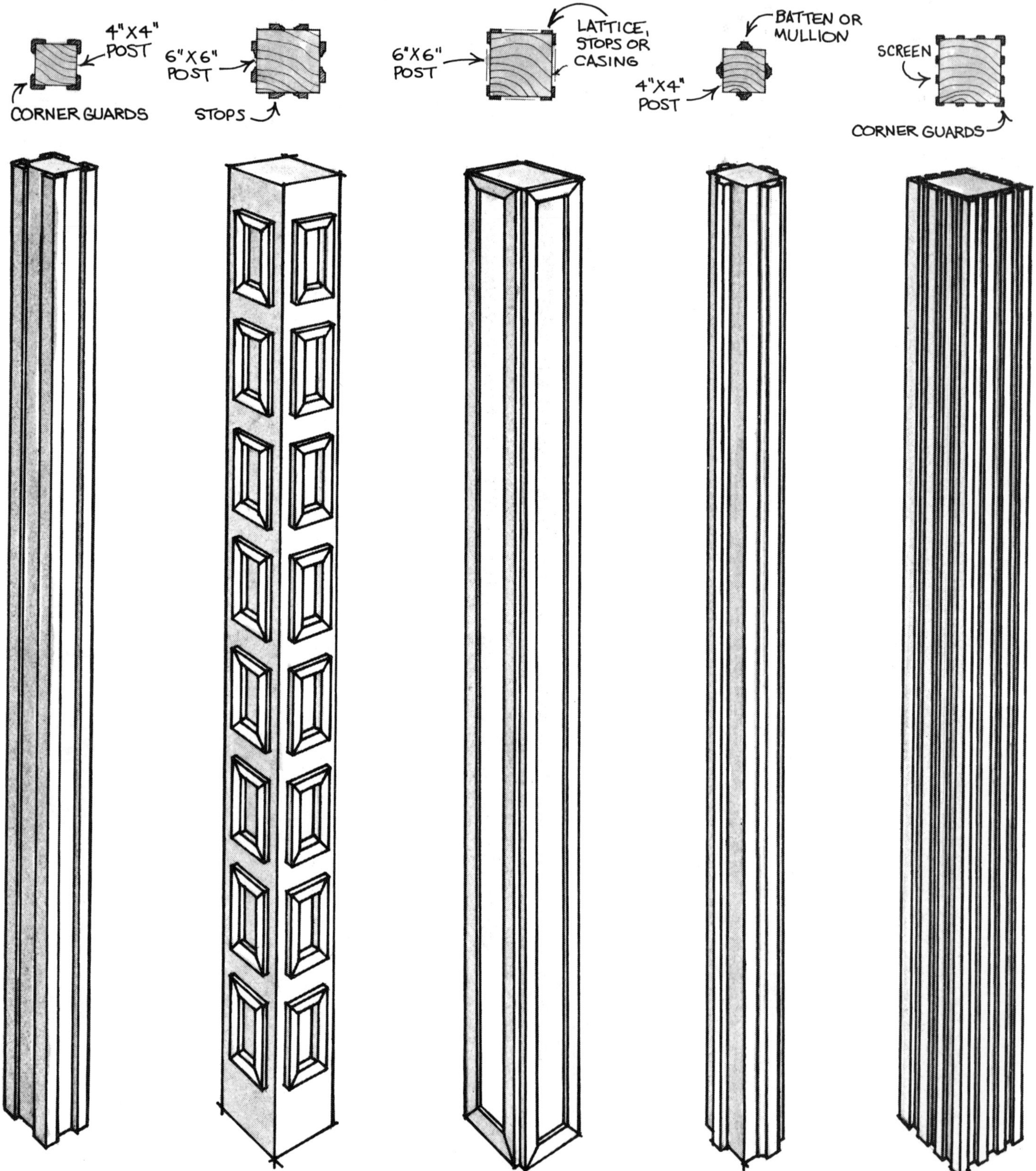

POT & PAN HOLDERS

Pity the harried housewife who has to fumble through her cupboards trying to find the right pot. None of that annoyance is necessary when it's so easy to build a pot and pan rack with wood mouldings.

A long strip of crown moulding with dowels for hanging the pots has the advantage of being just about any length you want, depending on the area it's to go in and the number of pans you want to hang. The moulding should be backed by a strip of S4S to give the unit rigidity and added thickness to support the dowels. Variations of this idea using different moulding patterns are shown. Besides dowels, various hooks are also available for hanging. If you decide to use dowels, make sure the size fits the pot and pan handles. An above-the-moulding shelf may also be added.

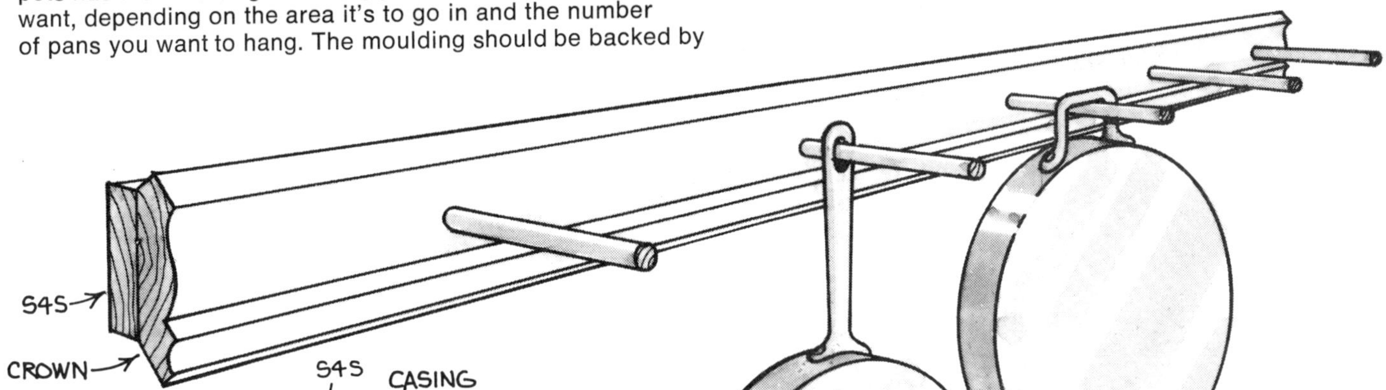

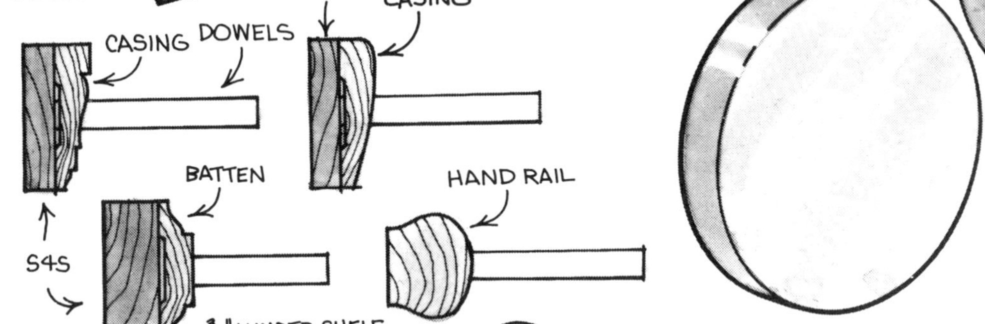

If you have enough space, multiple rows of moulding can be used to create a whole wall of pots and pans, such as the series of hand rails illustrated below. They should be spaced close enough together so the lower edges of the pans do not mar the wall.

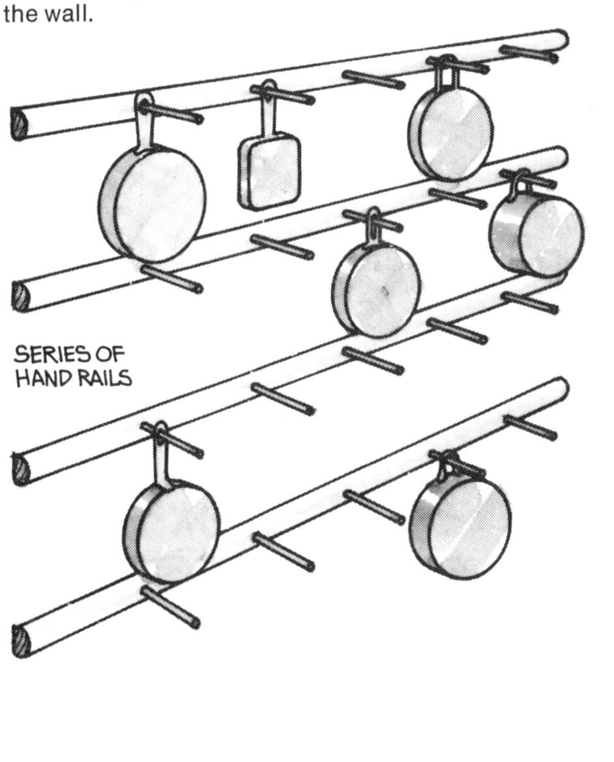

Because of the round top on picture moulding, it makes a natural hanging surface when used with the proper size hooks. A rack is easily put together merely by fastening rows of picture moulding onto two or more S4S uprights. The S4S also allows enough space between the wall and the hooks. Attaching any of the illustrated pot and pan holders to the wall is accomplished quickly and securely wih countersunk heavy duty lag screws.

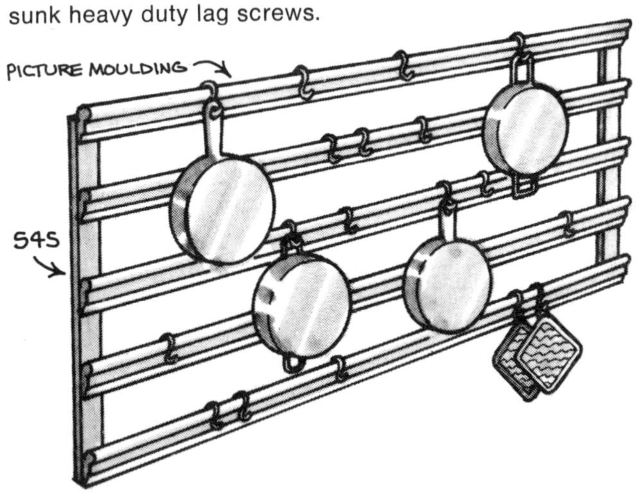

RECIPE BOOK HOLDER

How much easier it is to cook if the cookbook isn't covered with flour or other assorted ingredients. A recipe book holder can solve that problem simply by keeping the cookbook up out of the way. With this rack, the book sits on a ledge made of drip cap attached to a square. A series of dowels are inserted through the square to form the front legs and the upper back which holds the book. Another series of short dowels are inserted into the square to form the back legs. The ends of the dowels are then attached to the rounds to finish the job. This rack should be washable, so stain or paint accordingly. It's also a good idea to have a piece of plexiglass cut to the appropriate size to set over the cookbook when it's in use. This prevents the pages from getting splattered and also acts as a weight to keep the pages from flipping.

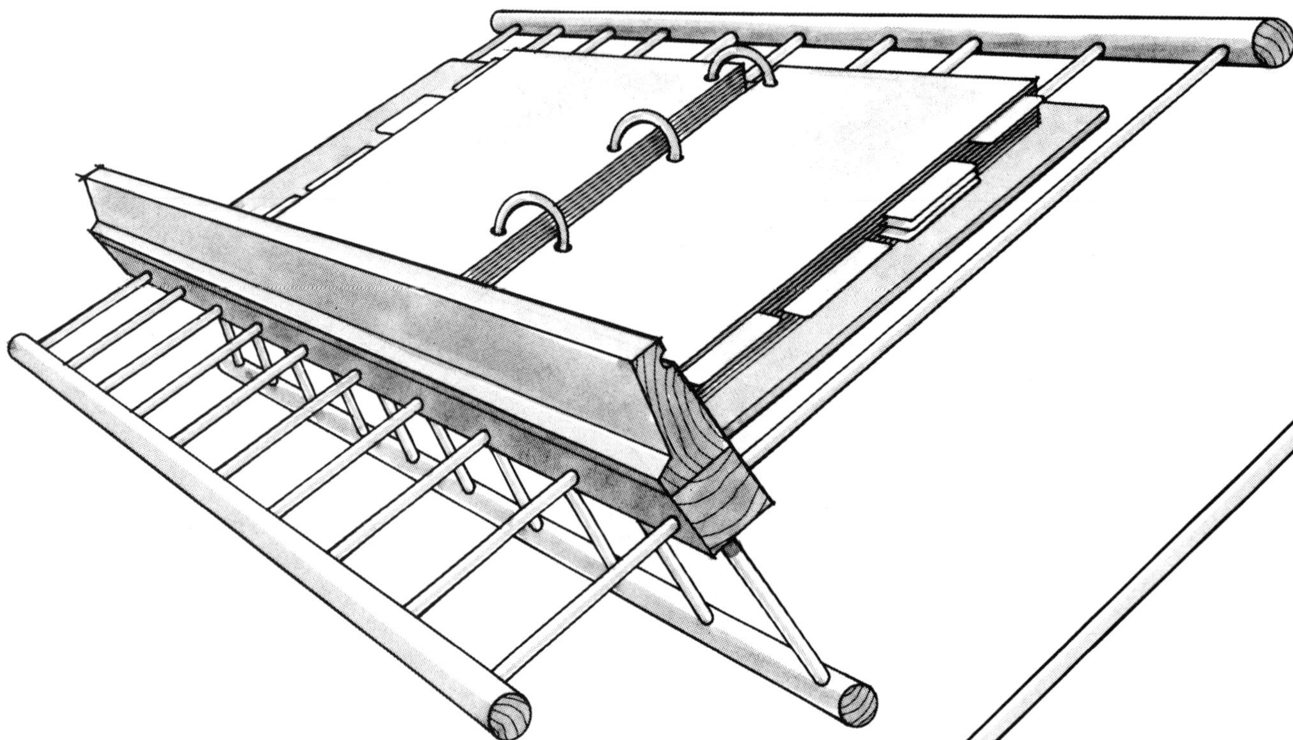

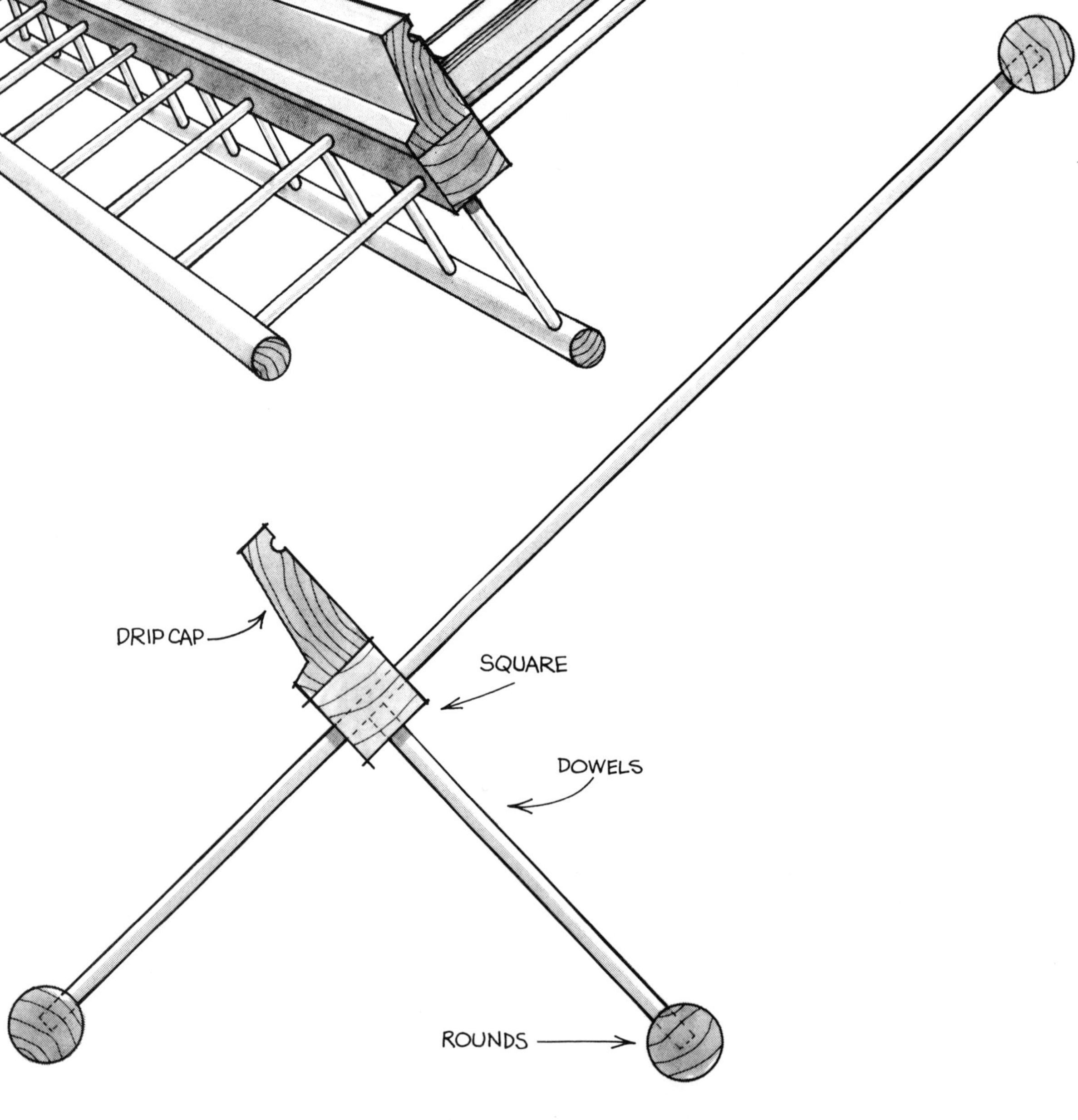

RECORD RACK

Making your own rack for your record albums takes about as long as it does to play one of the albums. Two pieces of hand rail connected by ¼″ dowels make up the base of the rack. Then holes are drilled about every two or three inches apart along the length of the hand rails in which the dowels are inserted to hold the records in place. Cut the dowels to about half the height of the albums. Depending on the size of your record collection, this record rack can be made anywhere from one foot to ten feet long. That oughta dazzle your record listening friends. Again, paint or stain it to match your decor.

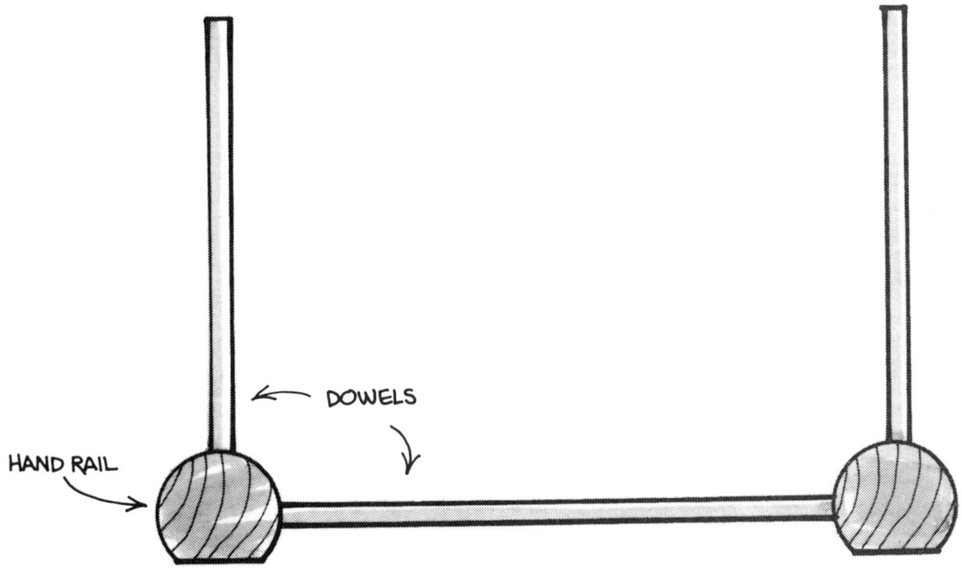

ROOM DIVIDERS

These handsome, easy-to-build room dividers offer a simple but decorative means of separating rooms, breaking up space, screening areas or directing traffic. Try the one that best fits your decore. You'll be surprised how fast it goes up.

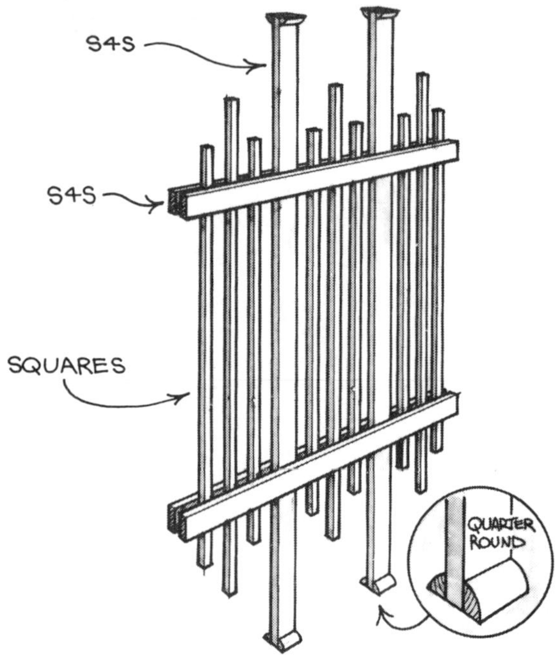

This easy to construct design is made up of varying lengths of vertical square moulding sandwiched between two rows of horizontal pieces of S4S moulding. S4S is also used for the upright supports which run from floor to ceiling. Short pieces of quarter round can be used to attach the unit at the top and bottom.

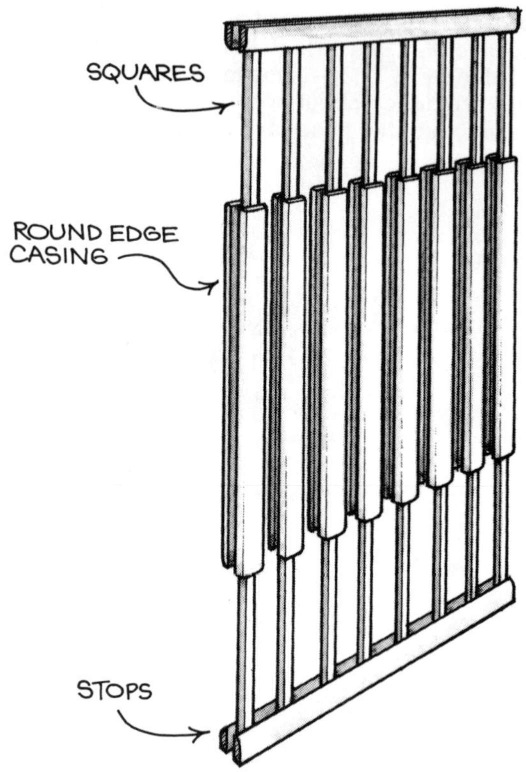

This one is just as simple. It's merely vertical square mouldings sandwiched between shorter spans of round-edged casing and then attached to the floor and ceiling with round-edged stops.

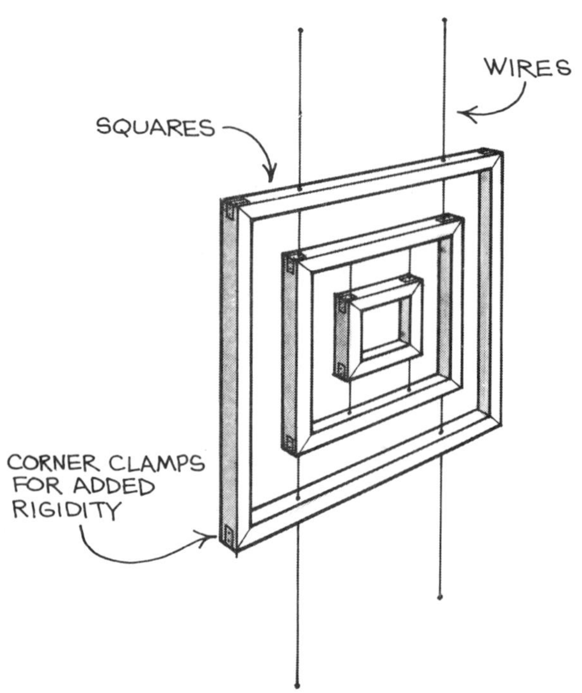

This floating square design is constructed of a square within a square within a square, which appropriately is made up of moulding squares. The structures are supported from ceiling to floor by wires or ropes attached to eye bolts. When nailing the mitered corners together it's a good idea to also attach corner clamps to give added rigidity.

An evenly spaced grouping of vertical rounds contained within a simple framework of S4S moulding makes this divider one of the easiest to build. The amount of "see-through" this design allows depends on the number and the closeness of rounds you choose to use. A good way to attach this unit to the floor and ceiling is by inserting the two supporting rounds into wooden closet pole holders or wooden macramé rings which may be purchased at most home improvement centers.

ROOM DIVIDERS

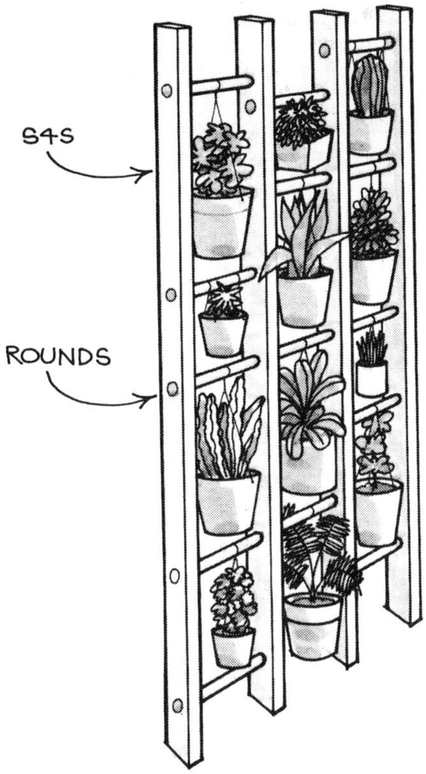

This wall of greenery is merely four vertical lengths of S4S connected by moulding rounds spaced at random distances, on which hanging planters are suspended. A simple but impressive looking divider.

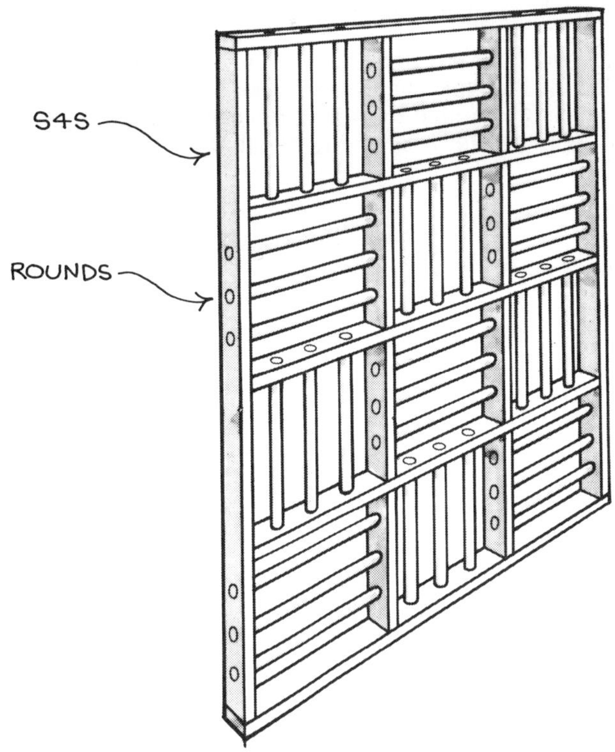

A framework of S4S modular squares make up this room divider. Inside each square three moulding rounds are inserted which alternate vertically and horizontally from square to square. The total width of the room divider depends on how many modules you choose to use.

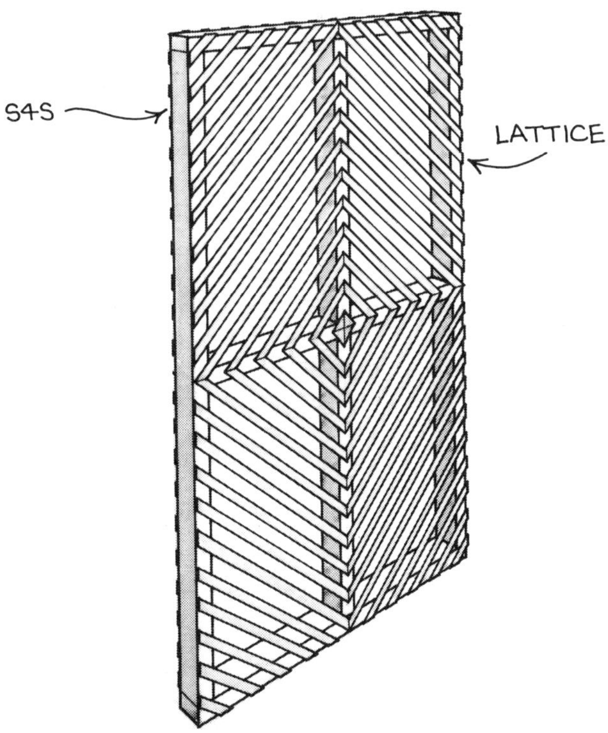

Because of its many uses around the house, both inside and out, it's easy to understand why latticework has been a favorite for so many years. This floor-to-ceiling room divider is put together simply by nailing lattice to the S4S frame in a diamond shaped pattern starting from the center of the inside cross pieces. All the lattice cuts are 45° miters. Complete the other side of the divider in the same way.

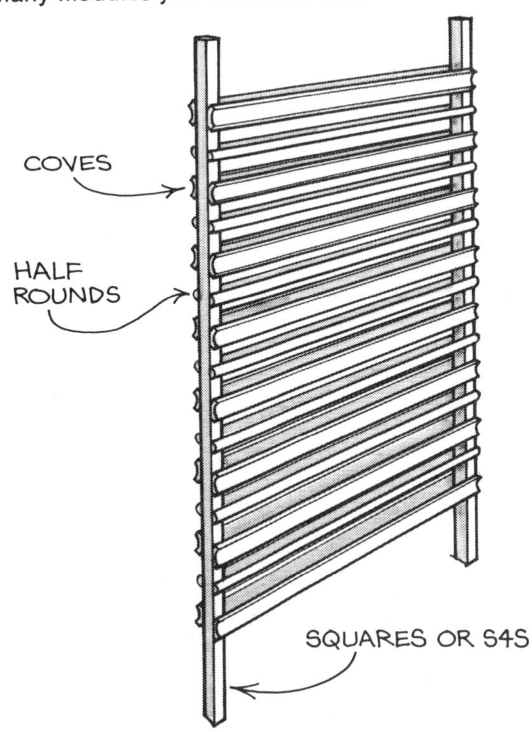

These horizontal rows of alternating cove and half rounds supported by two uprights of S4S or squares give a low contemporary look to your room. Besides these patterns, screen moulding, batten, mullion, lattice or casing can also be used. Again, both sides of the divider are done the same way.

ROSETTES

The dictionary defines a rosette as an ornamental arrangement, suggesting a rose. While the patterns shown here might not suggest roses, they do make excellent ornamental arrangements that have many uses. Their most common function is as decoration trim for doors, cabinets, furniture, shutters, ceilings, walls, room dividers, fences, headboards, planters and boxes. They can be plain and simple, using one piece of moulding mitered into a four piece square, or they can be very ornate, using a combination of many patterns and mitered into squares, hexagons or octagons. The moulding patterns pictured here are just starting suggestions. It's fun to design your own, so give it a try.

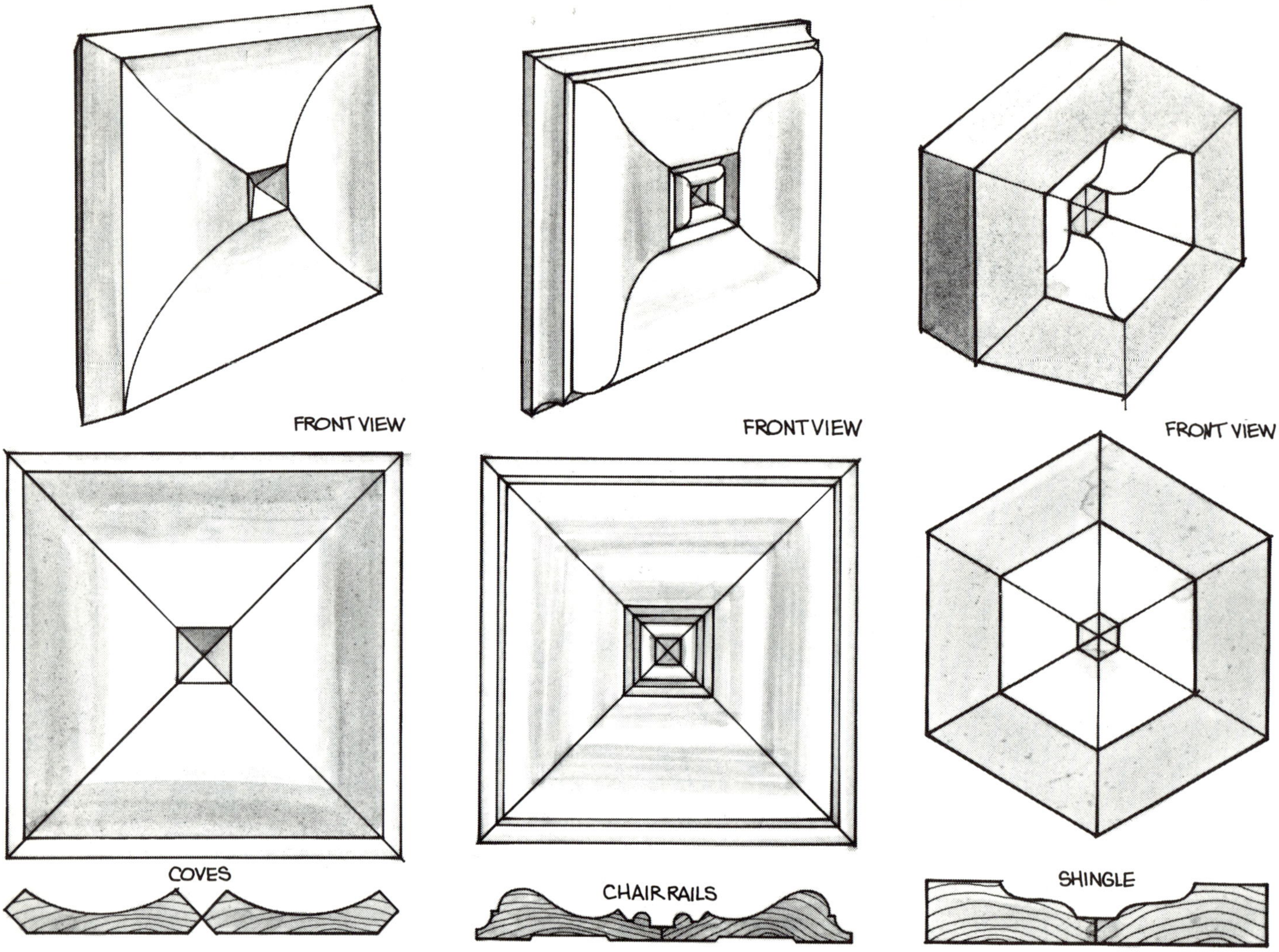

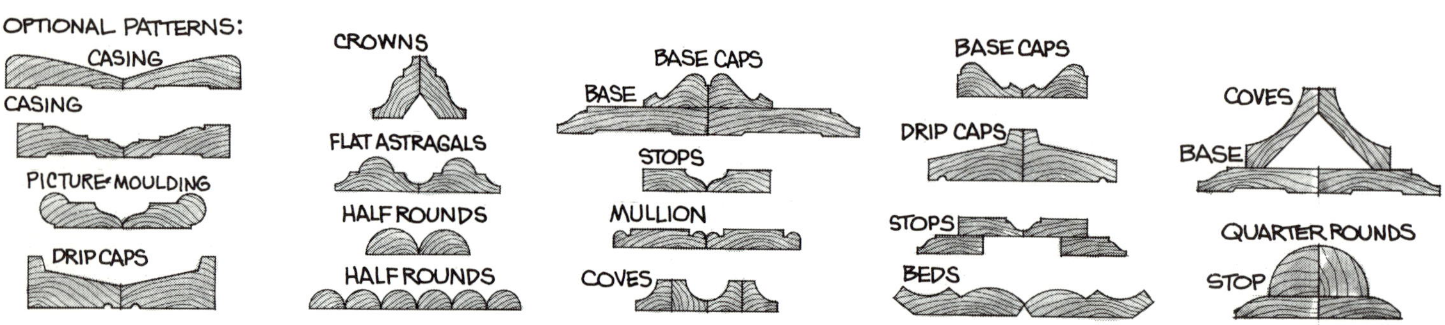

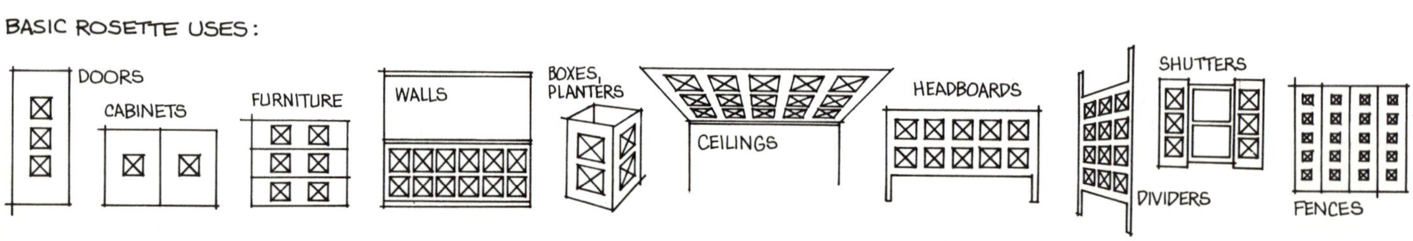

SERVING TRAYS

Always a practical item to have around the house, serving trays have many uses. All these shown here, can be built quickly with a minimum of materials.

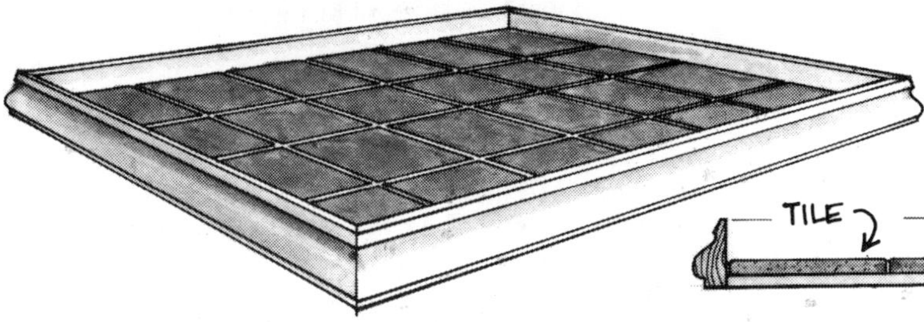

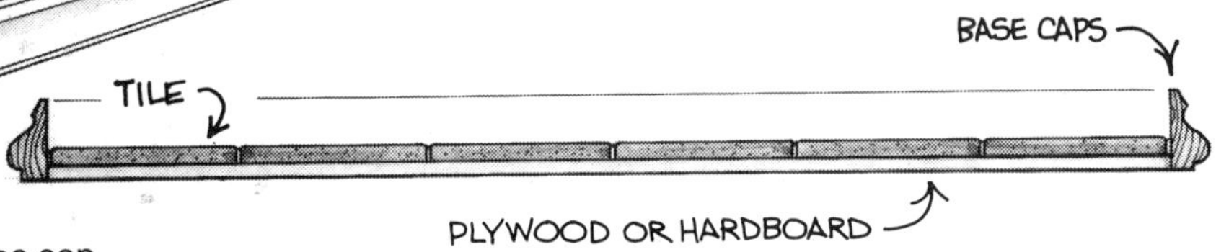

This tray is basically a 3/8″ plywood base with a base cap frame around it. Tile, in whatever size, shape and color you choose is then glued to the plywood. Nail the base cap frame to the plywood with brads. Care must be taken to keep from splitting the moulding.

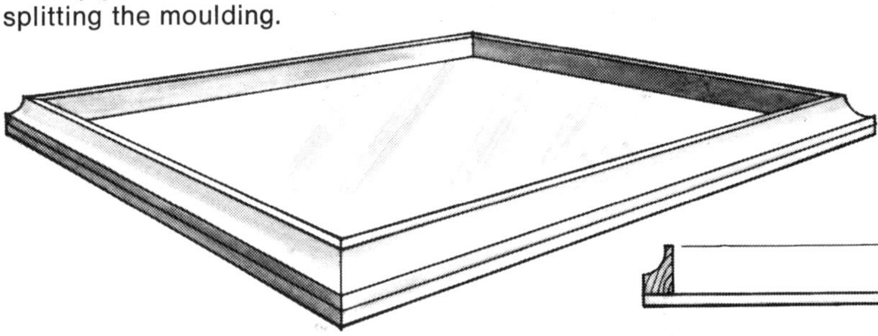

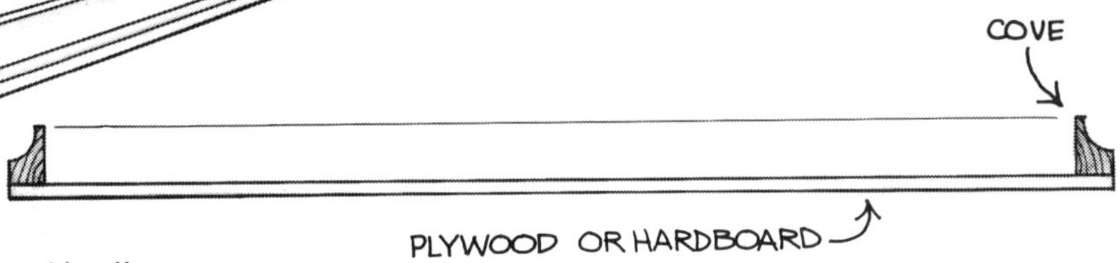

A frame of cove glued and nailed to a piece of plywood is all there is to this one.

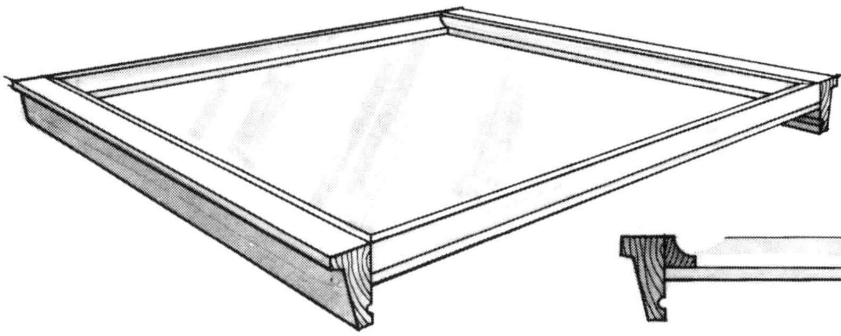

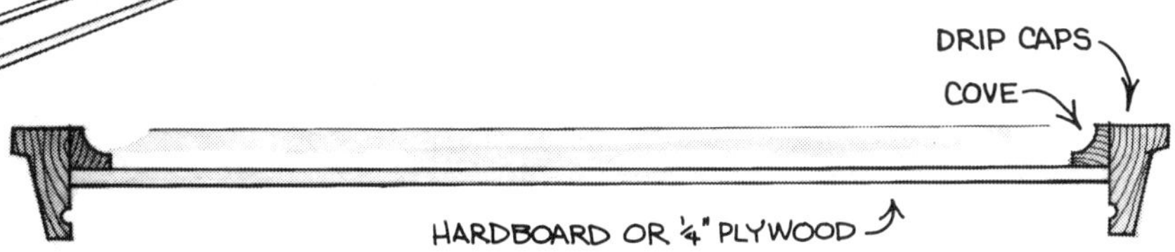

Another frame of cove, but turned in the opposite direction this time, is used on this tray. Drip cap moulding is then attached to each end to form tray handles and legs.

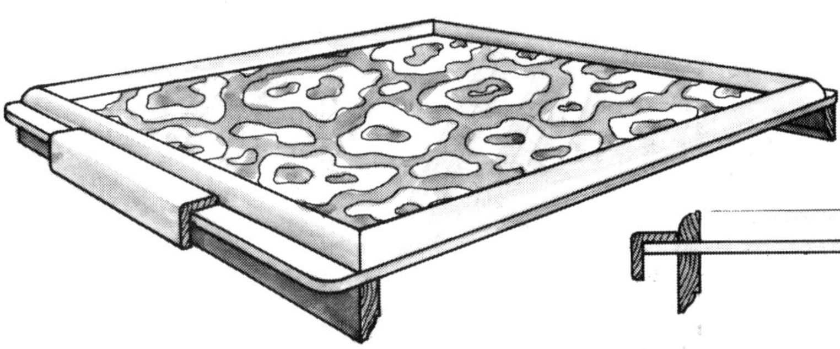

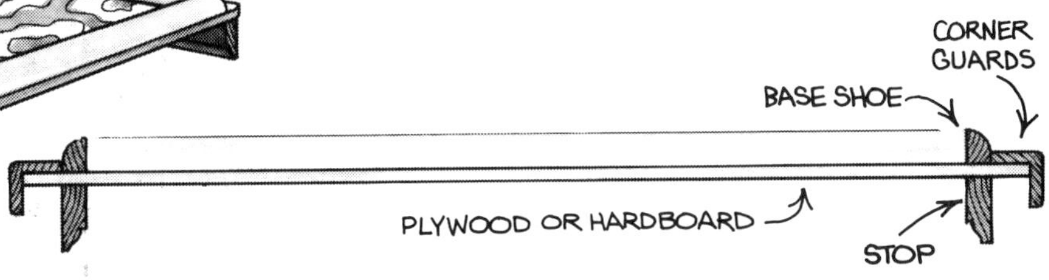

In this case, the frame is made from base shoe or quarter round. The hardboard or plywood bottom extends past the frame at each end to allow room for a handle of corner guard. Round off the corners of the exposed plywood. Legs of stop moulding are nailed to the bottom. Attach the legs before the frame is nailed in place. Use a high gloss paint on these trays for easy cleaning. To make them even more cheery, glue down a sheet of vinyl wall covering, also to be done before tacking the frame in place.

SCREENS

Screens basically have the same function as room dividers. They separate rooms, break up space, screen areas and direct traffic. Screens, however, are portable and expandable (foldable), thereby giving them more versatility than room dividers.

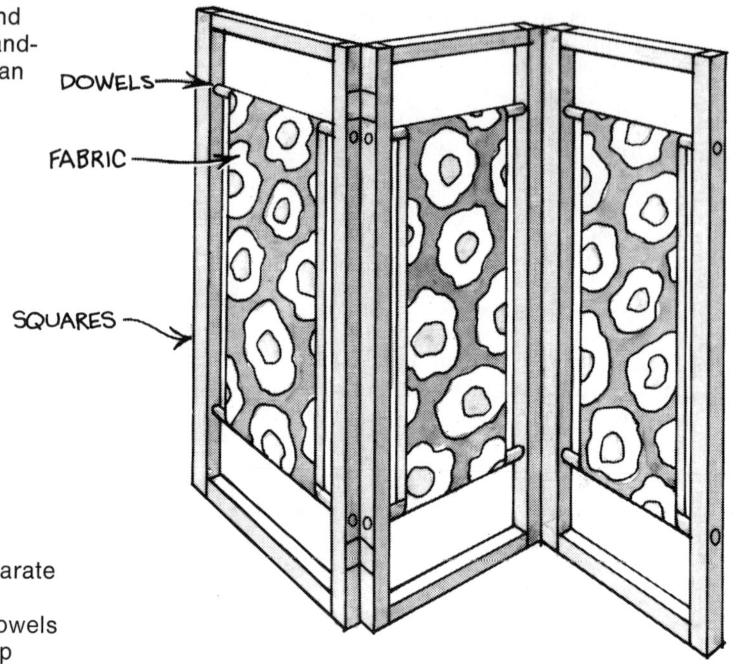

To build the first screen pictured here, make three separate frames of moulding squares. The height and width will depend on where it will be used. Next, drill holes for dowels in the side moulding about six inches or so from the top and the bottom, then cut six dowels of the proper length, which is the width of each frame. Fabric or canvas is to be added to this screen, but it's a good idea to do the painting or staining first. After you've constructed the three frames and cut the dowels to size, do the painting. Next measure the fabric accurately and sew it into a loop. Put the bottom dowel into the fabric loop and insert the dowel into the hole. Repeat the process for the top dowel. The fabric should be just short enough so it's stretched tight when both dowels are in position. Now hinge the three frames together. Remember, for a screen to stand up, each end must fold in opposite directions, so one set of hinges connects frames one and two on one side and another set of hinges connects frames two and three on the other side.

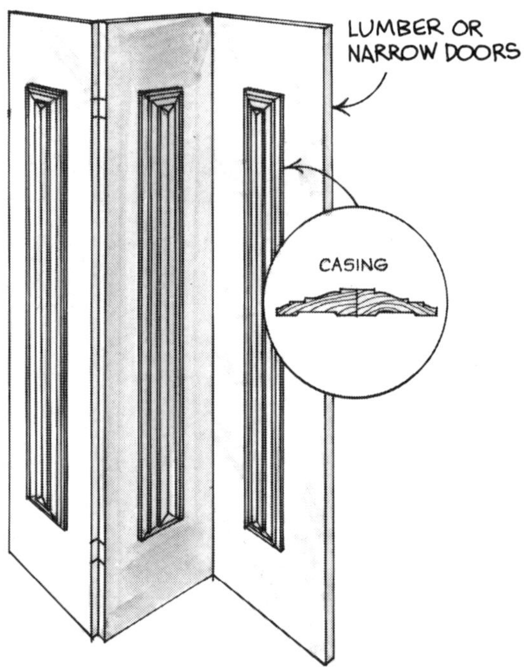

This screen is made up of three pieces of 1″ x 12″ lumber cut to the length you want, or three plain narrow doors. Centered within each panel is a decorative rectangle of casing mitered at the corners. Do both front and back of the screen in the same manner.

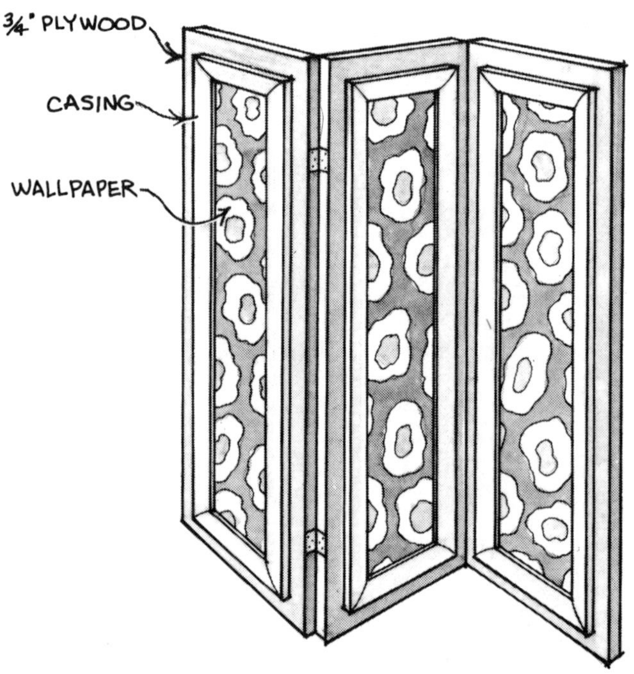

Three pieces of ¾″ lumber on which a framework of casing is attached, make up this lively screen. After cutting the lumber to size, add a rectangle of wallpaper or vinyl wall covering to each panel, then paint the exposed lumber background. Next, the moulding is measured, cut, painted and nailed in place. Hinge in the usual way. Besides casing, stop moulding, base caps, half rounds, shingle and coves can be used.

SCREENS

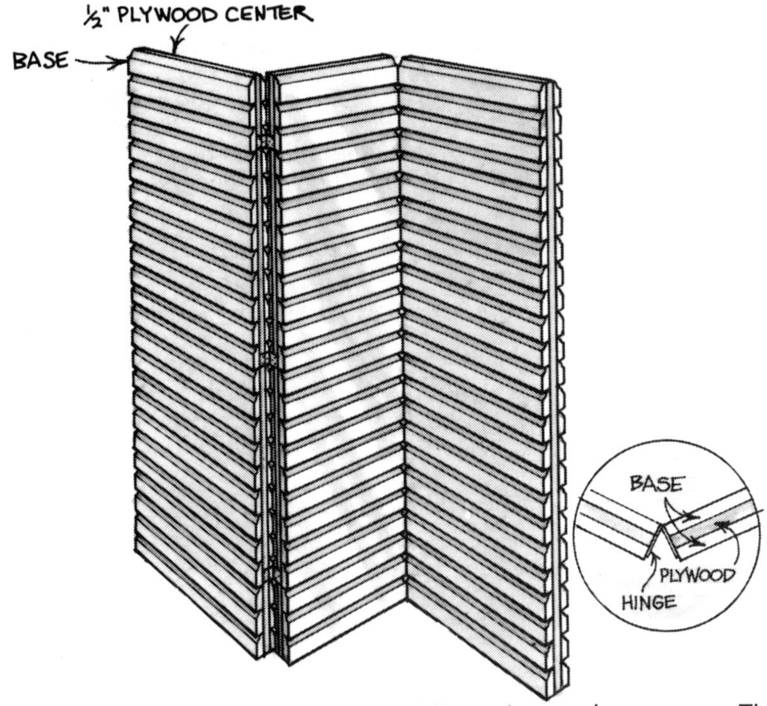

Here's a good way to make your moulding salesman happy. It also makes a very handsome screen and is easy to build. Solid rows of base moulding are attached to both sides of three plywood panels. For a smoother finish, first cut the pieces in rough lengths, attach to the plywood and then do the final cut. Sand away the rough spots and stain or oil to a handsome richness.

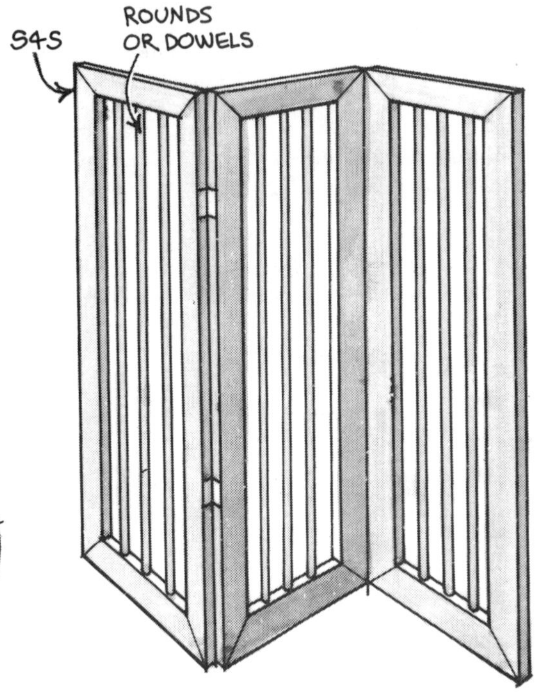

Three panels of S4S framing and rows of dowels is all there is to this screen. Drill holes about an inch into the bottom and top cross pieces. Connect the sides and bottom S4S, slip in the dowels and then add on the top S4S. The mitered corners of the frame are fastened together with lag screws in pre-drilled holes from the top and bottom so they don't show.

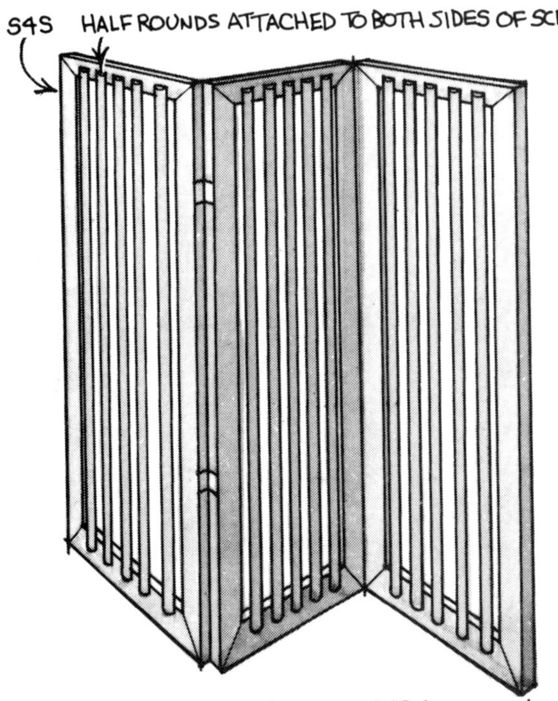

This screen has basically the same S4S framework, except instead of using dowels, half rounds are tacked to the front and back of the screen. The size and spacing depends on how much see-through you want or don't want.

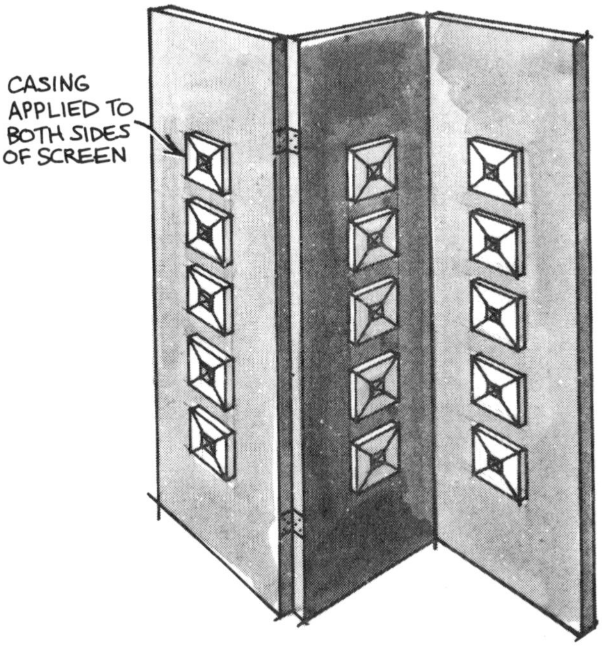

Three panels of lumber are again used in this screen, but in this case they're decorated with rosettes. For more detailed information and samples of rosettes, see page 88.

SEWING & YARN BOXES

There seems to be no end to the variety of uses for boxes. Here are two more to please the creative heart of any seamstress, whether she sews, crochets, knits, darns, stitches, weaves or macrame's. The kind of yarn, thread, needles and assorted sewing supplies will determine how big the container should be.

The first one featured here is just a square box of plywood or lumber. Each side of the box is decorated with a square frame or rosette of batten, mullion, panel moulding, stop or base cap. Base shoe around the bottom gives the box a more complete look. The lid is a piece of plywood edged with corner guard. Two pieces of short corner guard glued and tacked back to back is also used for the handle. This should be attached with brads from the underside.

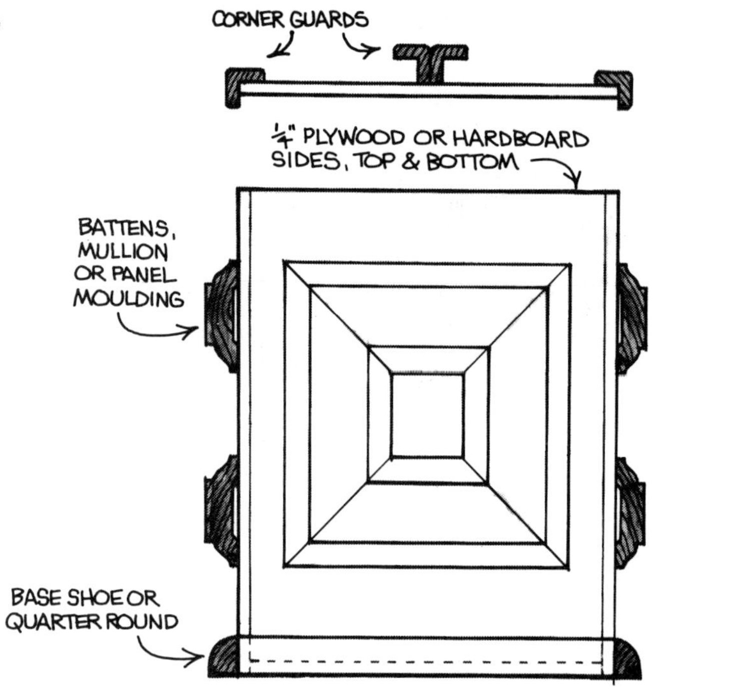

The next sewing box is made of ⅜" plywood or lumber with cove around the base. Use cove that does not have equal dimensions, so when it's applied to the lid, the inside frame of cove is laying flat and the outside frame of cove extends down over the box to form a lip that holds the lid in place. Since twelve corners are to be mitered here, accurate measuring is essential. Now paint it a sprightly color and finish off the job with a decorative wooden or ceramic knob. By the way, these containers also make excellent kitchen canisters. For other box ideas see pages 12 and 13.

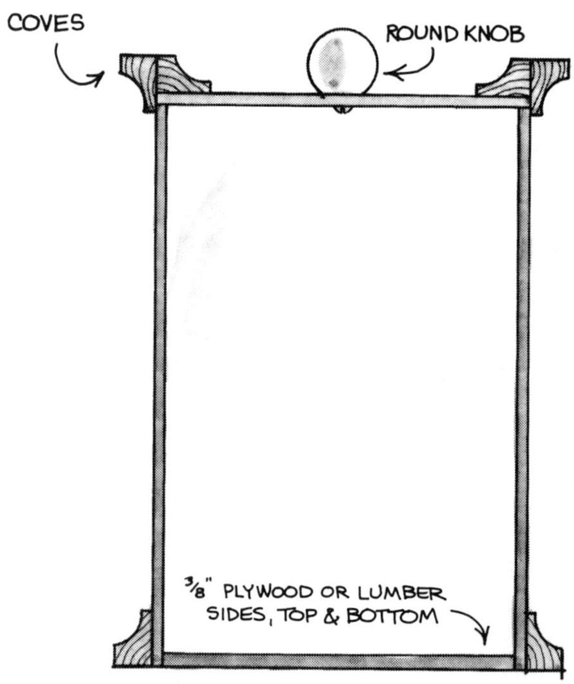

SHADOW BOXES

Who knows what lurks in the heart of a shadow box? Your friends will, when they see all your little treasures on display. Whether it's bottles, jars, toys, gadgets, numbers or letters, dried flowers, thimbles, sculpture, shells, photos or other miscellaneous what-nots, they'll all look at home when displayed in one of these handsome shadow boxes.

The first one is a 3/8″ plywood box with equal sized cubicles. The divider sections are also made of 3/8″ plywood. The whole box is framed with picture moulding, base cap or ply cap and mitered at the corners. Glue and nail throughout.
The smaller shadow box has sides made of shingle moulding. Stop, casing, base, drip cap or chair rail will also work. Use 1/4″ plywood for the back and lattice for the shelves.
The longest box is made with S4S or 3/8″ plywood sides and shelves. Lattice can also be used for the shelves. In this shadow box, the spaces are irregular, so arrange them in whatever way you find interesting. The whole box is then framed with corner guard. Use picture hangers for attaching them to the wall.

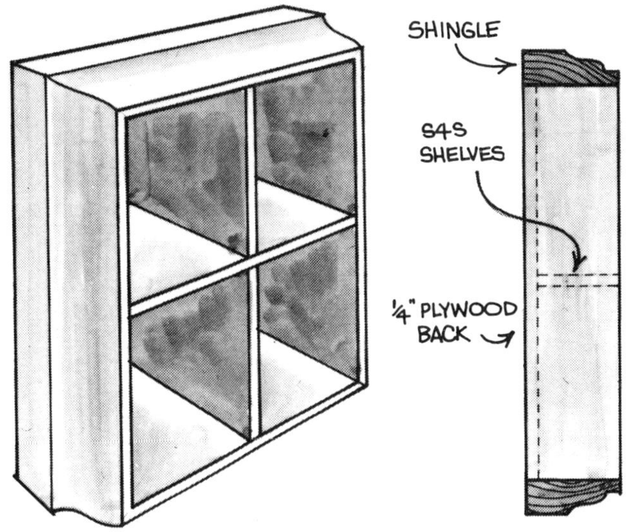

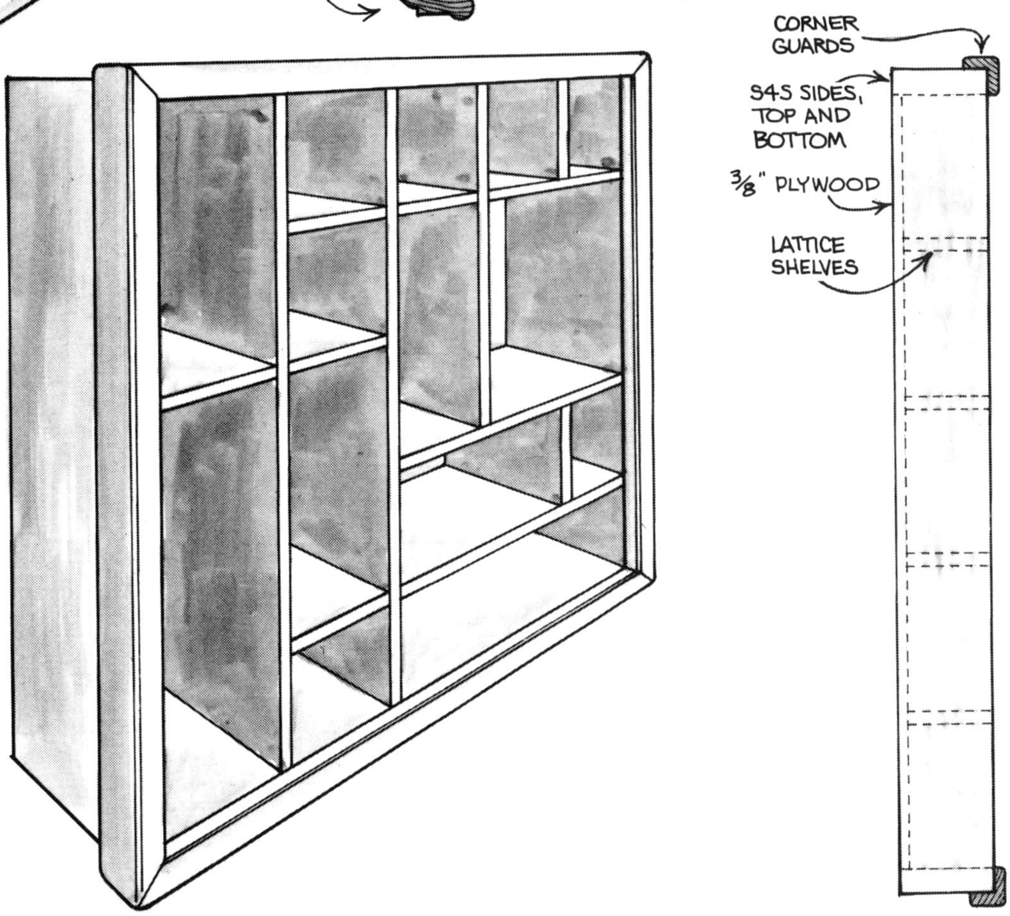

SHELVES

Shelves. Big, little, all sizes and shapes. They are a universal need for every room. Whatever kind you want, it's probably possible with wood mouldings. The examples shown here are but a sampling of what's possible. Take a look at these and then let your imagination take over.

The little shelf holding the flower pot is just a piece of 3/8″ lumber sitting on a piece of crown moulding that's been mitered at the corners. It might look a little complex, but it's just a simple 90° cut at each end and two 45° miters at the corners. Then cut a little back piece that fits inside which is used for attaching it to the wall with screws.

The shelf on the right is a length of flat stool or a stairway step. Half round is tacked on top at each end to act as a stop. Supporting the shelf are S4S braces with 45° miters. Holes are drilled near the ends of the S4S so lag screws can attach it to the walls. The front edges of the S4S are then caped with half rounds, which are also mitered. To attach the upper part of the shelf to the wall, use small shelf braces or counter sunk dowels.

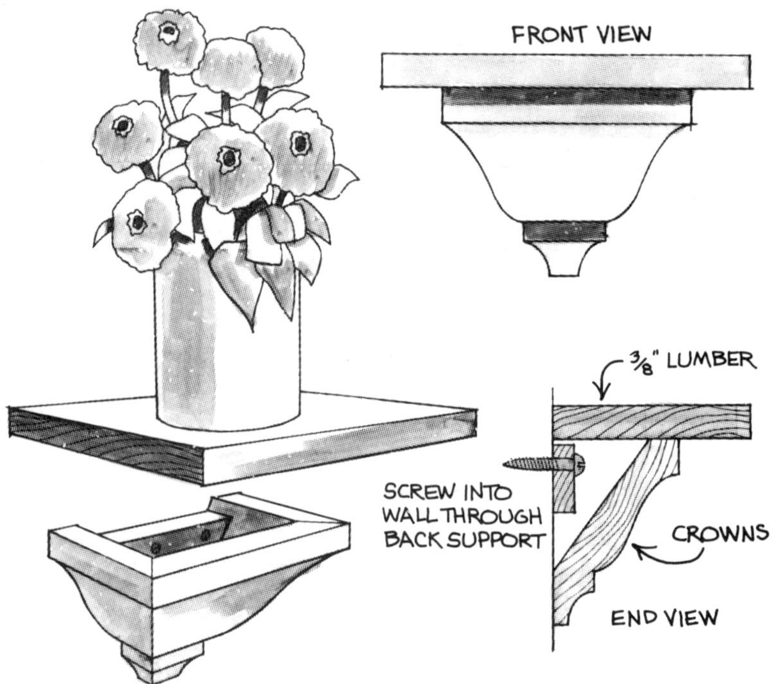

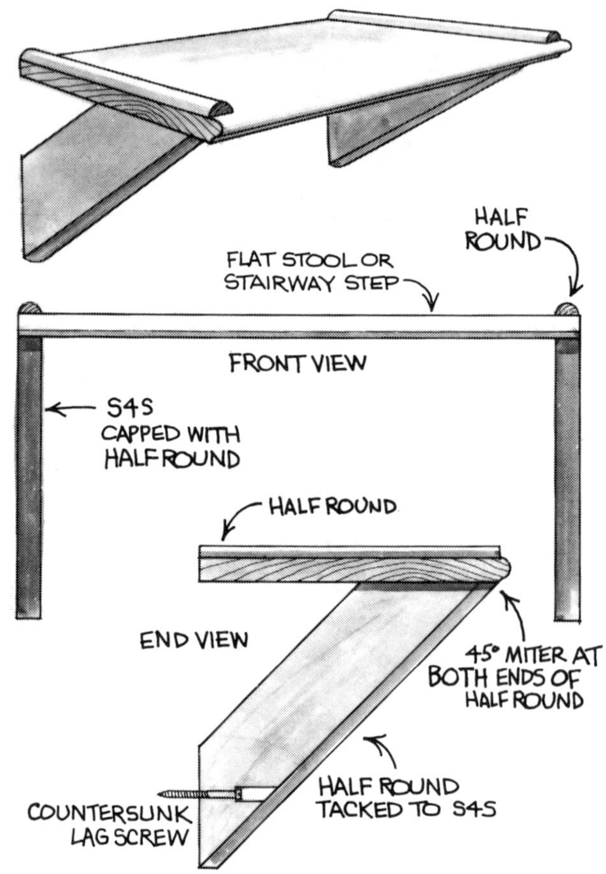

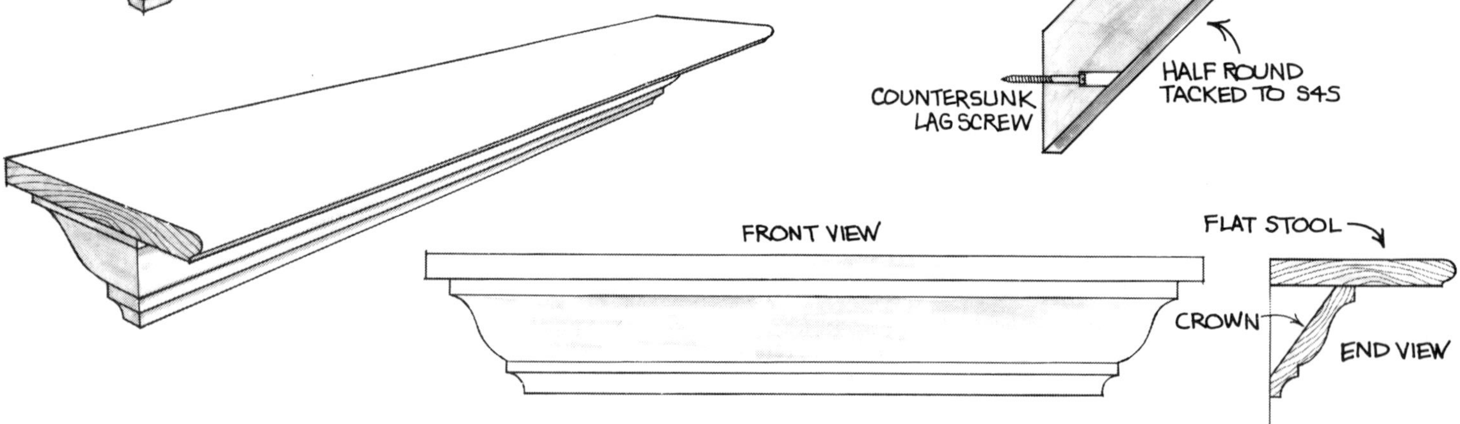

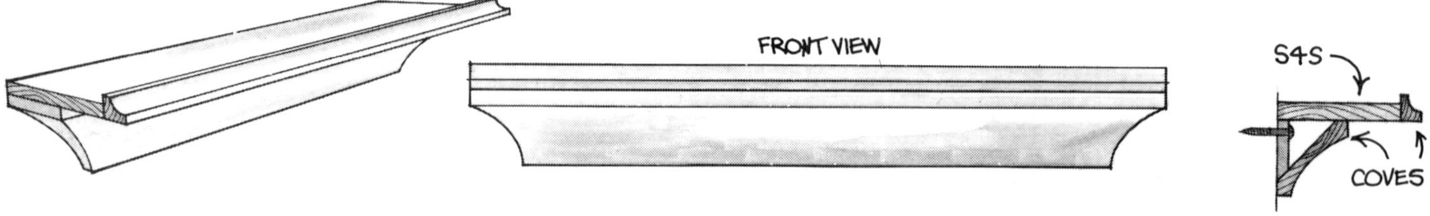

The long shelf in the middle is the same construction as the first small shelf, except it's long instead of short.

The bottom shelf is merely a long piece of S4S or lumber with a strip of small cove attached to the leading edge. The supporting structure is made of large cove mitered in the usual way. Add a couple of pieces in the back for attaching the shelf to the wall. After the supporting structure is connected to the wall, screw the top in place in pre-drilled holes.

SHELVES

Here's a popular shelf design you've probably seen in magazines, or stores. Now you can build a variation of your own with wood mouldings. First construct the ends and middle support ladders which are cross pieces of S4S sandwiched between uprights of round edge casing. The dimensions are left up to you. Drill holes for the rounds which hold the shelves. Glue and nail all the pieces together. If you would rather have the option of dismantling the shelves, leave the rounds unglued. Use 3/8″ safety glass cut to size at a glass shop for the actual shelves. Just their weight will hold them in place.

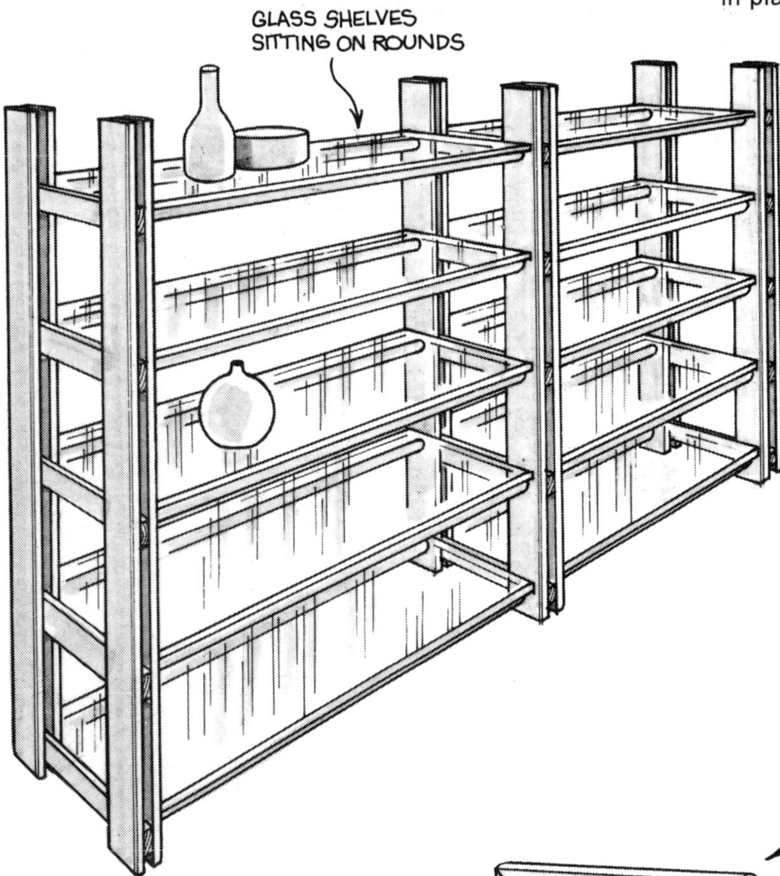

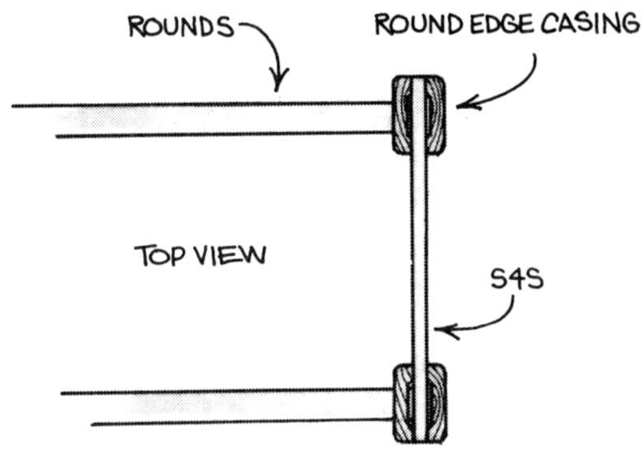

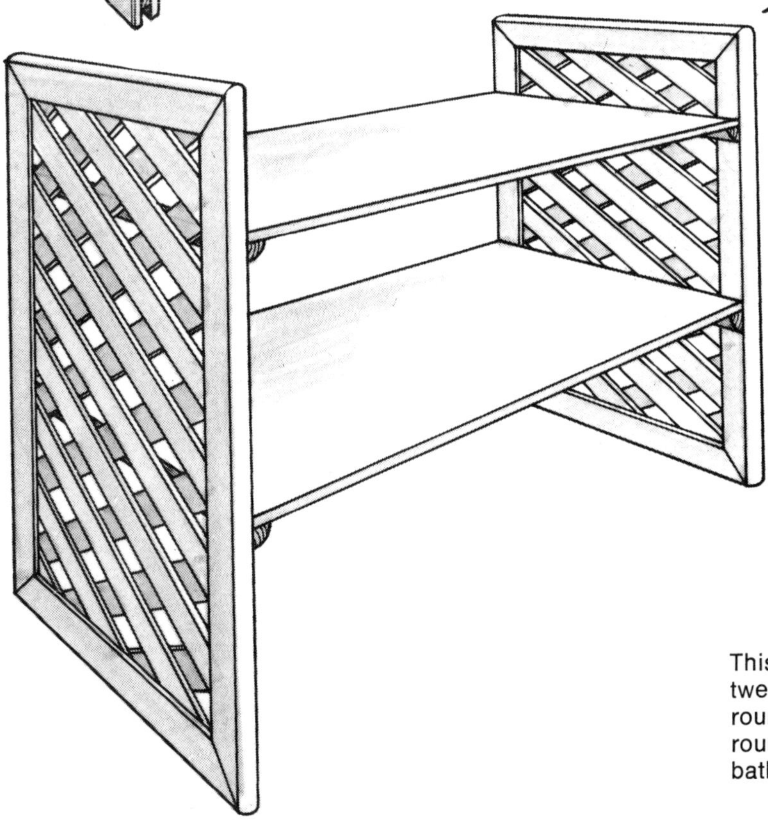

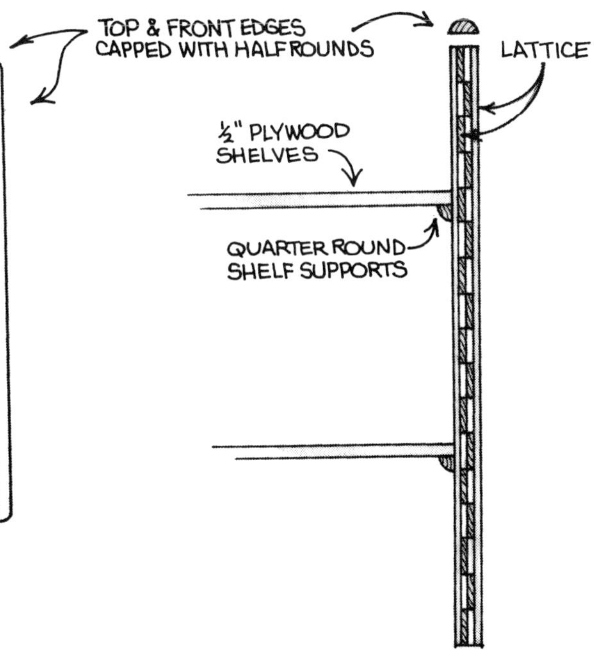

This shelf has ends made of lattice grids sandwiched between frames of lattice. The edges are capped with half rounds. The shelves are 1/2″ plywood which sit on quarter round supports. This shelf design works beautifully in the bathroom.

SIGNS

Whether you need a sign for your business, home, shop, school, club or organization, there's no need to go to an expensive professional sign company when you can do it yourself with wood moulding and still end up with a sharp, professional looking job.

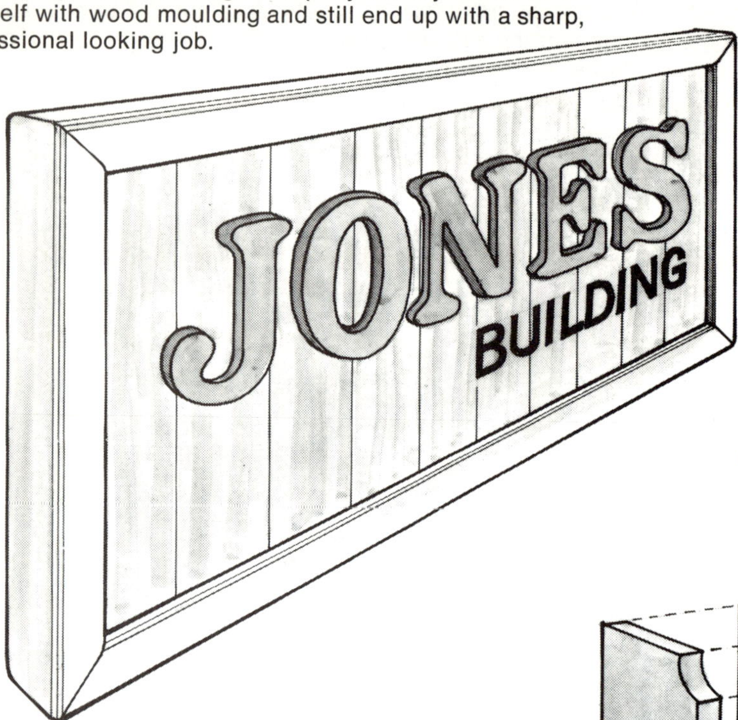

The simple-to-make sign on the left is constructed with either rough sawn boards, paneling or siding attached to a backboard of plywood and then framed with corner guards. The lettering can be wooden, plastic or metal letters, or self adhesive lettering. If the sign is to be used outdoors, paint or stain accordingly. If it's not too heavy, use picture brackets for hanging. Otherwise, use screws for bolting to the wall. Suspended signs can utilize hooks and chain for hanging.

The lighted sign on the right uses a standard fluorescent fixture attached to plywood or lumber backing. The box around it has S4S for the top and bottom. A screen moulding spacer strip is used, and then strips of cove are nailed to the outside. Colored glass, plexiglass or plastic is inserted in the spacer slots. S4S is also used for the end pieces and the corners are coped to match the coves. Self adhesive letters or numbers are then applied to the glass.

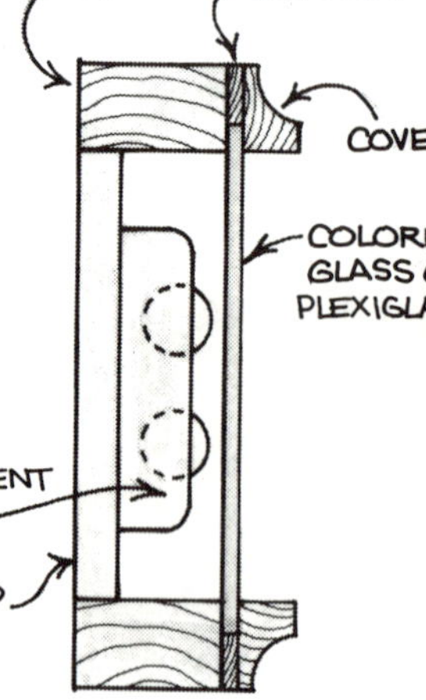

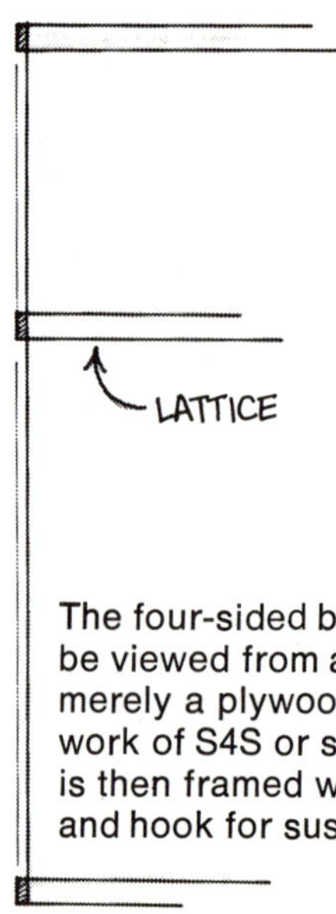

The four-sided box sign is meant to be viewed from any direction. It is merely a plywood box built on a framework of S4S or squares. The plywood is then framed with lattice. Use chain and hook for suspending the sign.

SIGNS

This sign extends out from a wall and is meant to be viewed from two sides. It also hangs by hook and chain. The sign itself is made of ¾″ lumber on S4S stock with a capping of S4S on the top and bottom. Drip cap is applied to both sides and the end. The drip cap corners are mitered. If a more traditional look is desired, use crown, bed, traditional casing or base cap instead of the drip cap.

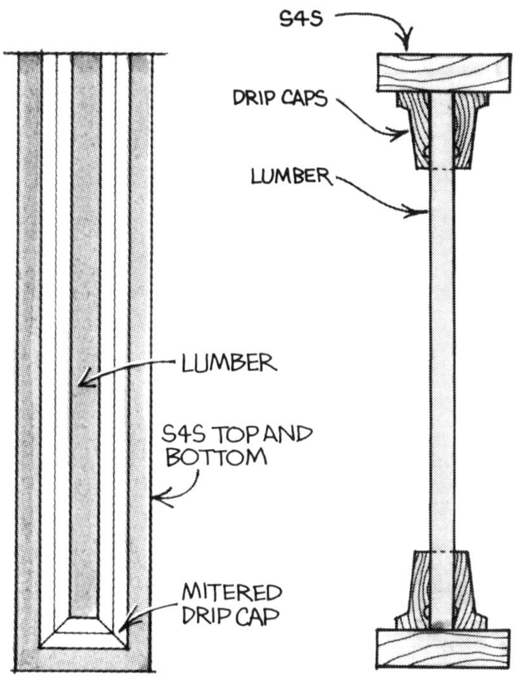

TOP VIEW

FRONT VIEW

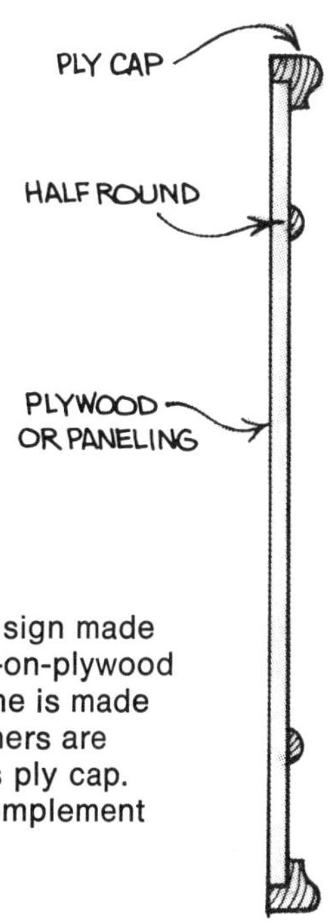

This is another simple wall sign made with a plywood or paneling-on-plywood background. The inner frame is made of half round. Octagon corners are optional. The outer frame is ply cap. Paint or stain the sign to complement its subject matter.

SHUTTERS

Modifying your existing window shutters or building new ones is just as simple as giving your door a facelift, however, in this case, it will give your home a whole new appearance. Picking the style, size, color, moulding pattern and layout is what will give your house that look of individuality. The most common mouldings used for shutters are lattice, batten, shingle, screen, half round, panel, casing, stops and mullion. Since shutters are exposed to the elements, use exterior plywood or lumber for the shutter backdrop and finish it and the moulding with the appropriate paint or stain.

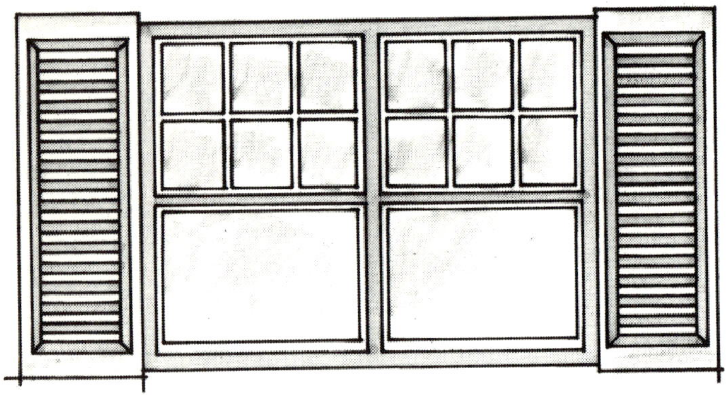

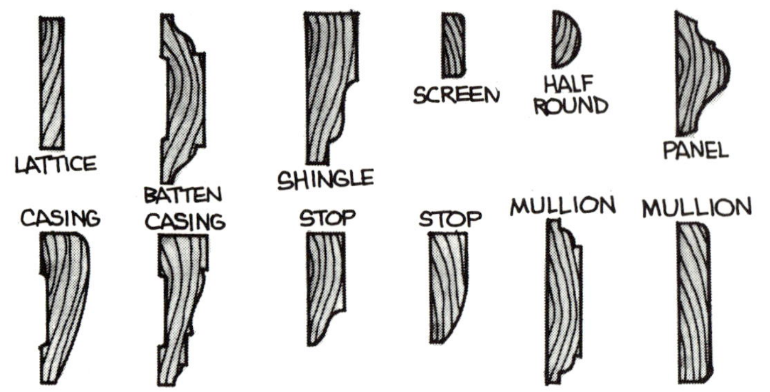

SPICE RACK

Here's a quicky project that's sure to please the creator of culinary delights in your kitchen. Keeping all the needed spices in one place also makes cooking that much more efficient. This rack can be built with as many shelves and as large as you want, depending on how spicey the cook is in your house. The sides and shelves of the rack are made of S4S stock and has either hardboard or ¼" plywood backing. Use strips of screen moulding to hold the spice jars in place. Where it's placed will depend on the size and shape of the jars to be used. The fit cannot be too tight or they'll be difficult to remove. The top of the rack should be left open and the side corners rounded off. Paint or stain it to match the kichen.

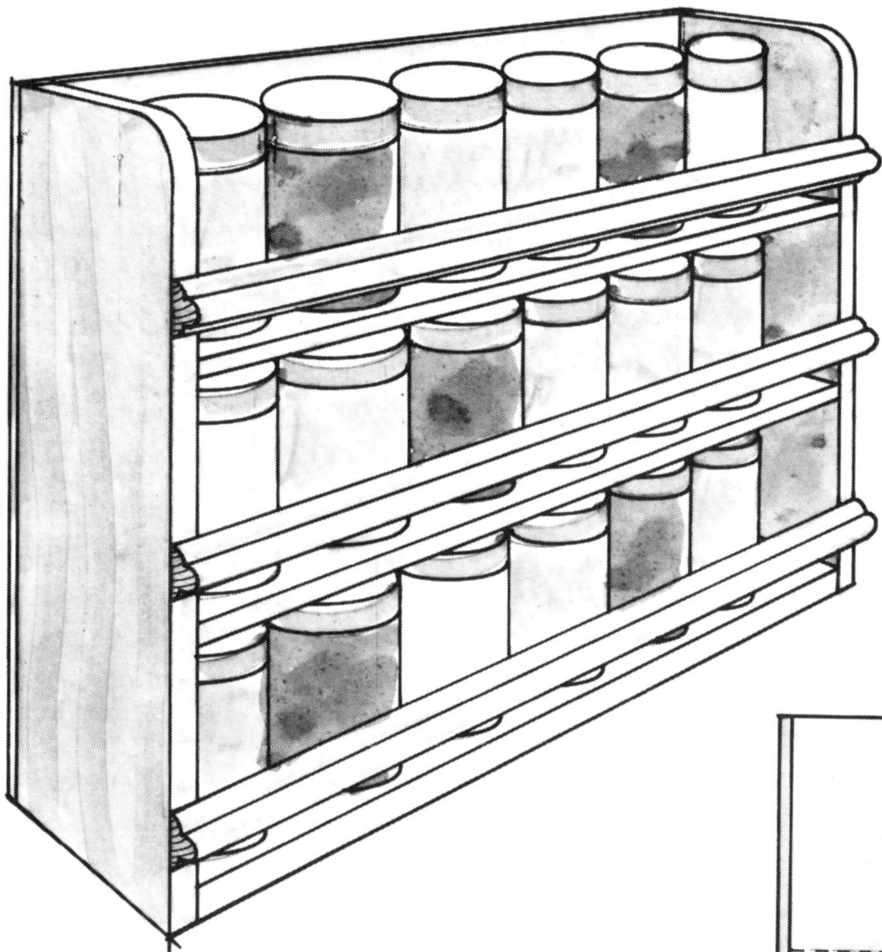

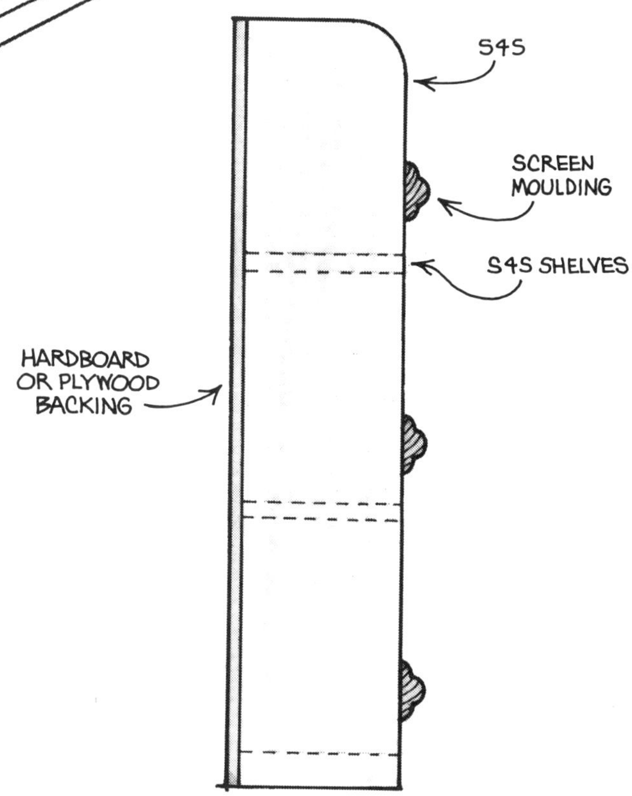

SPOOL HOLDERS

Whether the seamstress in your family has only a few spools for those occasional repairs or is a multi-spooled super sewer, these handy spool organizers will help keep the sewing simple. No more wasting time looking for that right color. All the threads are in sight. The number of spools these racks hold depends on you. Cut the moulding as long as you want and add as many shelves as you need.

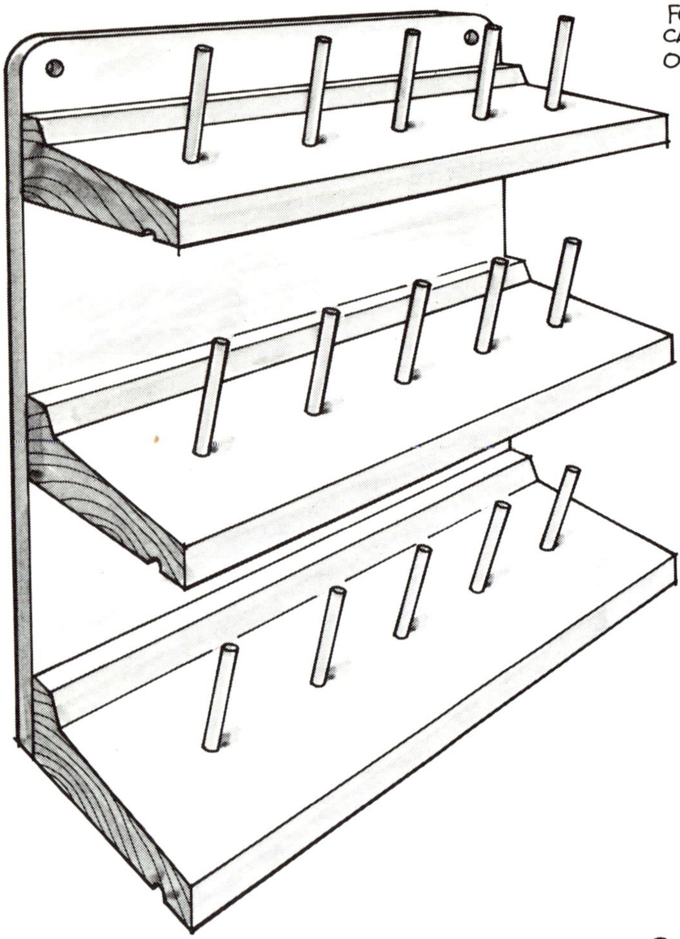

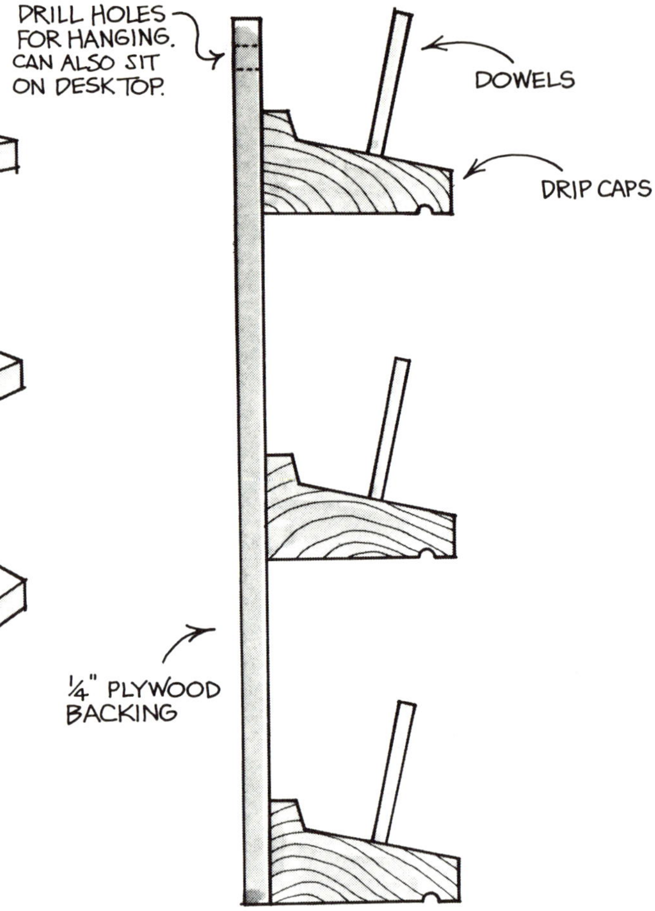

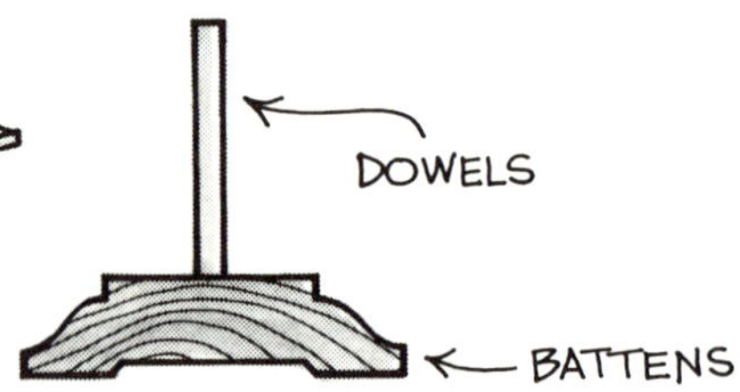

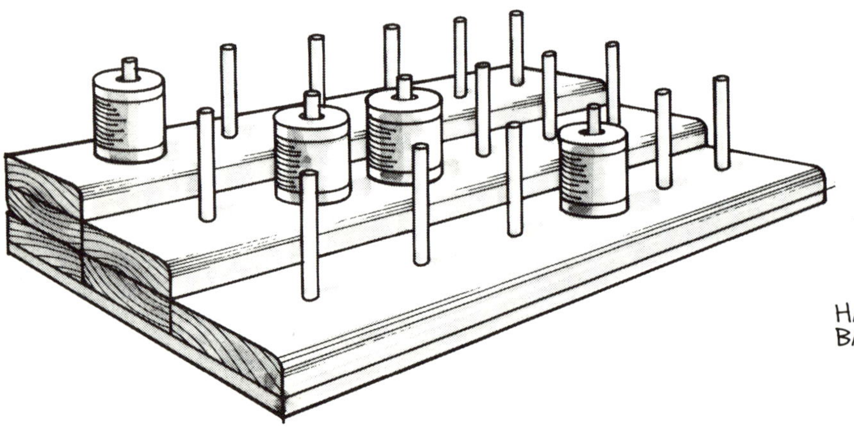

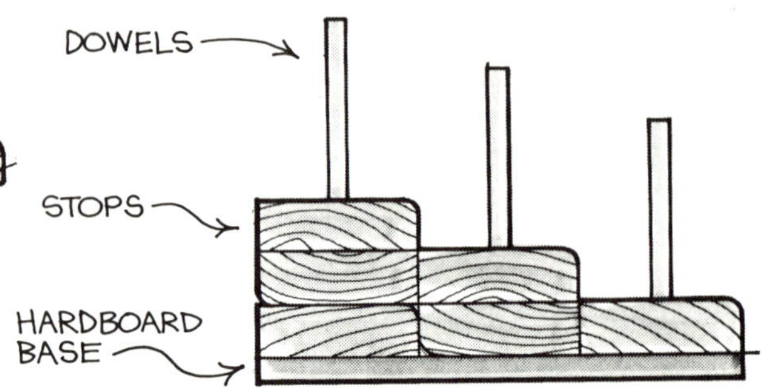

STAINED GLASS HANGING WINDOW/DIVIDER

This decorative window can either hang in front of another window or can be suspended as a room divider. All the glass is cut into rectangles and no lead or solder is used, so it is much simpler to make than most stained glass windows. An outer frame is made of large S4S stock and the inner frames are smaller S4S. Build all the framework to the size you want and nail securely. Do the uprights first, and then the cross pieces. The glass is to be held in place from both sides with small quarter round, so next, cut all the quarter round to the proper sizes. Miter the corners accurately. Then tack it in place on one side of the window only. Now is the time to stain or paint the frame. Cut the glass to the size panels you've selected and position the pieces in the window where you want them. Since this is to be an inside window, no putty is needed. Now tack in the quarter round on the remaining side. At this stage, it's a good idea to use a cardboard shield next to the glass so it will not be hit by the hammer. Use good sized screw hooks and chain for hanging.

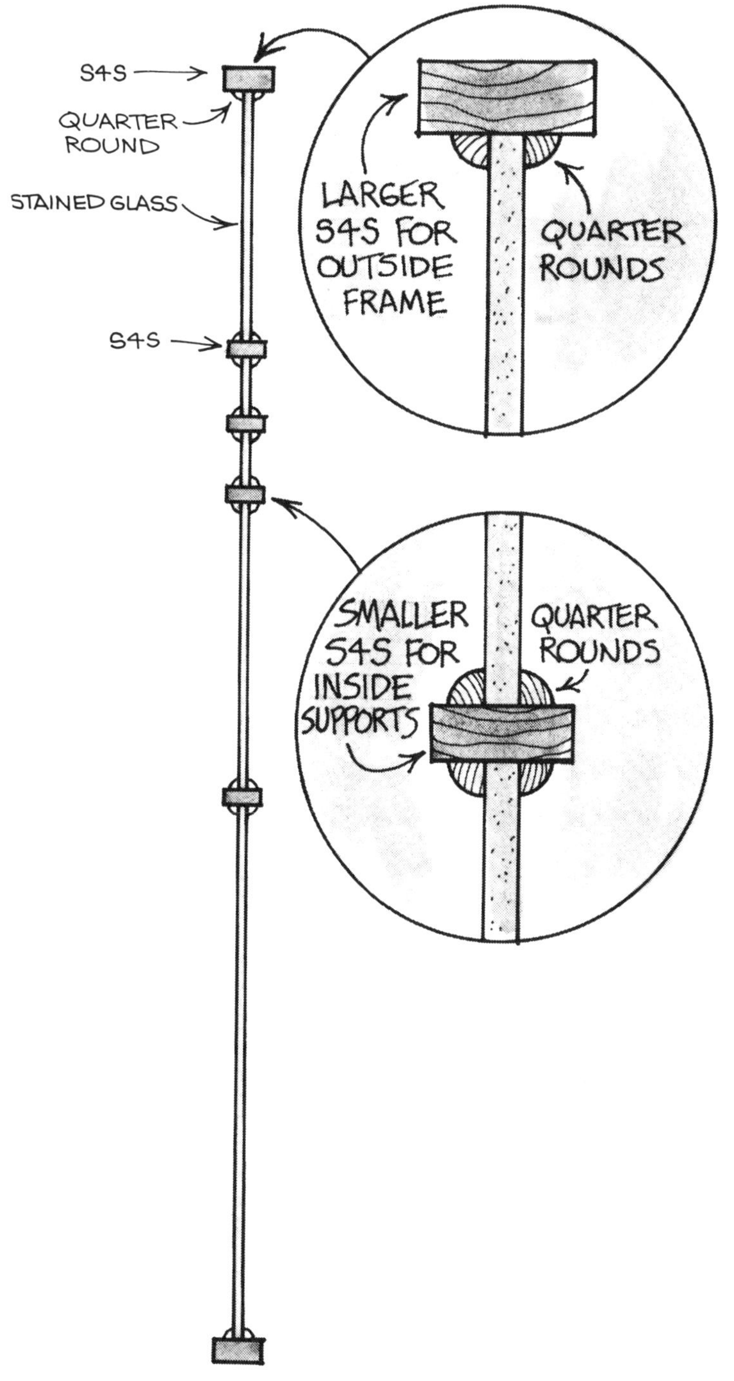

STORM WINDOWS

In these energy conscious days, a few cents spent now makes a lot of sense later. These do-it-yourself storm windows offer a good heat-loss solution to that winter time problem. And the expense is something you can live with 'til you get those factory made aluminum storm windows or the double-paned permanent windows.

To build these, all you need is S4S stock for the main framework and screen moulding or lattice for the tack-on frame. The plastic should be 8 or 10 mils thick. First, measure the house windows to be covered, then build your S4S frames accordingly. Either nail the ends together or screw into predrilled holes so the S4S doesn't split or use corner clamps on both sides of the frame. Next, cut the lattice to fit on top of the S4S. Paint or stain before putting on the plastic. Lay the plastic sheet over the frame, staple in place and trim off the excess. If the plastic is thinner than 8 mils, it will tear when stapled. The final step is to place the lattice over the plastic and nail in place. This makes it even more air tight, plus gives it a more pleasing finished look. How you attach your storm windows to your house windows will depend on what kind you have. Nailing them is the most common solution, but several other types of fastening cleats are available at your building supply center.

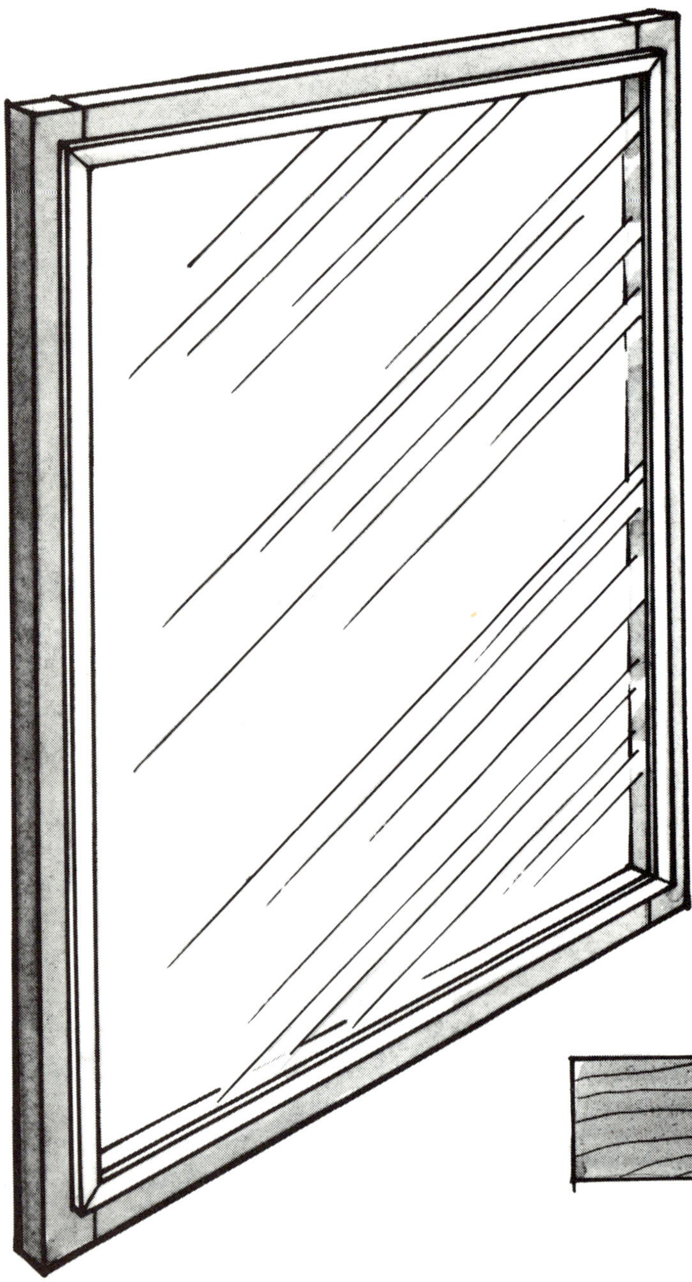

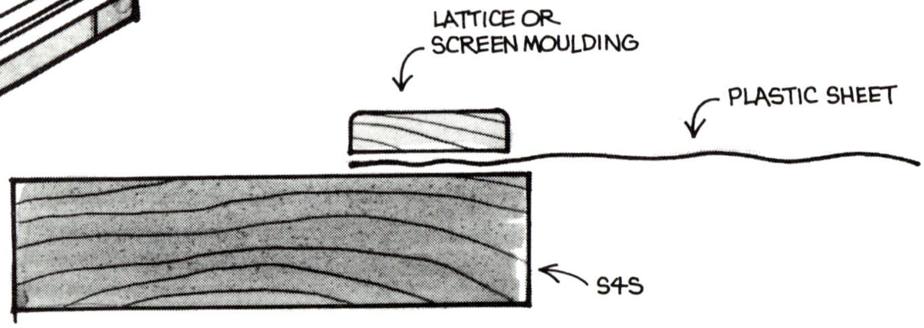

TELEPHONE RACK

If that shiny plastic telephone looks a little out of place in your traditional kitchen, build a rack of wood moulding around it and add a couple of shelves for that paper and pencil you always seem to need when you're on the phone. Use traditional base moulding for the sides and ⅜″ plywood for the backing. S4S or plywood can be used for the shelves. Measure your phone so the rack fits over it with a little extra space on the sides. More space is required at the top and bottom so the cover can be removed when necessary. There also needs to be cord room at the bottom. The phone is attached to the rack in the same way it's attached to the wall. Another alternative is to not have a backing on the rack and just slip the frame over the phone and attach the rack to the wall with small brackets under each shelf. If your kitchen has the modern look and you want a telephone rack, use ranch base instead of the traditional base and paint it a bright cheerful color.

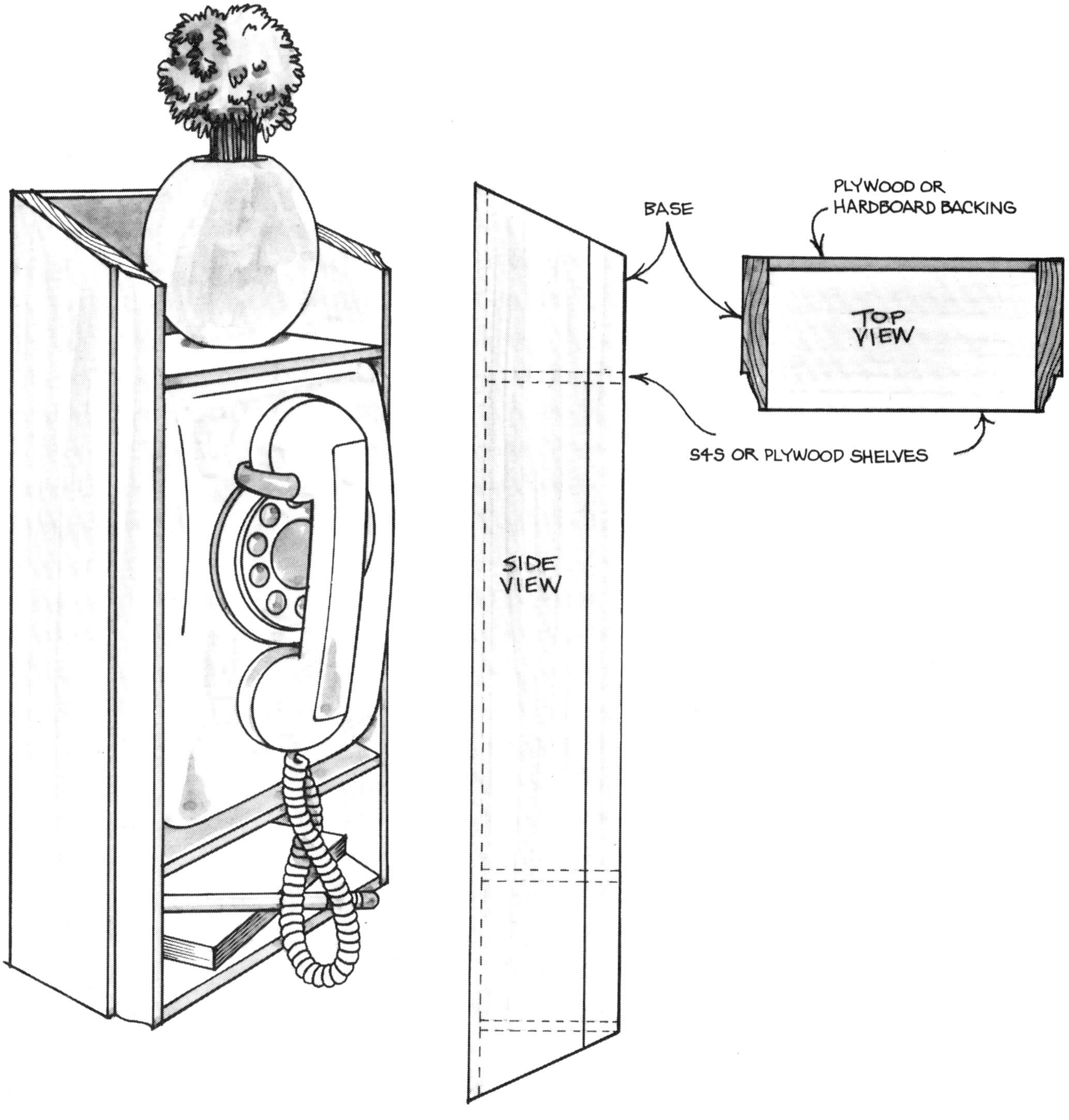

TIE RACKS

Now the man of the house can see all of his ties at a glance. No more wondering where his favorite one is. They're all right at his fingertips, whether he has six or sixty. You can probably make one of these handsome wood moulding tie racks in less than an hour. The one pictured here is a piece of simple base moulding with a double row of ¼″ dowels. Besides base, the tie rack can also be made from casing, hand rail, batten, shingle and crown moulding.

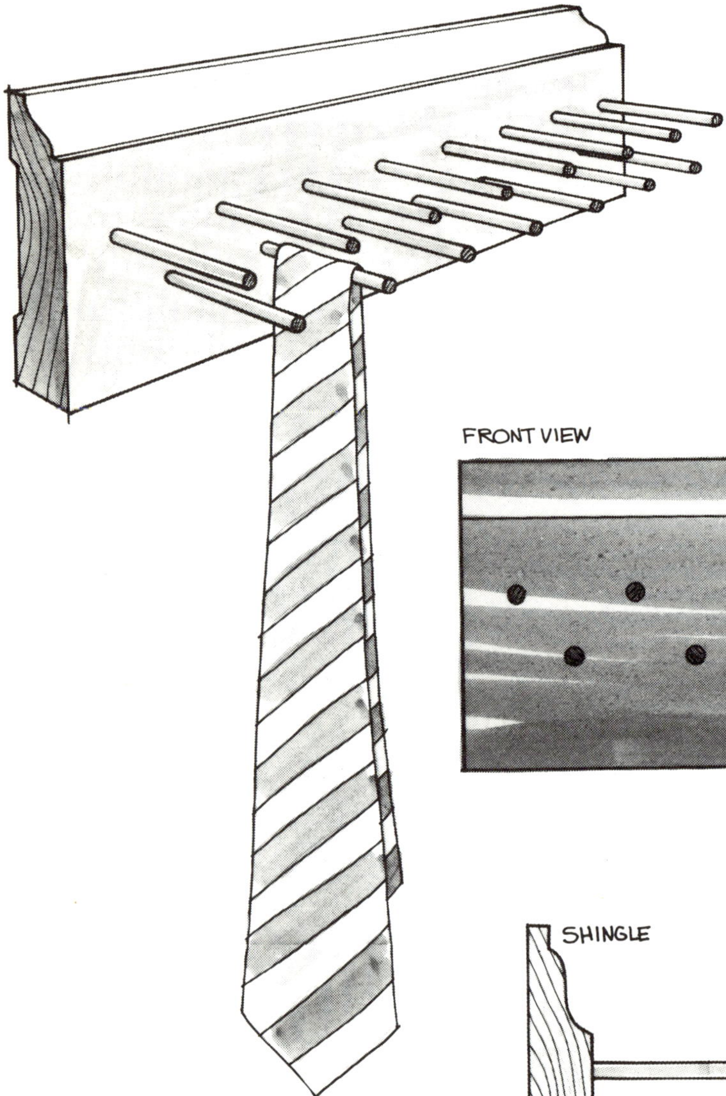

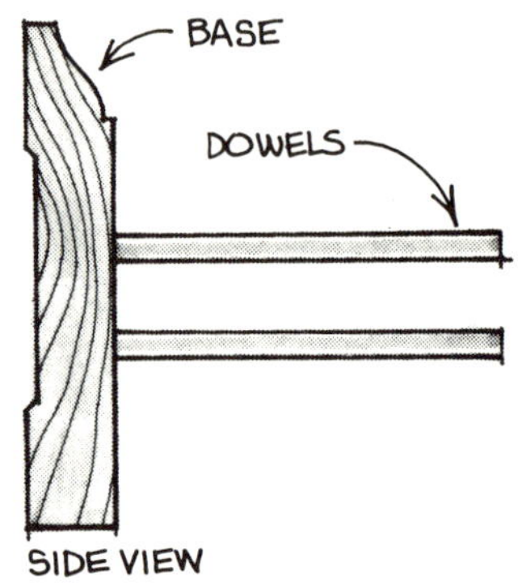

FRONT VIEW

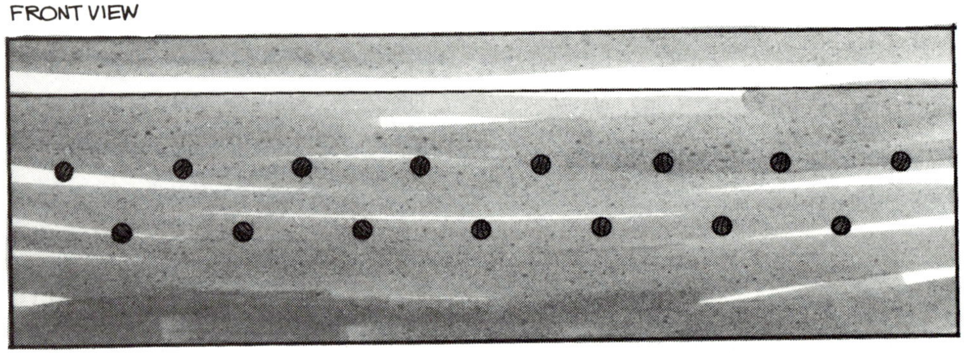

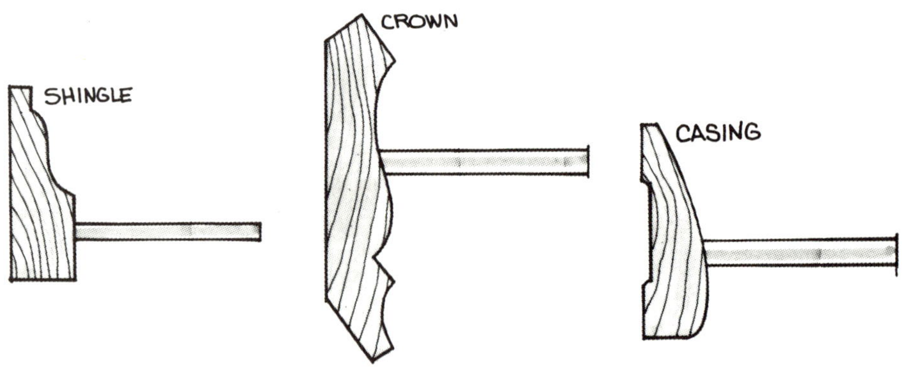

OPTIONAL PATTERNS:

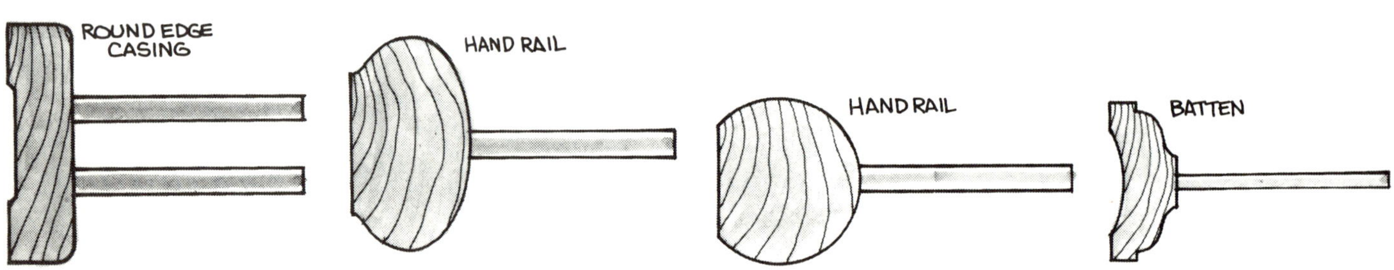

TOMATO PLANTER

Tired tomato plants get that way because their load gets heavy. As the tomatoes grow large, the plants droop and the crop ends up on the ground where they spoil. That can easily be prevented by giving them something to lean on.

This simple little fence, made of lattice nailed to quarter round legs, is a tomato lifesaver. Just put it around the young plants and soon they'll be keeping their crops where they belong.

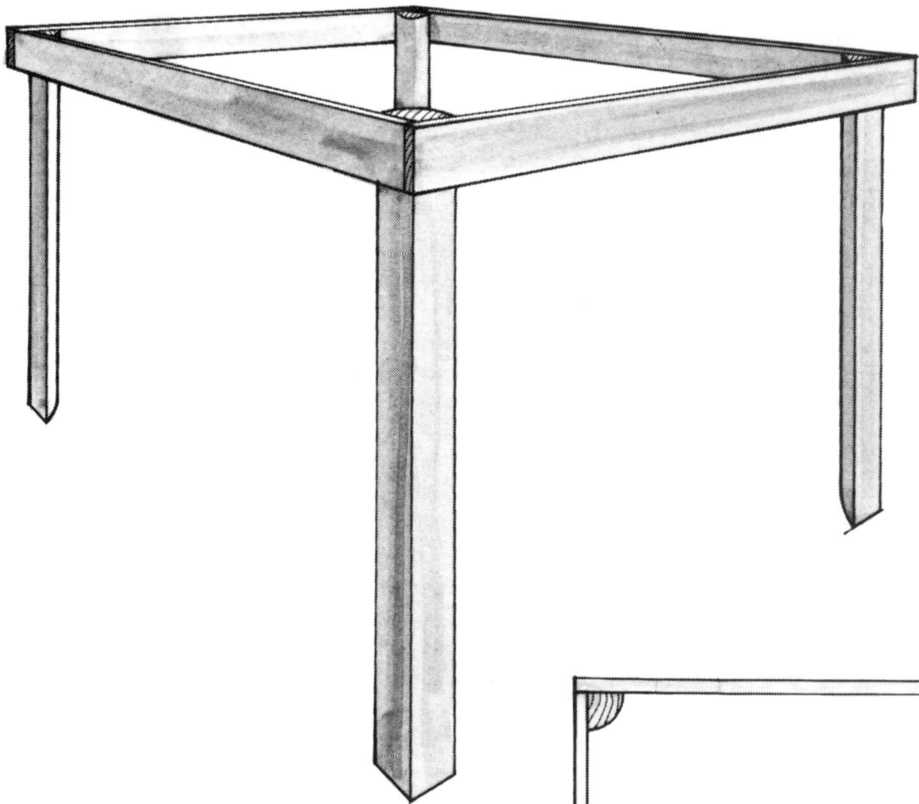

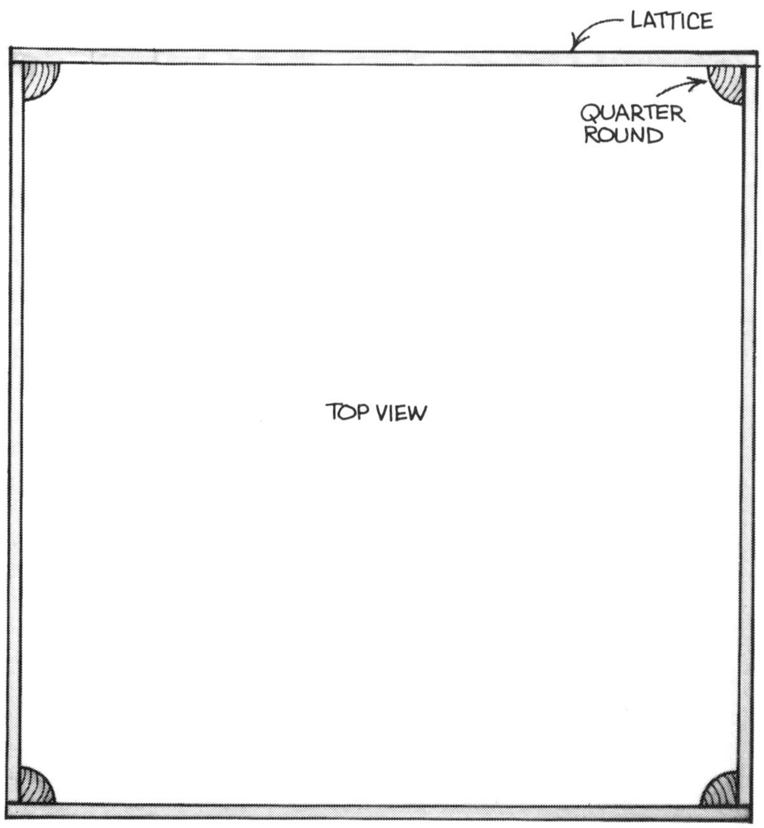

TOOTHBRUSH HOLDERS

A tooth brush holder performs a simple little job, but that doesn't mean it can't have character. When stained to show off the elegance of the wood, these easy-to-make little holders will add a dash of richness to your bathroom.

The first one below is a piece of S4S sandwiched between two coves. The corners are rounded off and ½″ holes are drilled. The number depends on the size of your family. The other holder is a piece of S4S with ⅝″ holes drilled about three inches deep. For decoration, a piece of chair rail can be added to one or both sides. Casing, base or base cap could also be used.

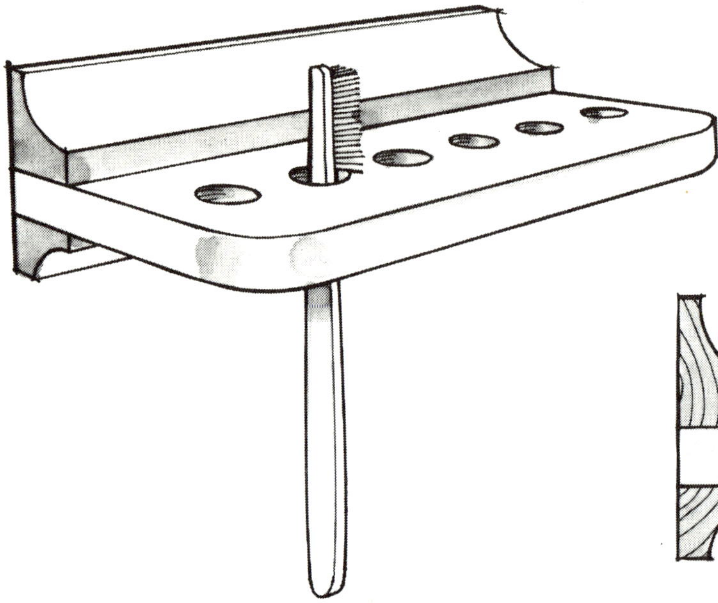

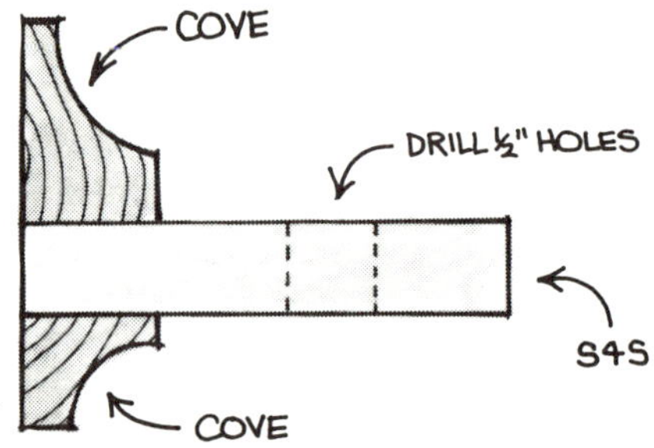

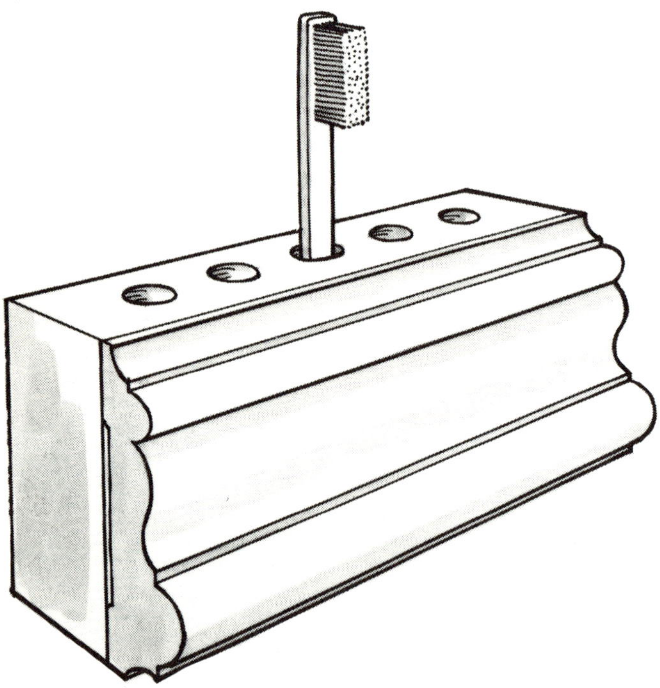

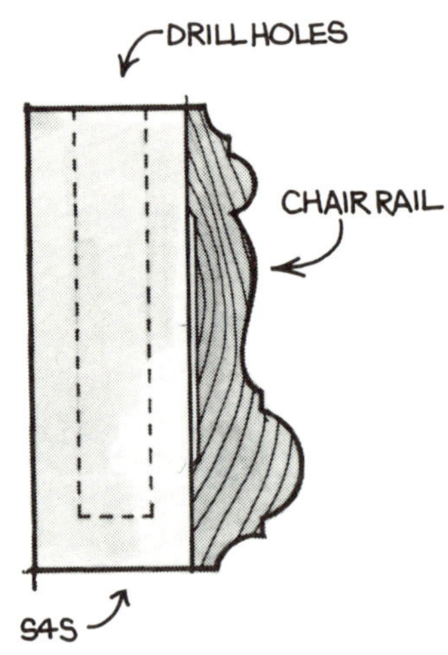

TOWEL RACKS

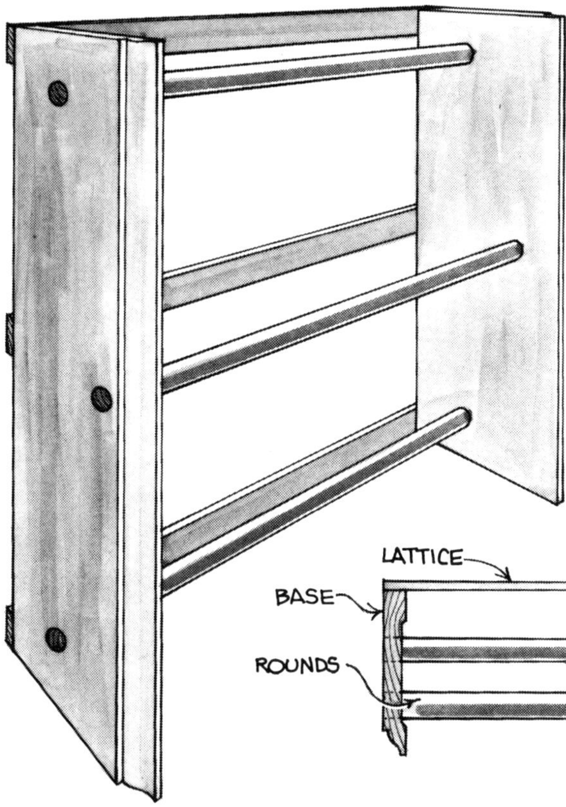

The towel rack on the left uses three moulding rounds between two upright ends of base moulding. Vary the placement of the rods so the towels will hang with space between them. Lattice is used on the back to attach the rack to the wall. The size of the unit depends on the wall space available.

The rack below is a strip of crown moulding with ½″ dowels about four inches long and spaced about four inches apart.

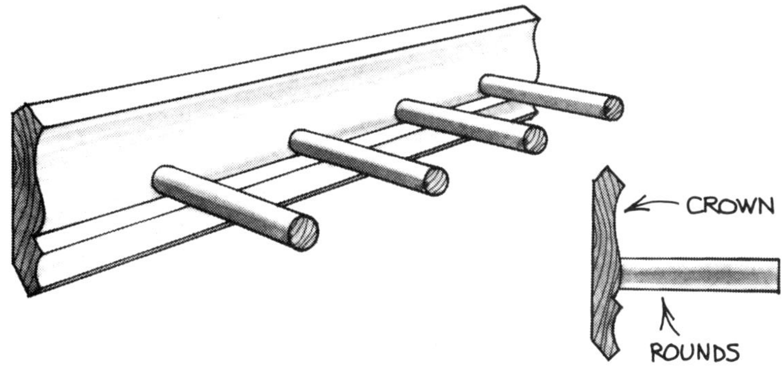

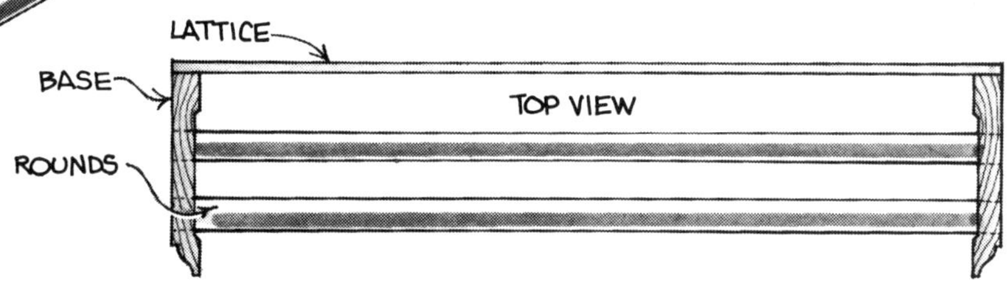

This towel rack uses square frames of drip cap spaced to form a center hole for the towel rod. Coves are nailed to the back pieces for attaching the rack to the wall. The length of these designs can be varied to hold towels, wash cloths or bathroom tissue.

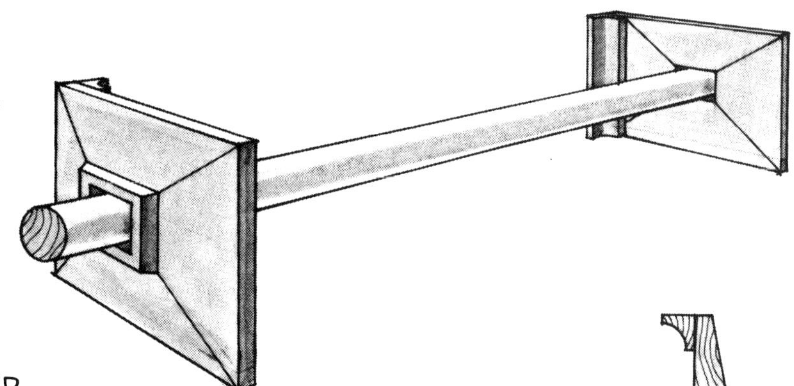

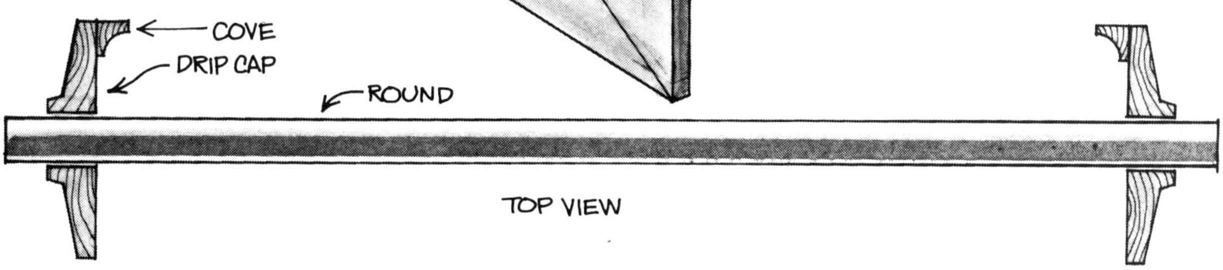

The rack to the right uses two pieces of casing back to back on each end. The inner casing is notched in back to hold a lattice strip for attaching the rack to the wall. The foreward edges of all four casings are coped in half circles to hold the towel rod which is glued in place and secured with screws in pre-drilled holes. When inserting the screws, clamp the casings together to prevent separating.

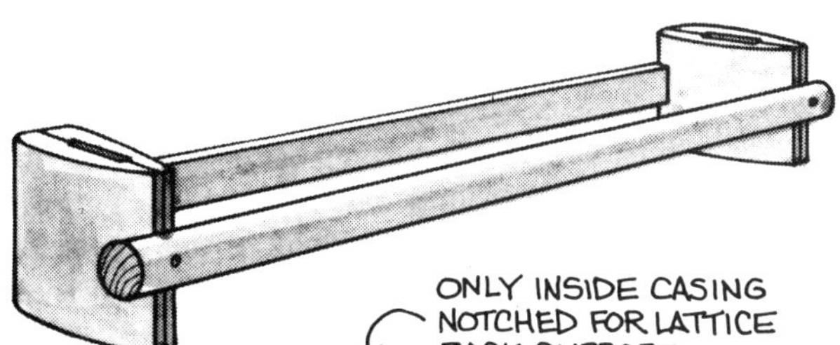

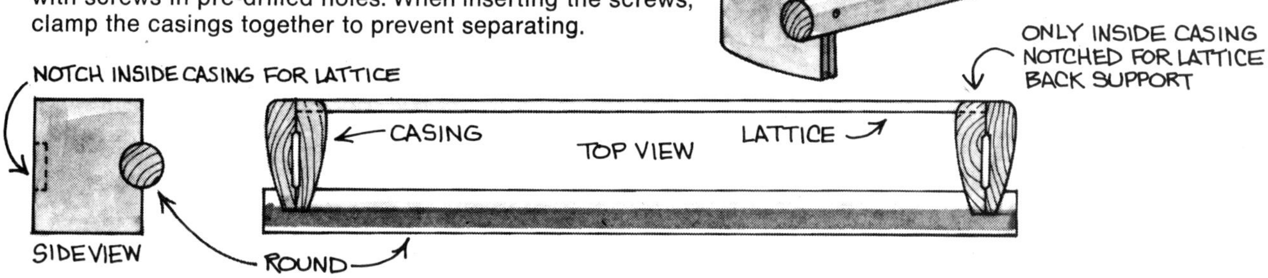

TOYS

Making toys yourself will bring out the kid in you. It's surprising how much fun a few hours of toy making can be and how much fun it'll bring to others. Most of these toys you've probably seen in toy stores or hobby shops before, but now try making variations of your own. Some short lengths of basic wood mouldings, a sharp saw, some glue and a little imagination is all you need to become an architect, builder, mayor, pilot, boat captain or zoo keeper.

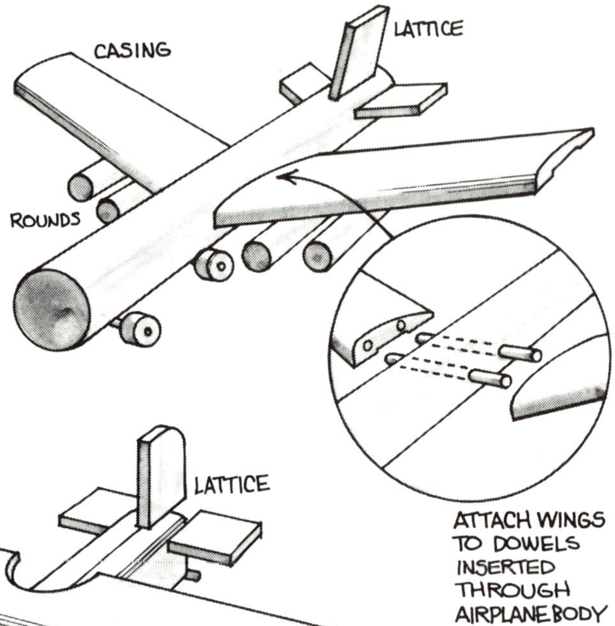

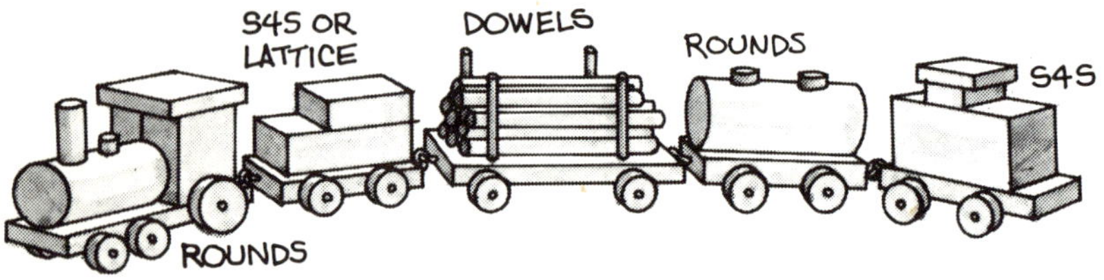

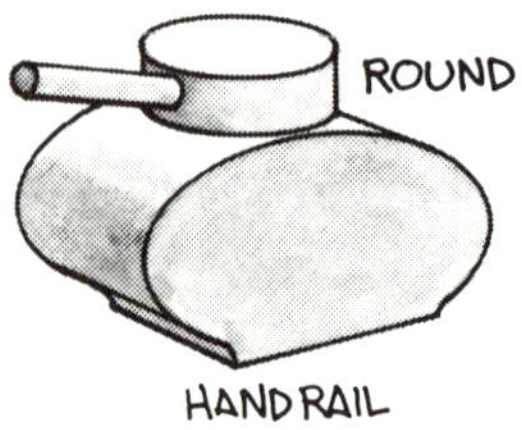

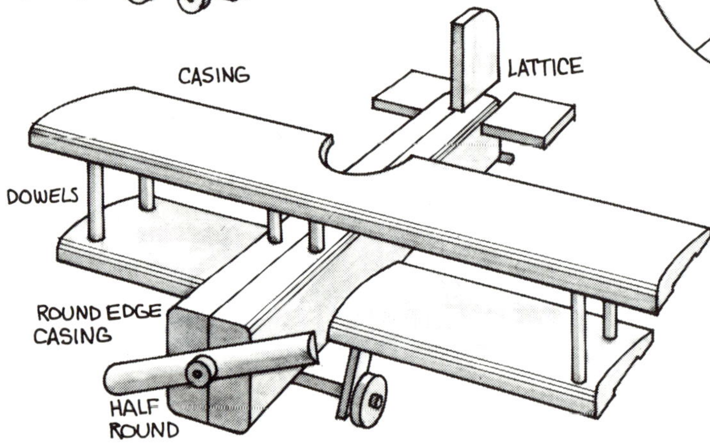

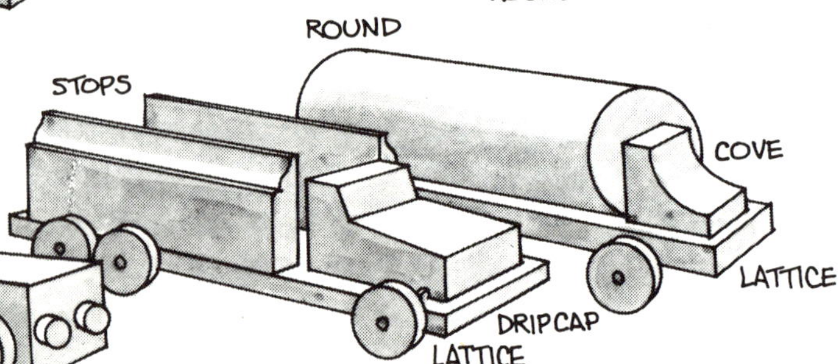

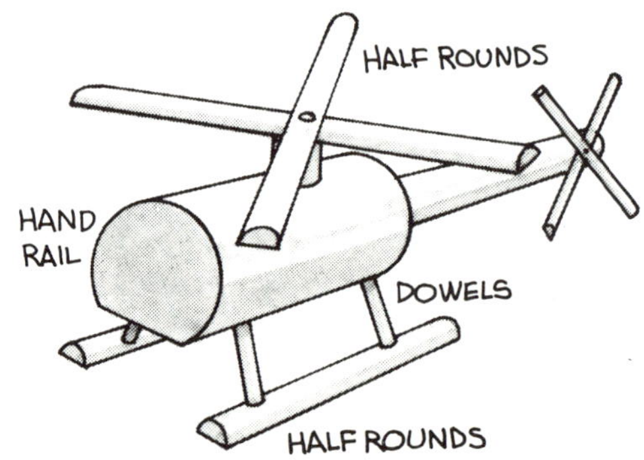

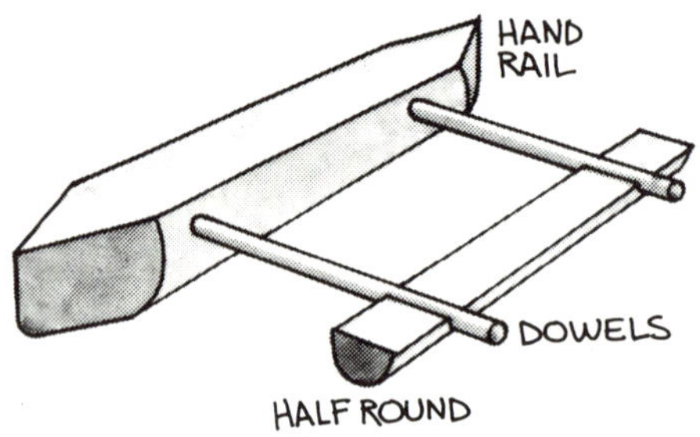

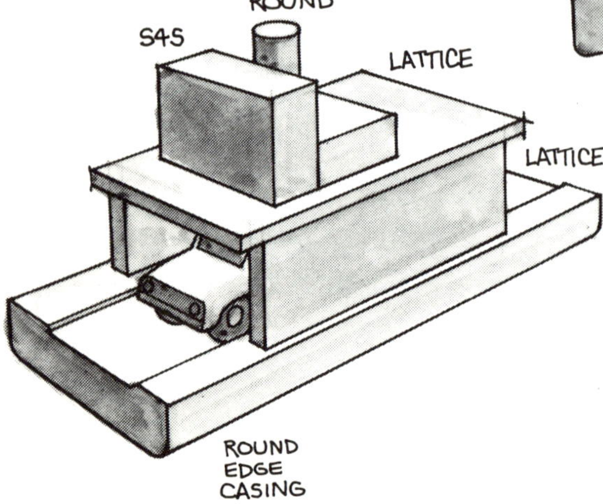

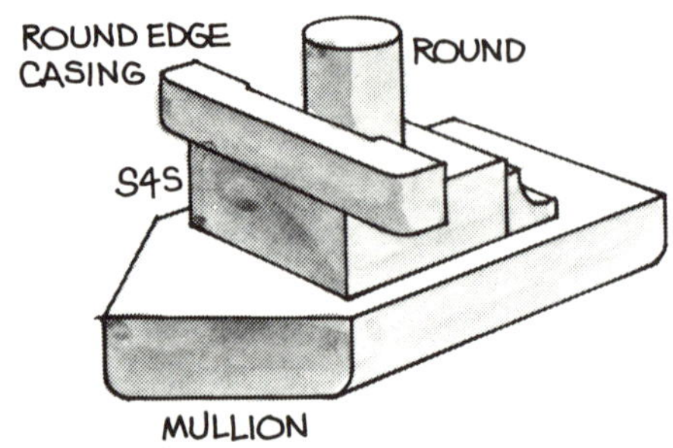

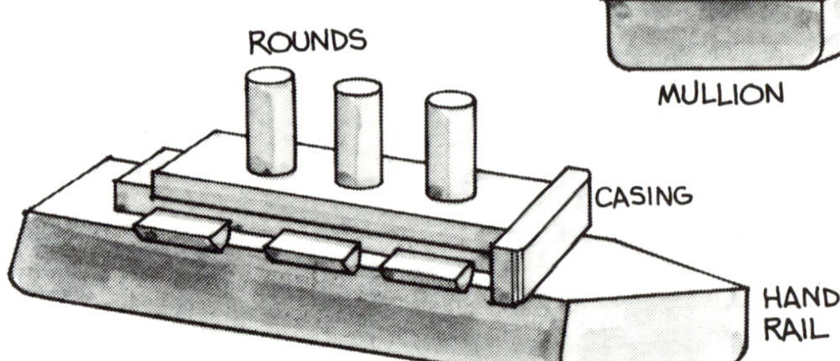

TOYS

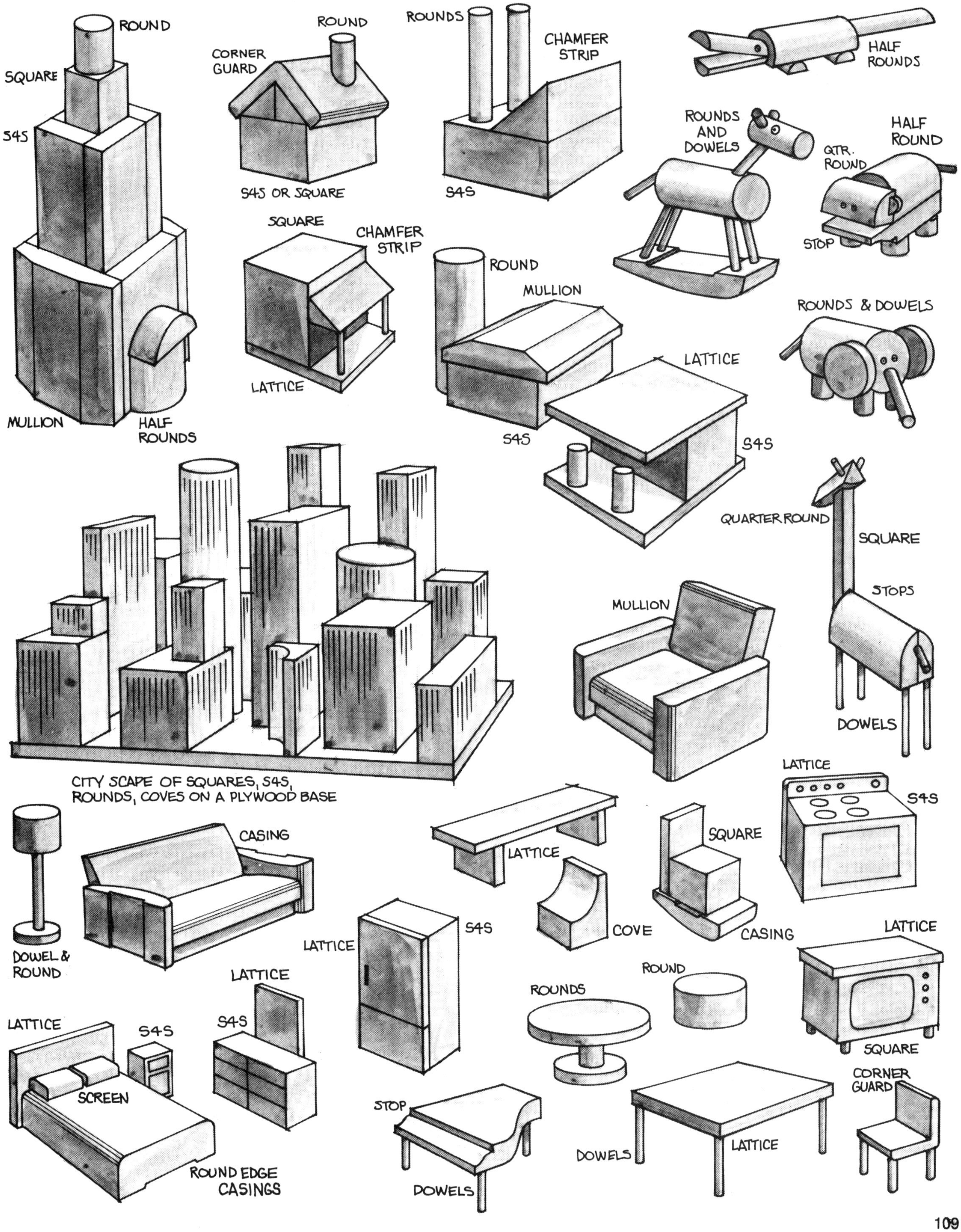
ROUND
SQUARE
S4S
MULLION
HALF ROUNDS
ROUND
CORNER GUARD
S4S OR SQUARE
ROUNDS
CHAMFER STRIP
S4S
HALF ROUNDS
ROUNDS AND DOWELS
HALF ROUND
QTR. ROUND
STOP
SQUARE
CHAMFER STRIP
LATTICE
ROUND
MULLION
S4S
LATTICE
S4S
ROUNDS & DOWELS
QUARTER ROUND
SQUARE
STOPS
DOWELS
MULLION
CITY SCAPE OF SQUARES, S4S, ROUNDS, COVES ON A PLYWOOD BASE
CASING
DOWEL & ROUND
LATTICE
COVE
SQUARE
CASING
LATTICE
S4S
S4S
LATTICE
LATTICE
LATTICE
S4S
S4S
SCREEN
ROUND EDGE CASINGS
ROUNDS
ROUND
STOP
DOWELS
DOWELS
LATTICE
SQUARE
CORNER GUARD

TRELLISES

A few pieces of lattice or S4S is all it takes to make a home for your climbing plants. Most trellises are used outdoors, but they can be used indoors, too. A handsome wood trellis with an exotic bit of greenery makes an excellent room divider or wall decoration.

The trellises on these pages are made of lattice nailed to a framework of S4S or squares. These designs can also be used indoors and out. Paint or stain according to where they'll be used.

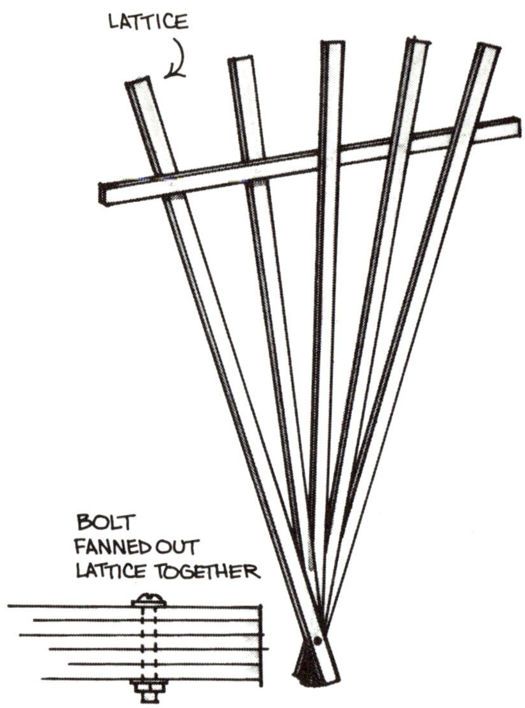

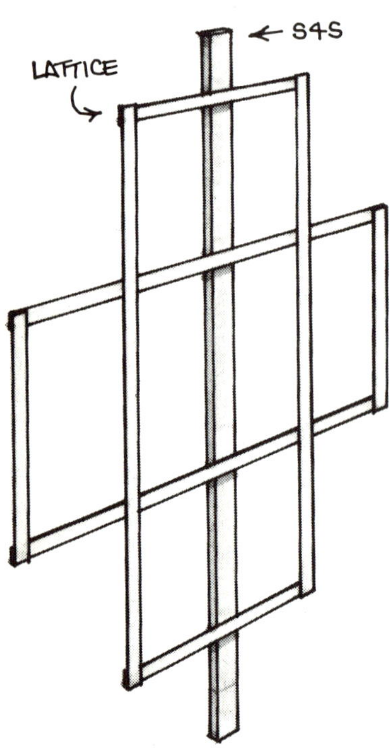

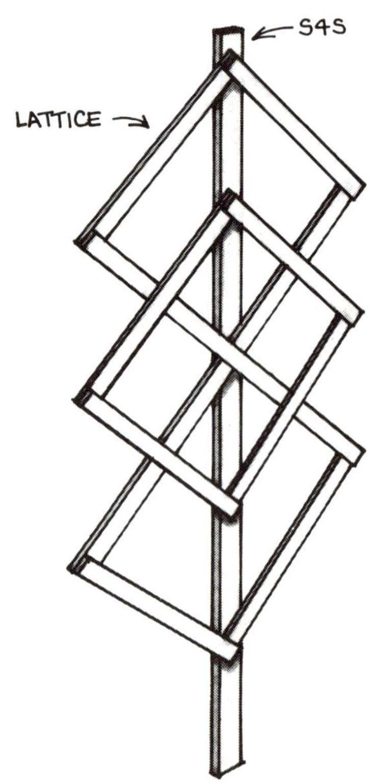

TRELLISES

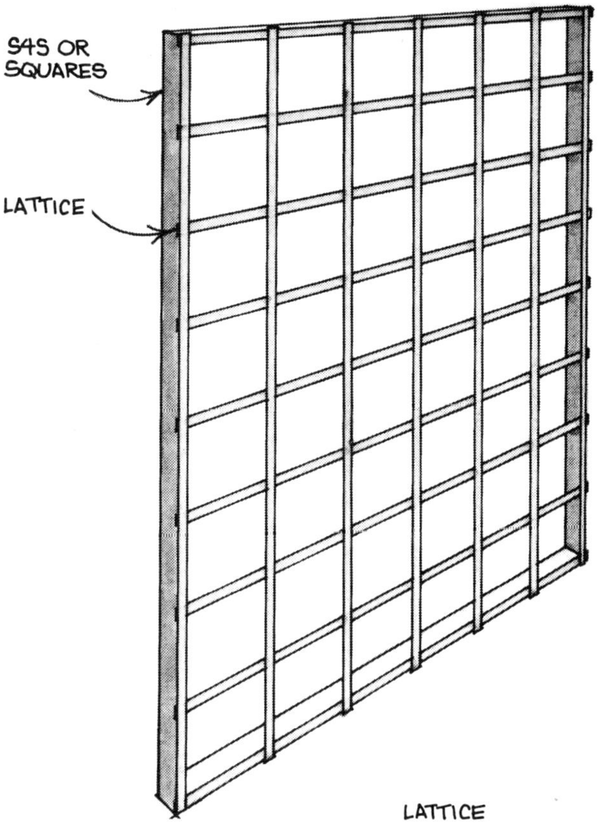
S4S OR
SQUARES
LATTICE

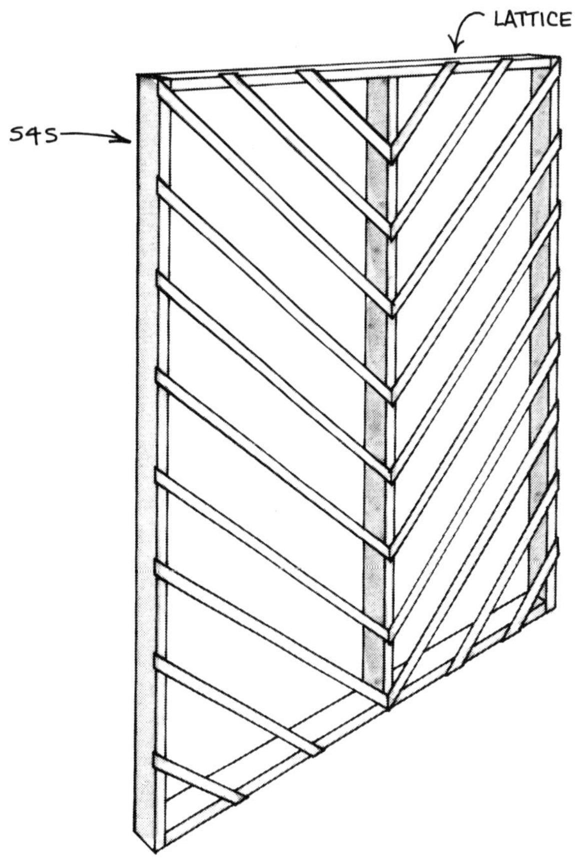
LATTICE
S4S

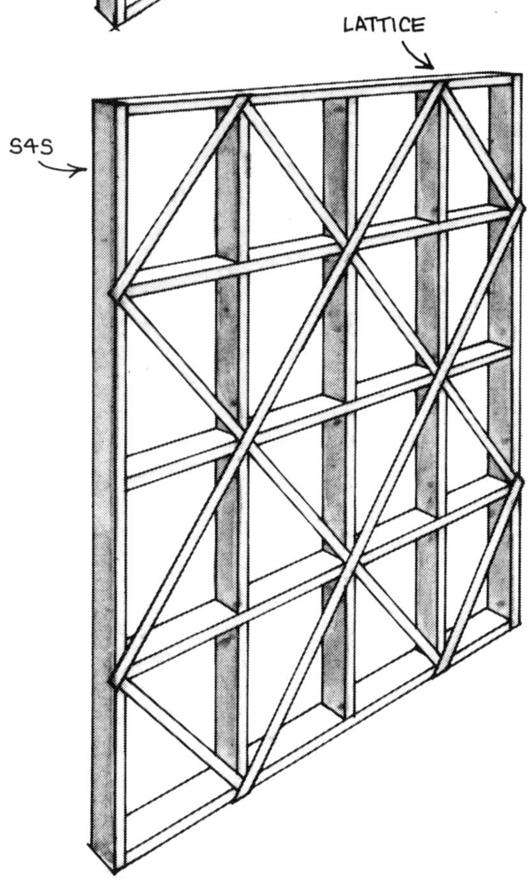
LATTICE
S4S

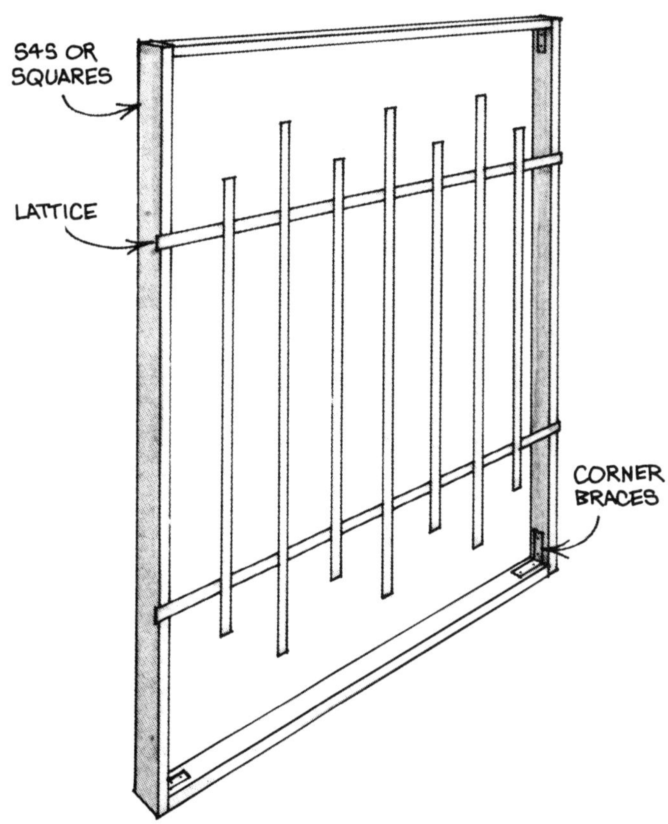
S4S OR
SQUARES
LATTICE
CORNER
BRACES

TRIVETS

Originally, trivets were tri-pods that held hot cooking pots. Now, any kind of rack or pad that holds hot dishes is called a trivet, but something with such a basic function as surface protection need not look ordinary. Building them with mouldings will add the warmth of wood to your table. The simplicity of the designs below are self-explanatory. After gluing, stain or oil to the richness you desire.

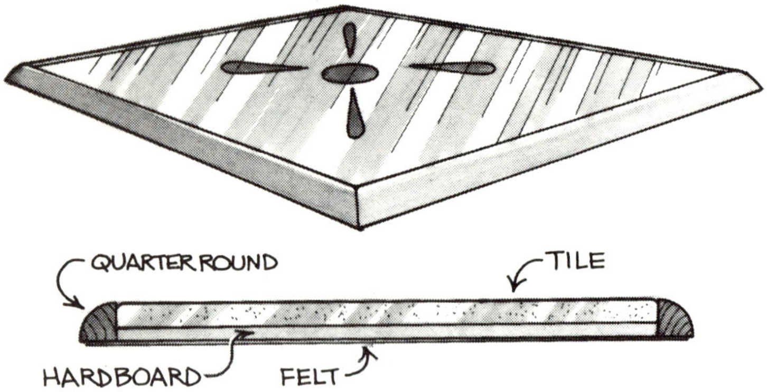

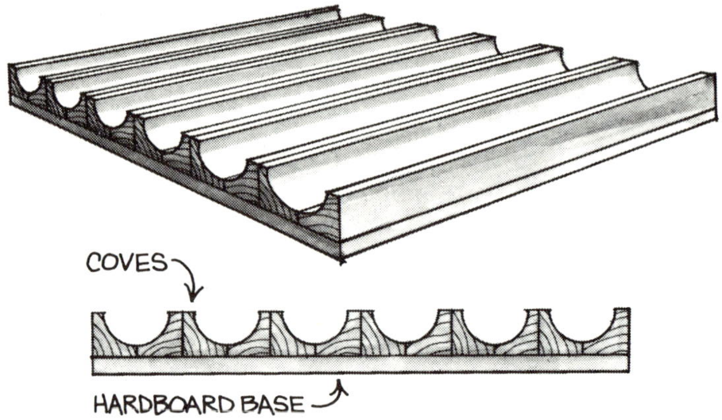

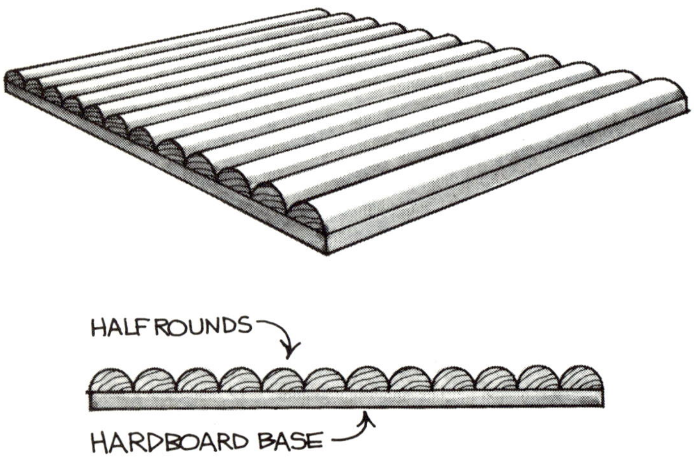

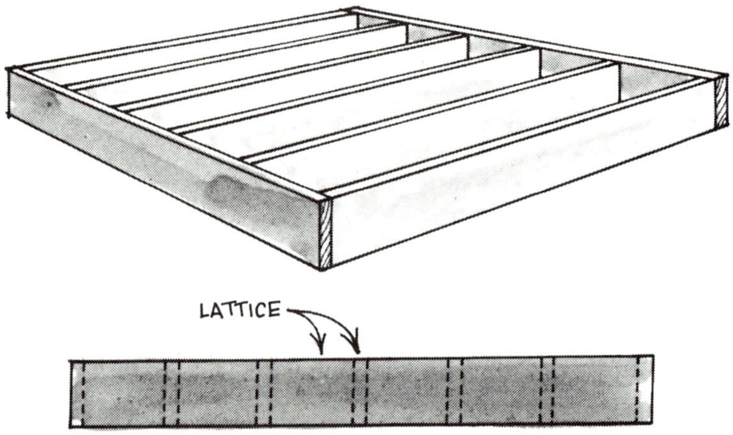

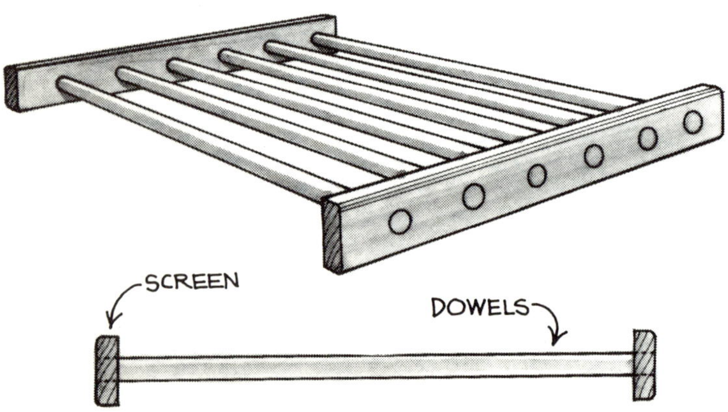

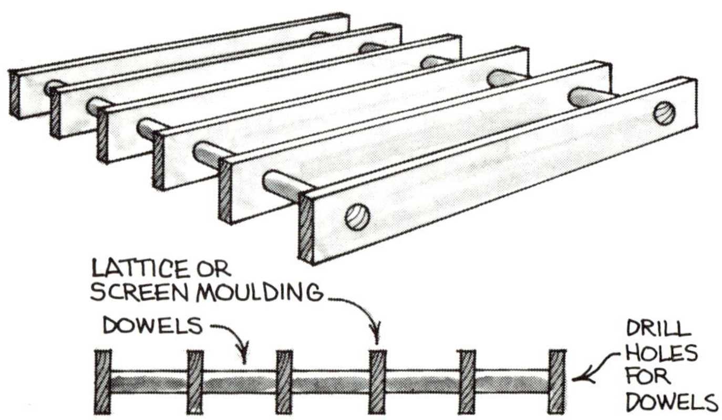

UMBRELLA STANDS

Umbrella stands are probably not as predominant in most homes nowadays as they once were, but if you can use one, why not build it yourself? The four handsome umbrella containers illustrated on this page are all simple plywood boxes covered in various ways with wood mouldings.

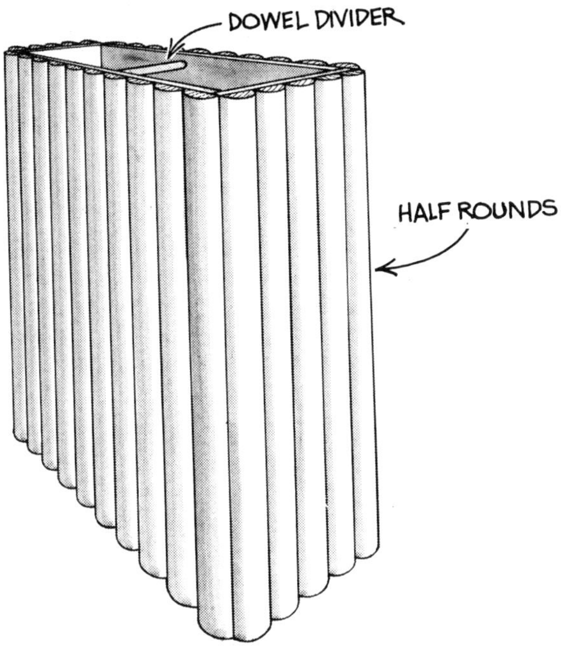

The first one uses vertical strips of half rounds. Small coves, battens, astragals or mullions can also be used. A dowel is inserted into holes drilled in the plywood sides to act as a divider.

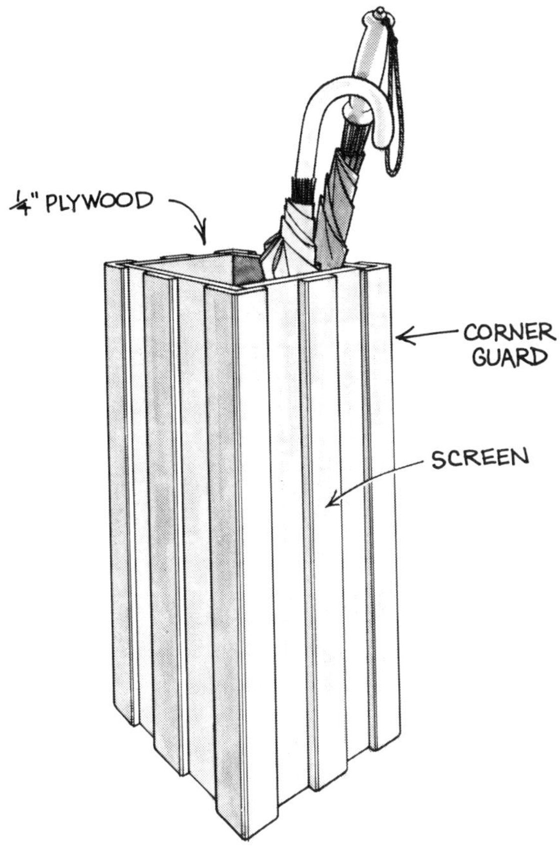

This stand uses corner guards at each corner with screen moulding in between.

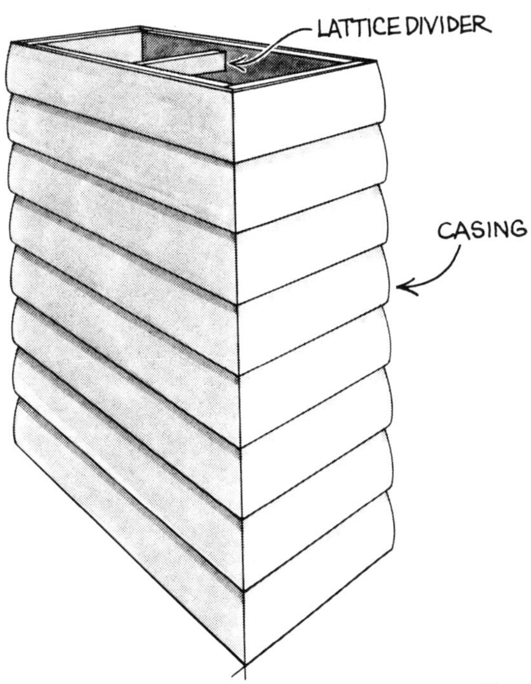

This box is covered with horizontal rows of casing. The corners are mitered, so good measurements are a must. Either traditional or ranch casing can be used. A lattice divider strip is nailed and glued in place.

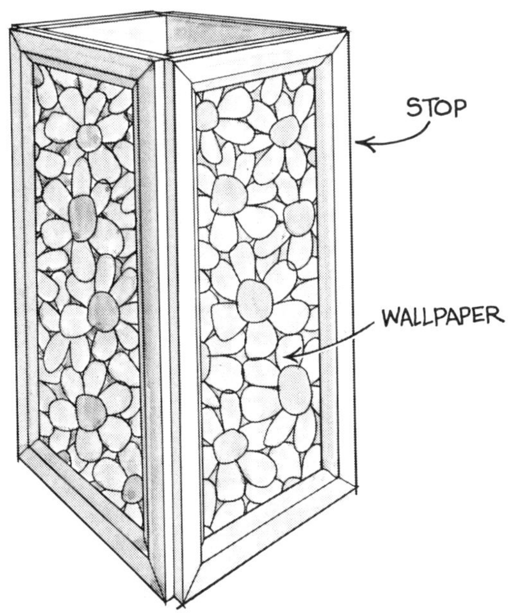

Here, the plywood box is covered with either wallpaper or vinyl wall covering and then framed with stop moulding. Base cap, small cove, casing, screen, lattice, mullion, shingle or drip cap can also be used.

UTENSIL RACK

Keeping your kitchen utensils in view where they're handy to get at is the purpose of this charming kitchen rack. This particular design has the look of a traditional early American kitchen, and is built with a combination of crown moulding and mullion on a background of S4S and traditional stop. Use square bend screw hooks to hang the utensils. Staining it maple, oak or walnut will also add to the desired appearance. If a more contemporary look is needed, use a large cove instead of a crown and ranch stops instead of the more ornate patterns. A bright paint job will also add to the modern look.

A variey of other utensil rack possibilities are shown below.

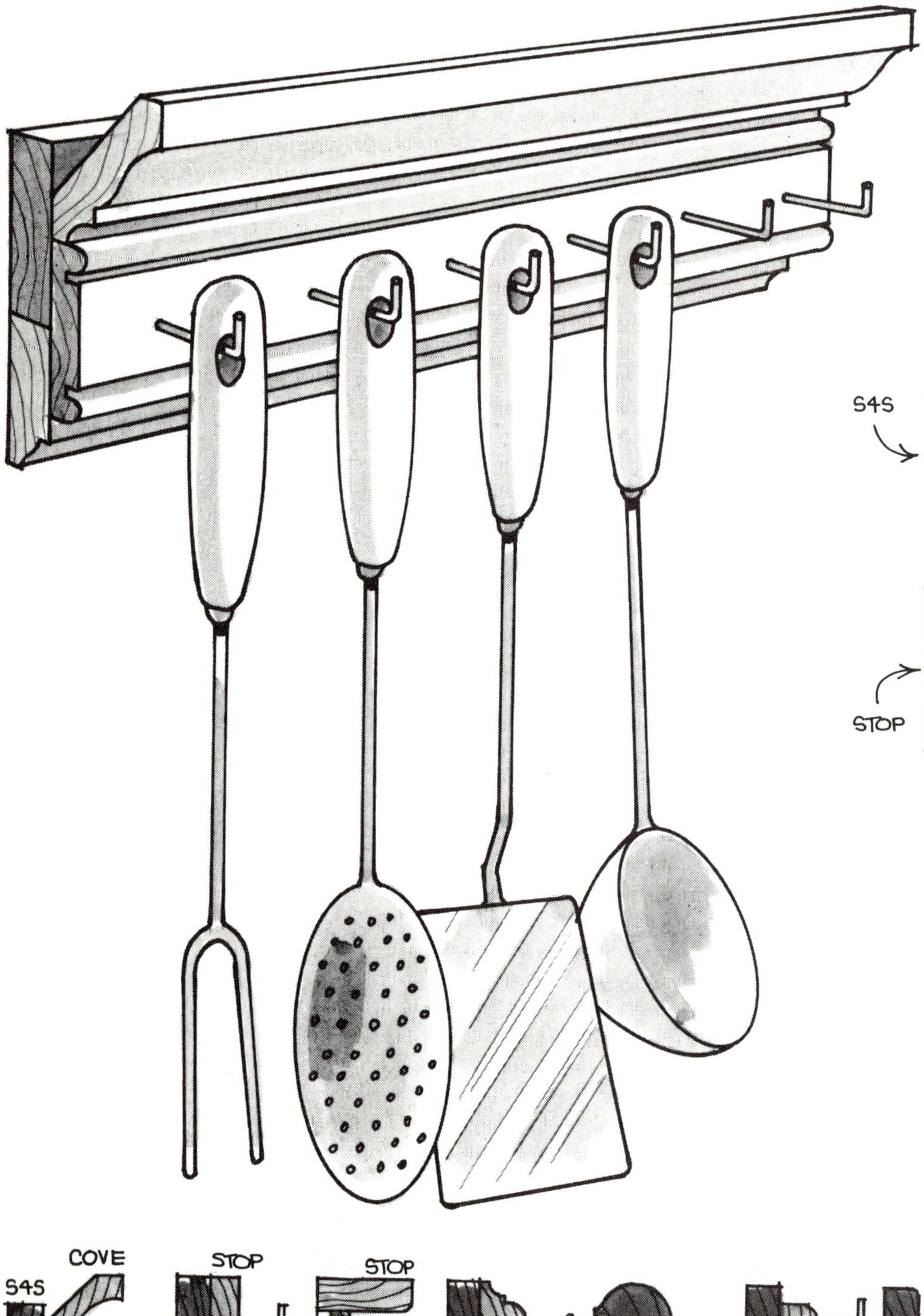

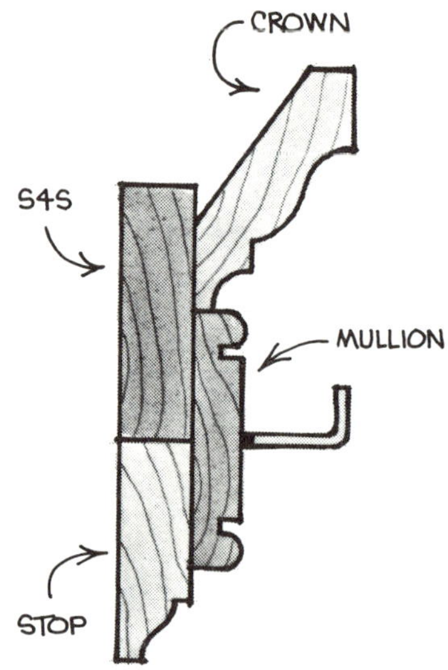

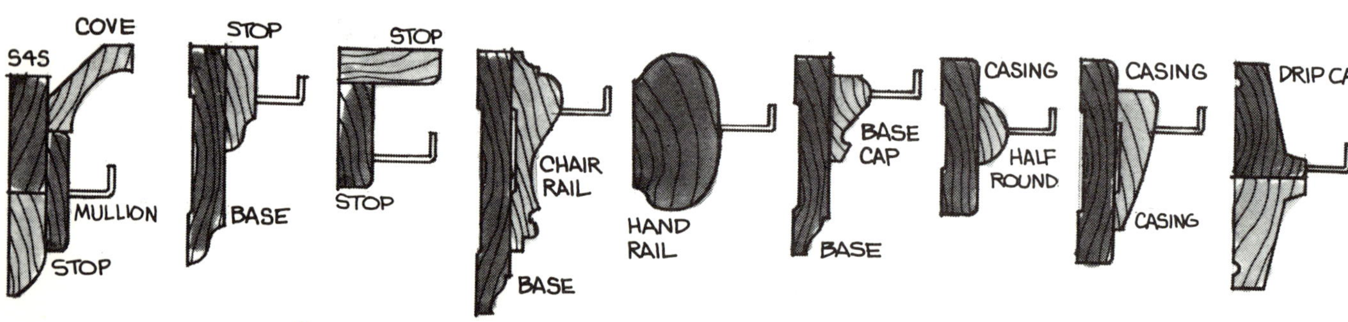

VALANCES

A valance is a bordering structure used to cover the top of draperies, disguise or redirect lighting or hide a shower rod. Although used for many things, they are simple to build.

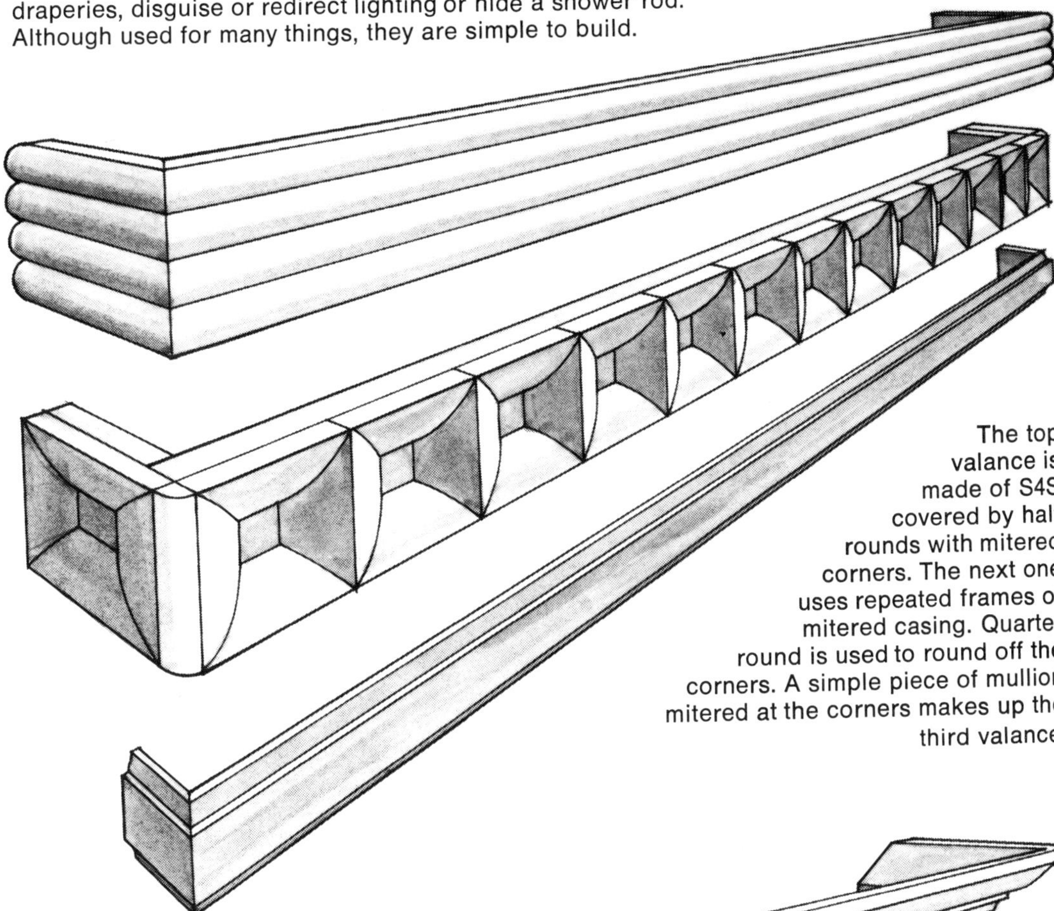

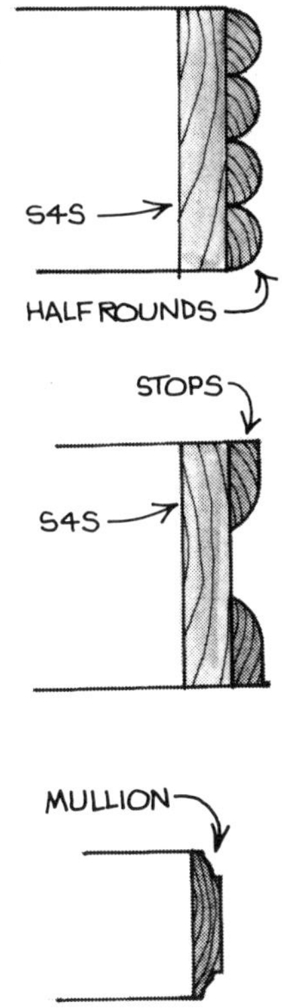

The top valance is made of S4S covered by half rounds with mitered corners. The next one uses repeated frames of mitered casing. Quarter round is used to round off the corners. A simple piece of mullion mitered at the corners makes up the third valance.

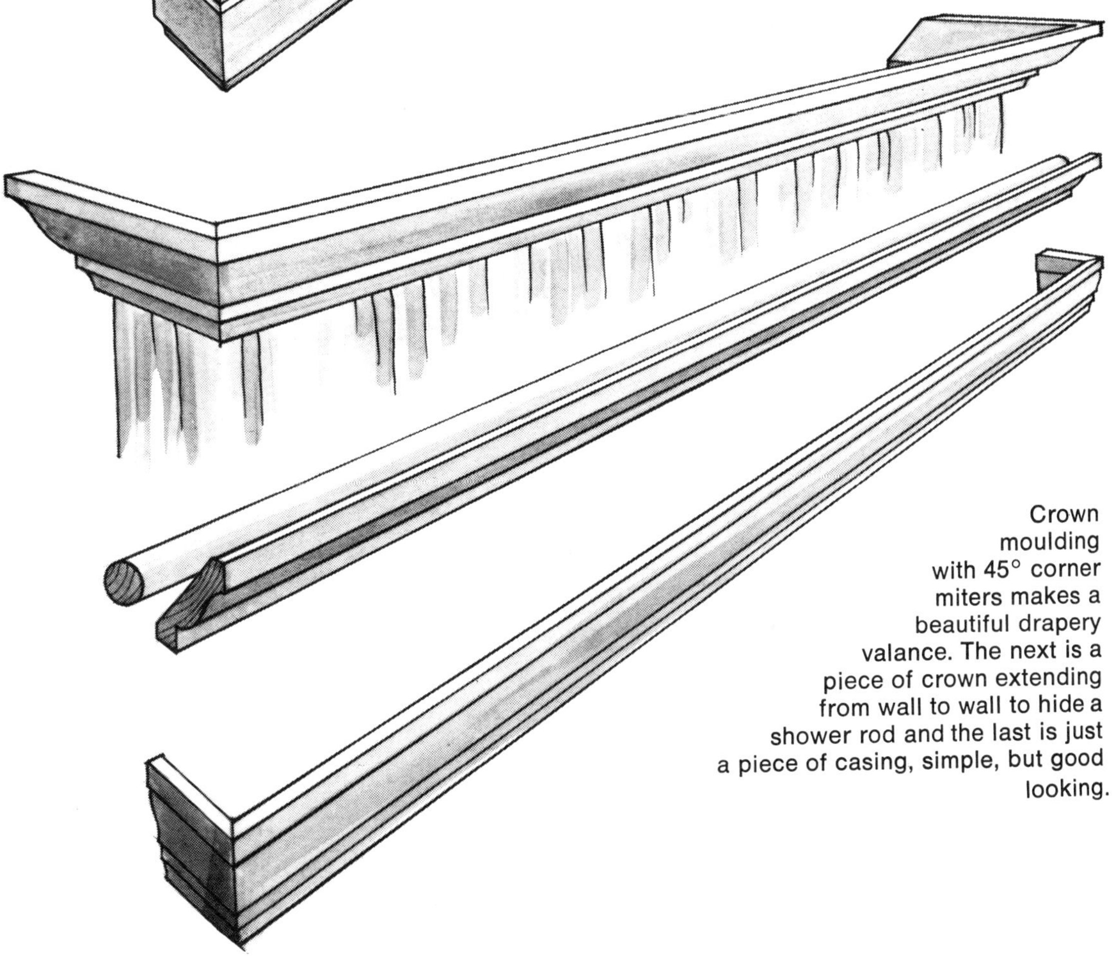

Crown moulding with 45° corner miters makes a beautiful drapery valance. The next is a piece of crown extending from wall to wall to hide a shower rod and the last is just a piece of casing, simple, but good looking.

WALL GRAPHICS

Dress up your walls with individuality. You can add excitement, color, elegance, whatever you want with the three-dimensional uniqueness of wood mouldings. Whether it's a large free-flowing form, a sunburst of color, a floor-to-ceiling stylized tree, a collage of patterns, a giant super-graphic of wood or a quiet panel design of oil stained elegance, wood mouldings can do it for your walls. The wall graphic samples pictured here are meant to be only appetizers to get you started on your own imaginative wall creations.

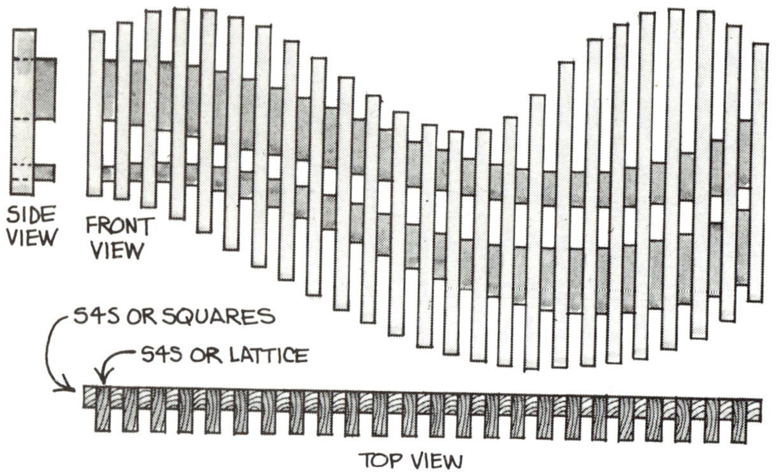

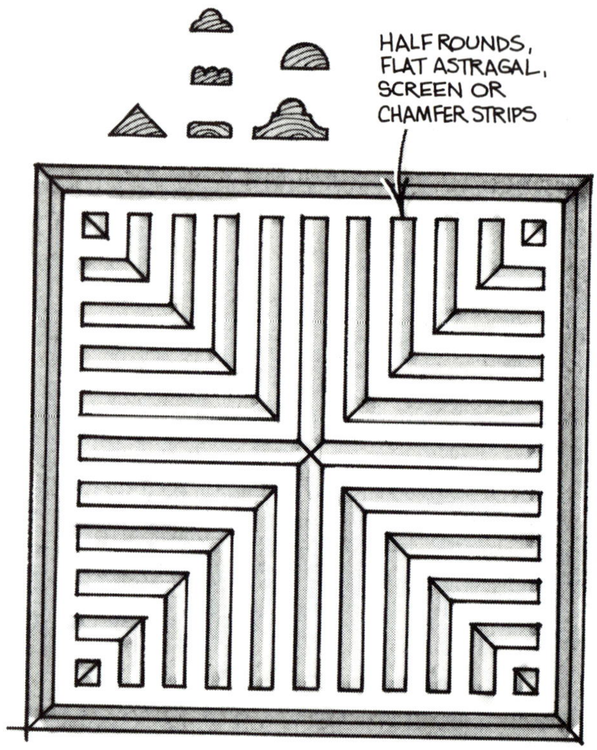

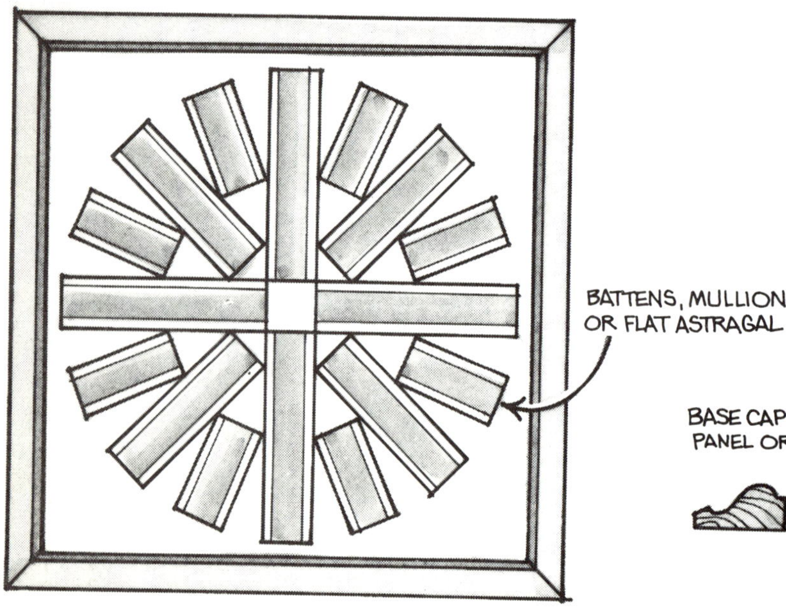

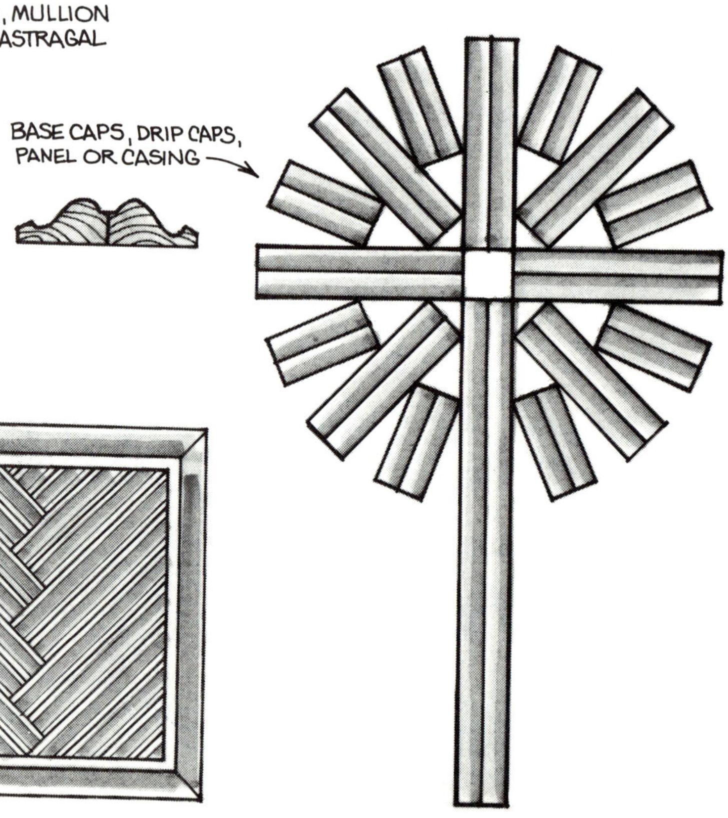

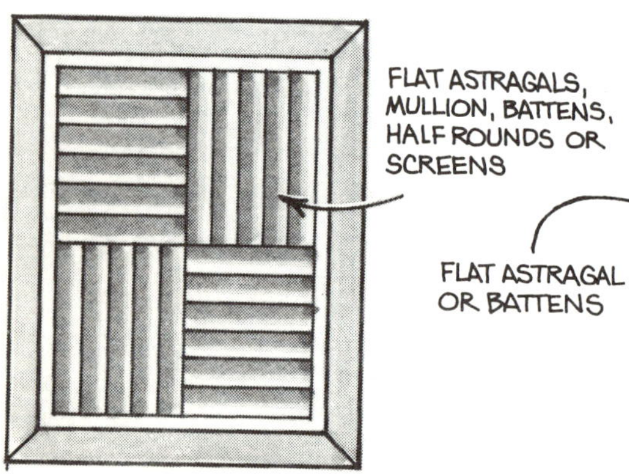

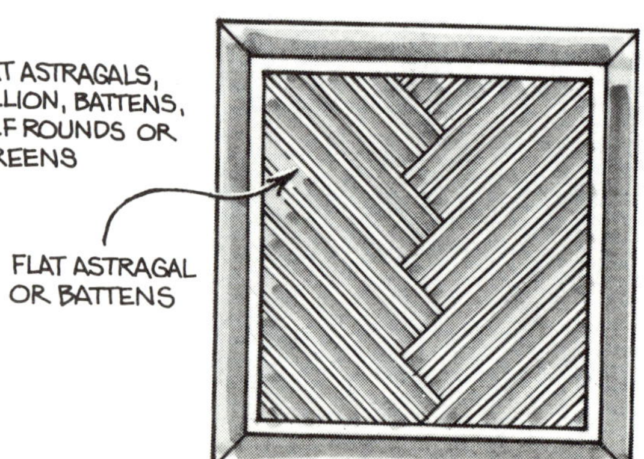

WALL GRAPHICS

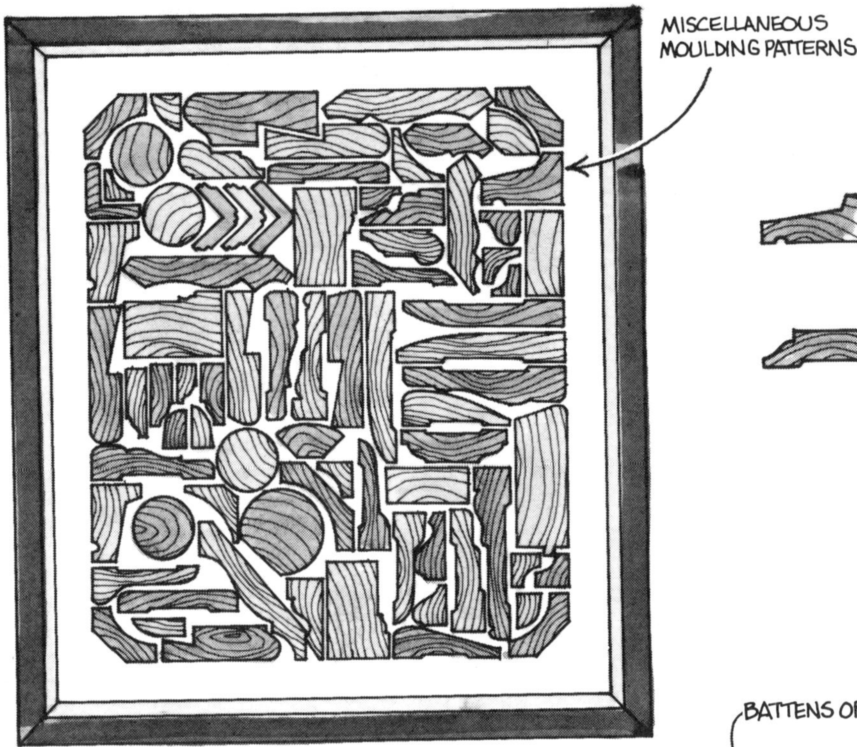

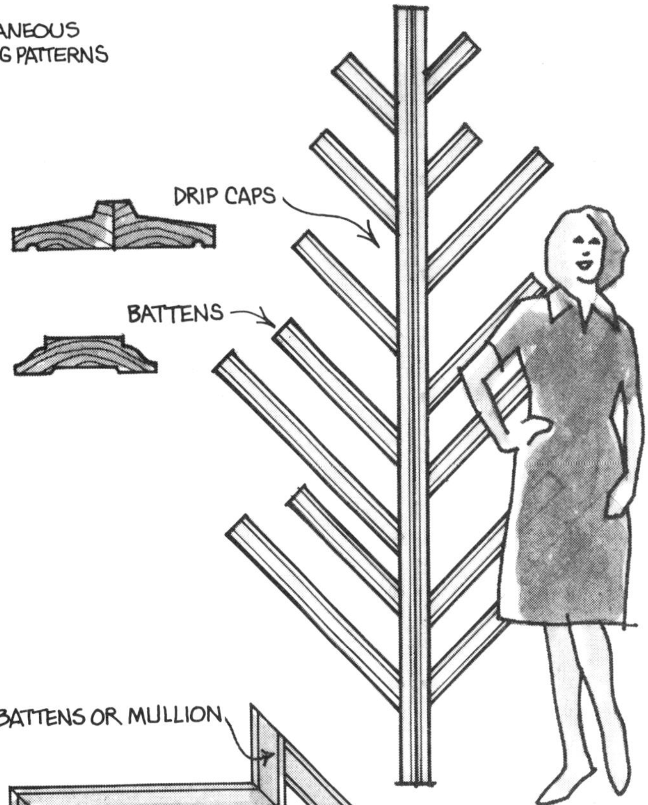

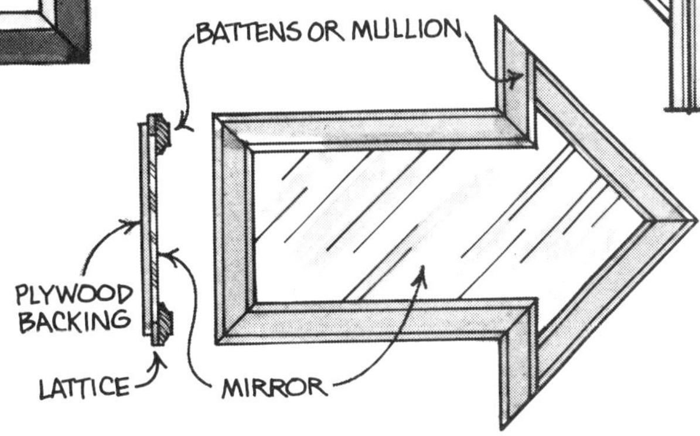

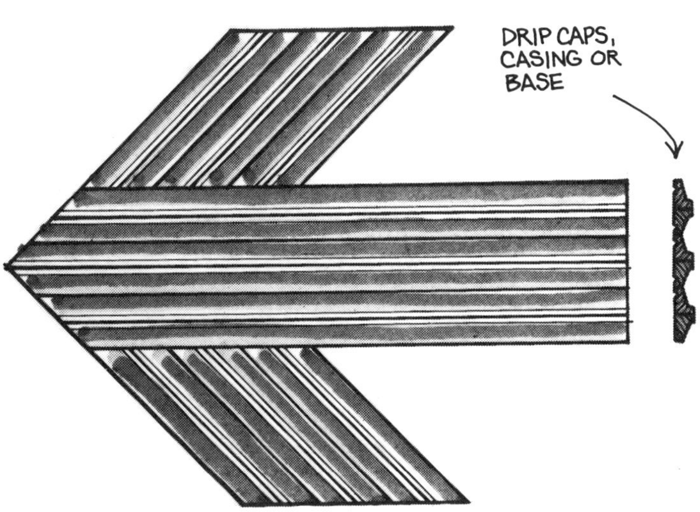

WALLS

How easy it is to transform an ordinary wall into a wall of uniqueness, beauty, charm, color, dimension and excitement. You name it. Want it to have the quiet purity of colonial elegance? The rustic charm of the old west? The dazzle of the disco scene? Whether it's the look of yesterday or today, the endless uses of wood moulding can give your wall the character you want. The next few pages illustrate a wide selection of wall treatments, ranging from classic simplicity to richly accented variations. Also pictured below are a variety of chair rail possibilities, plus base variations.

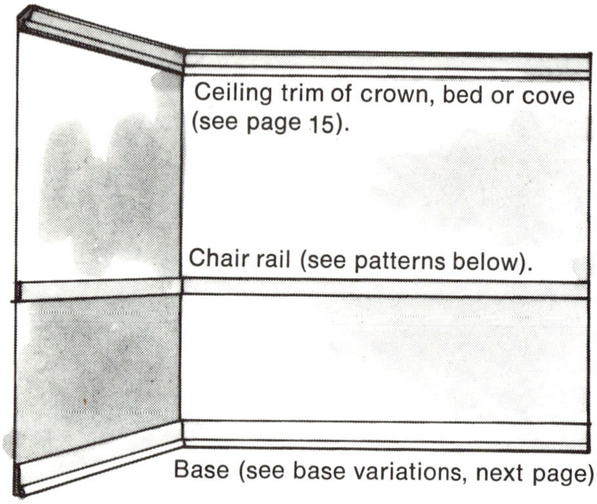

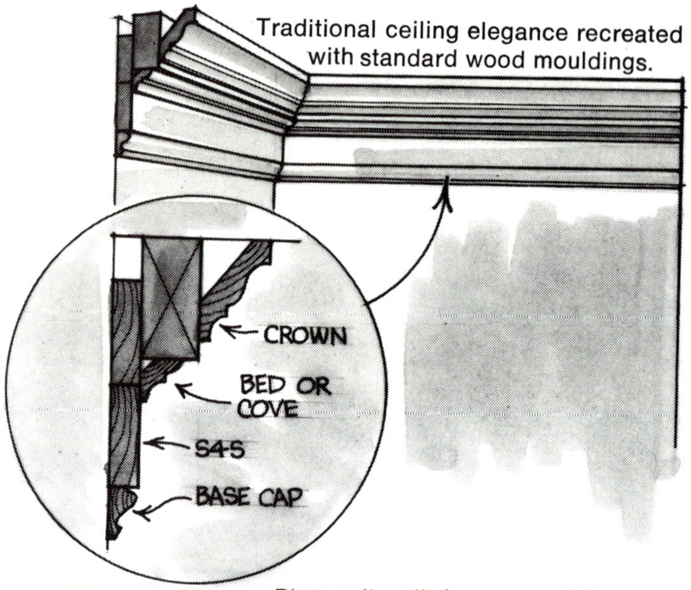

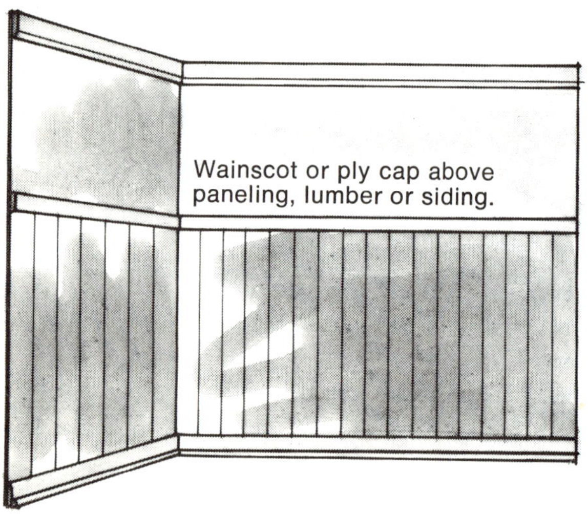

Plate rail wall (see page 79). Lower half can be paneled, wallpapered or painted.

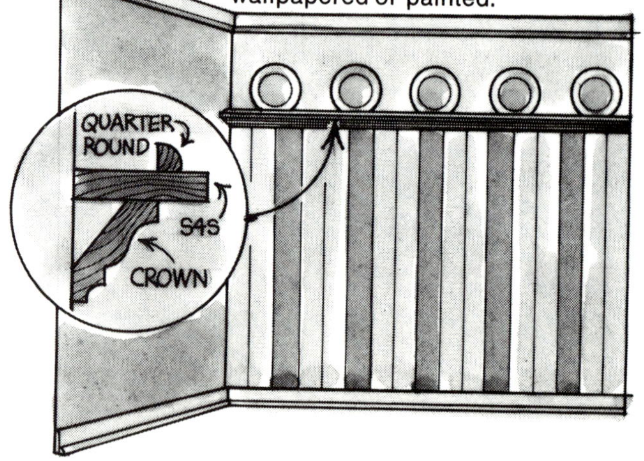

CHAIR RAILS

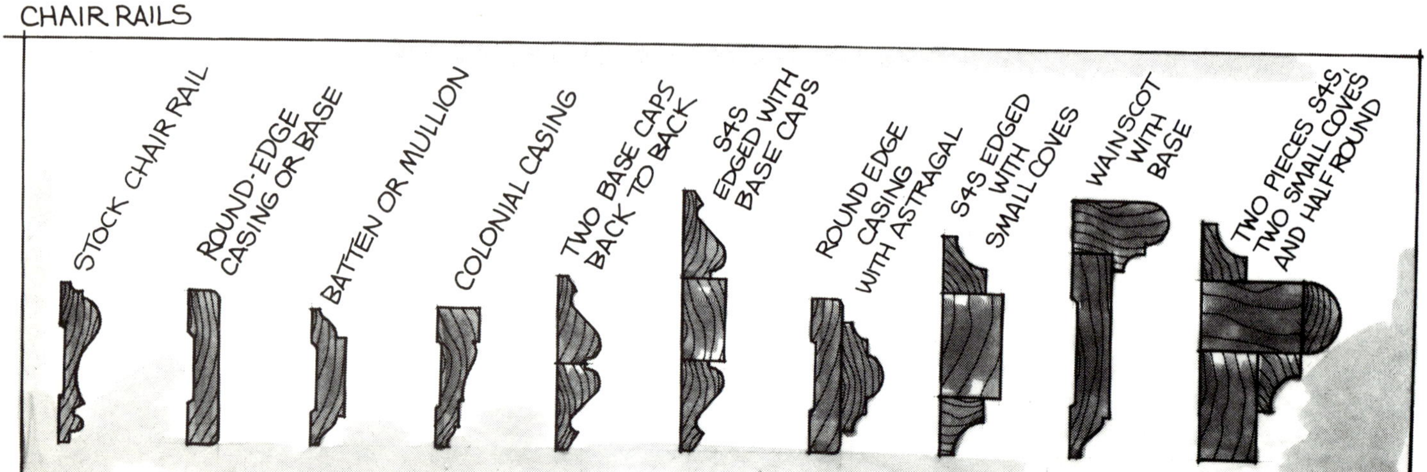

WALLS

Vertical strips of lattice or screen moulding.

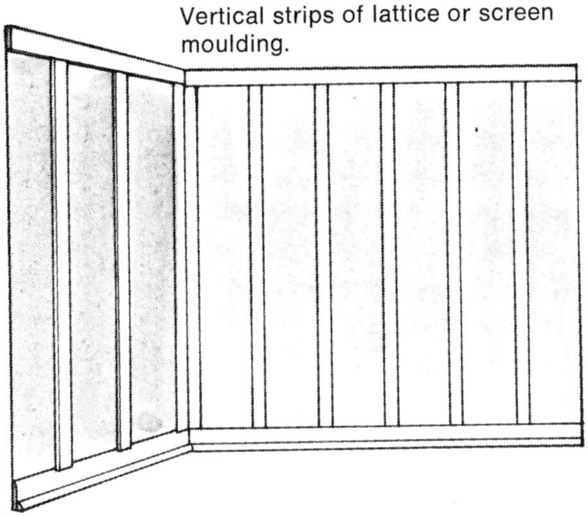

Varied rectangles of lattice.

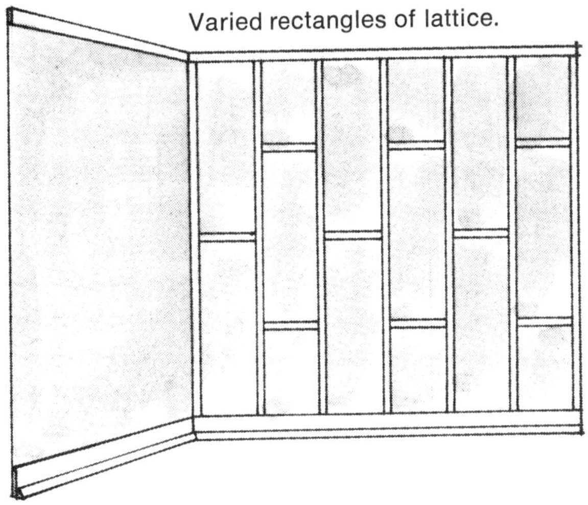

Square frames of moulding above chair rail. Can also reverse it and put them below.

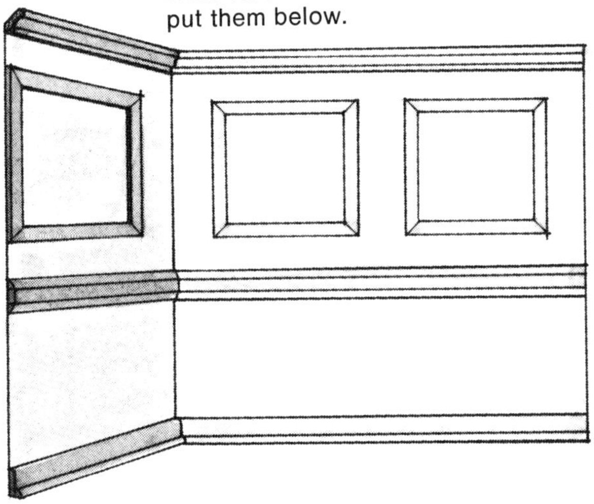

Vertical frames of lattice, screen, casing, ply cap, stop, batten, half rounds, panel, shingle or mullion.

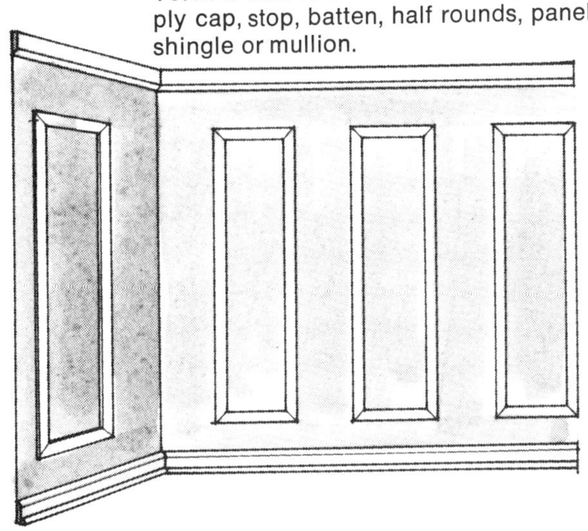

BASE VARIATIONS

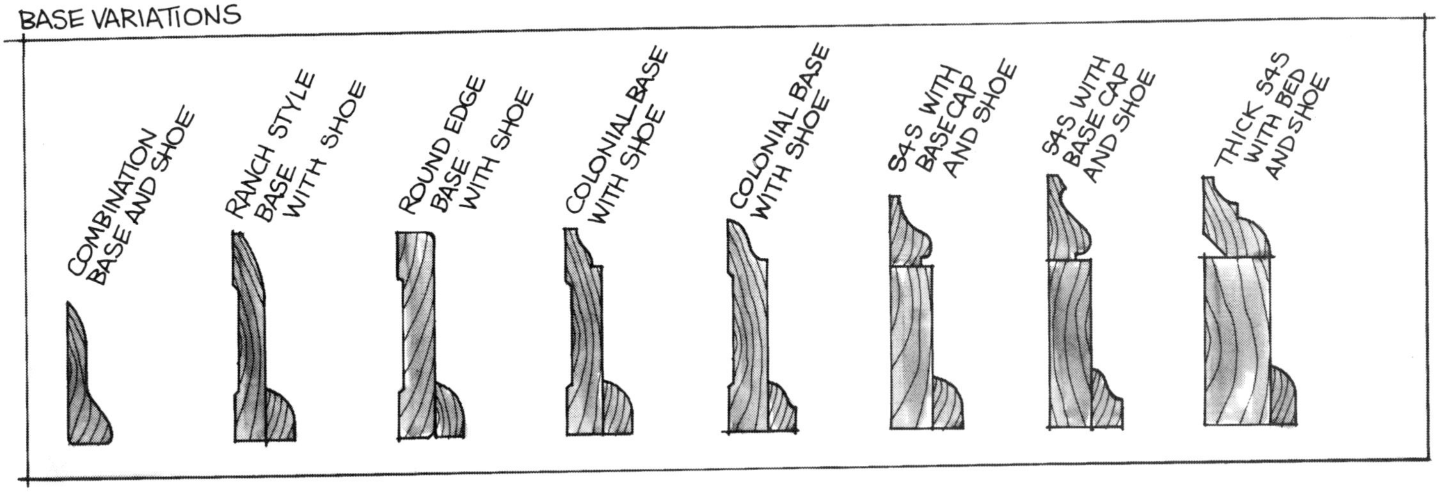

WALLS

Horizontal frames of lattice, casing, ply cap, stop, batten, half rounds, screen, panel, shingle or mullion.

Alternating sizes of moulding frames of lattice, screen, casing, ply cap, stop, batten, half rounds, panel, shingle or mullion.

Square frames applied above and below chair rail.

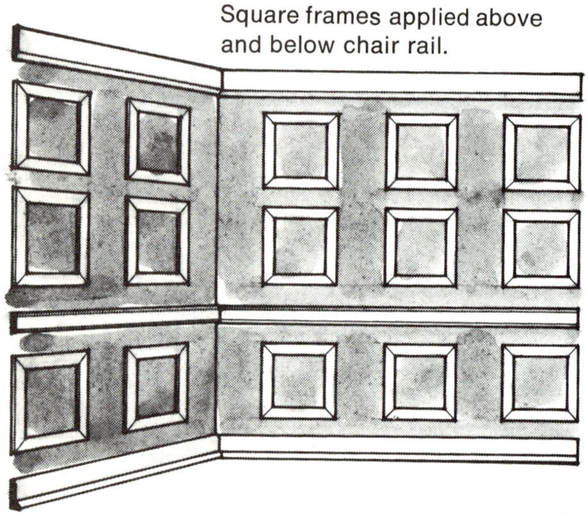

Varied sizes of frames applied above and below chair rail.

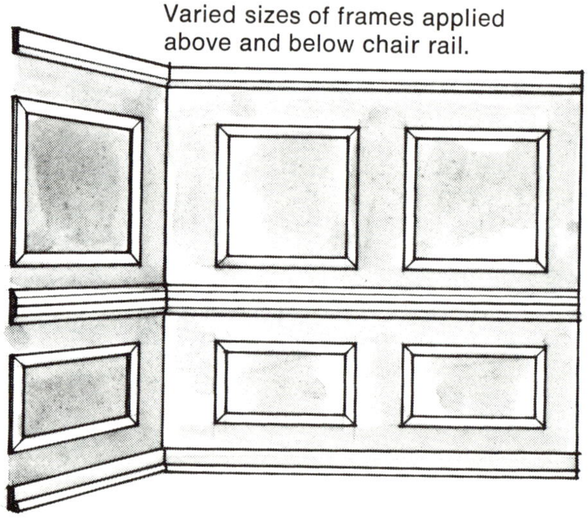

Alternating sizes of vertical frames using base or chair rail moulding.

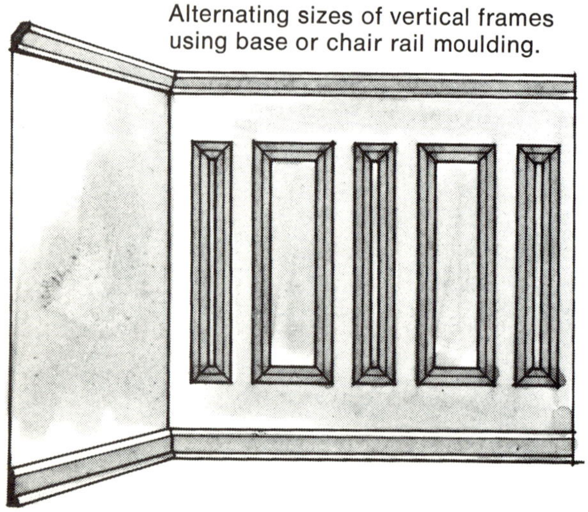

Lattice, ply cap, batten or casing, configured around wallpaper or vinyl covering.

WALLS

Vertical strips of lattice, mullion, S4S or batten.

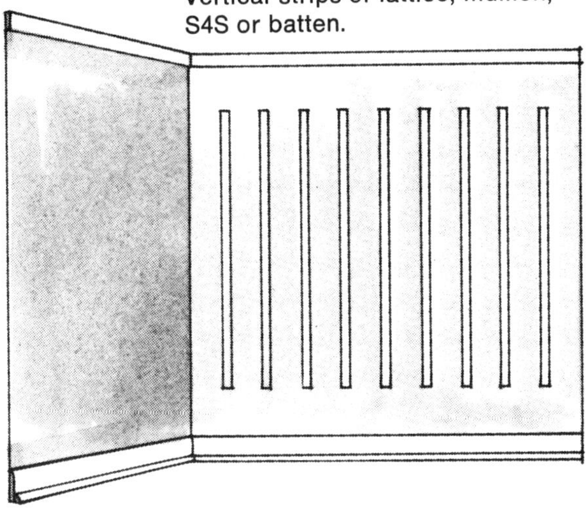

Varied rectangles of S4S to form shallow shelves, with each rectangle receiving a different wall treatment.

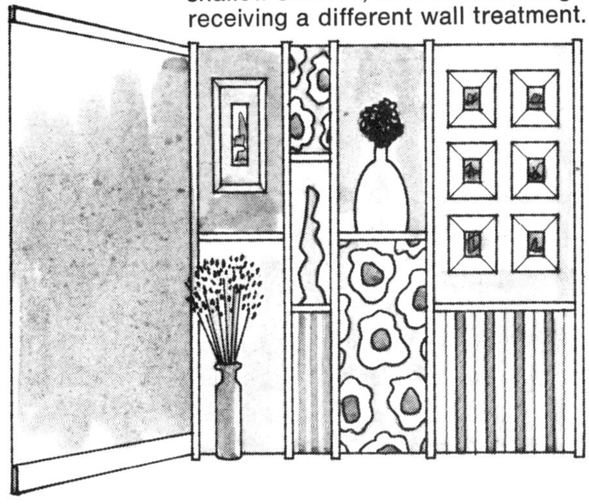

Varied sizes of diagonal rectangles of lattice.

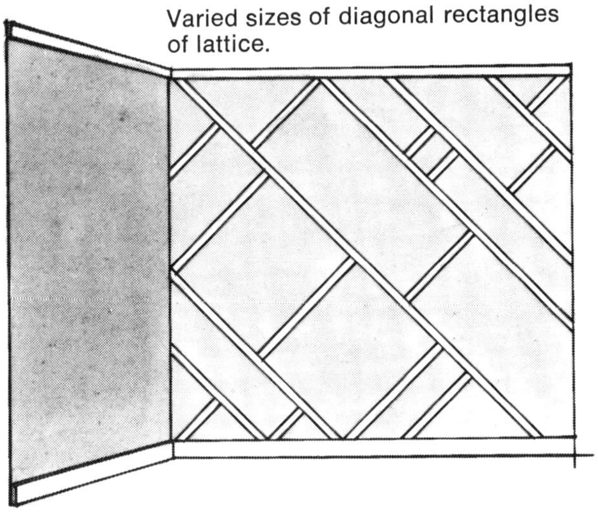

Vertical rows of half rounds, battens, coves, casing or drip caps.

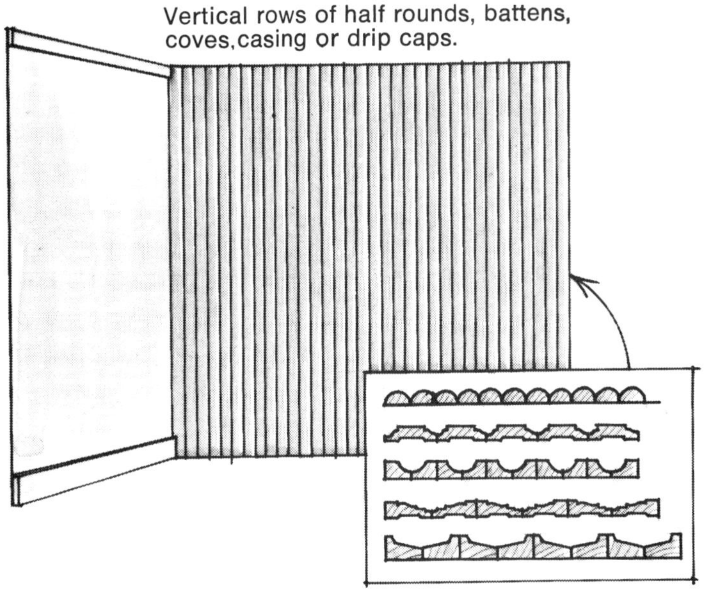

Alternating vertical and horizontal squares of moulding squares and S4S.

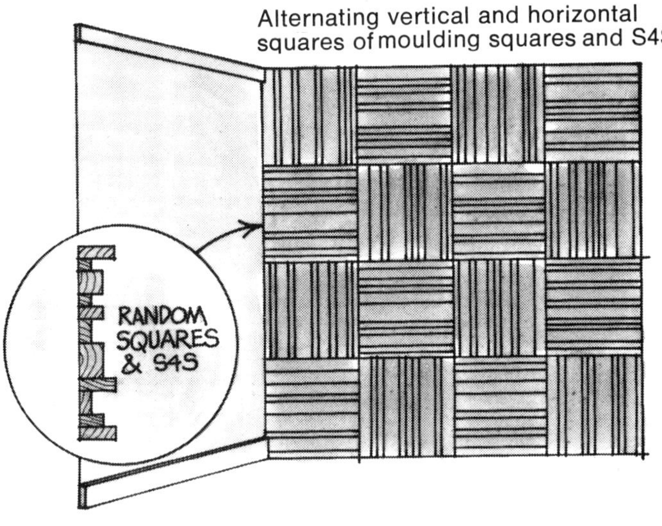

Horizontal rows of base caps, drip caps, cove/S4S/cove or stops.

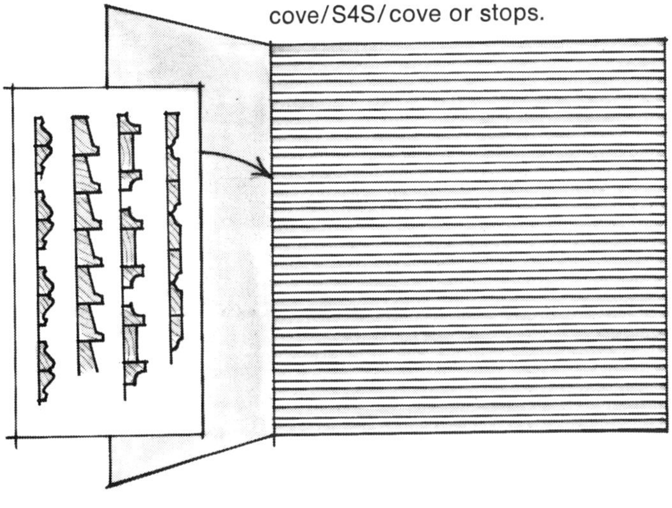

Rows of varied sized frames of moulding.

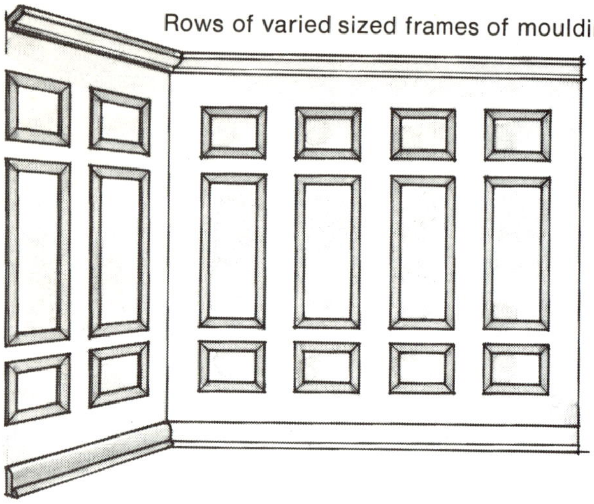

Diamond shaped patterns of lattice.

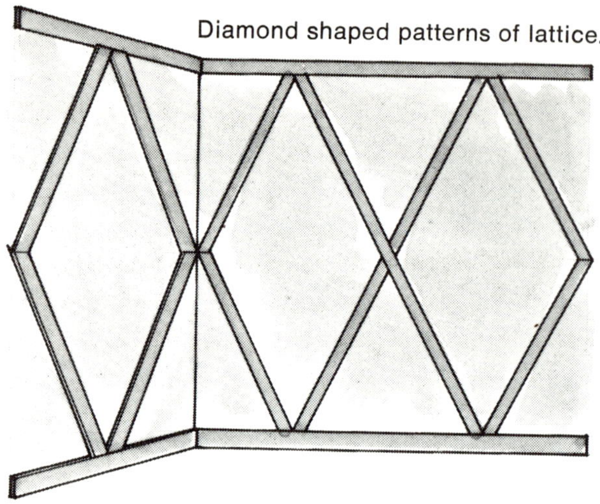

S4S as super graphics
or used to frame mirrors.

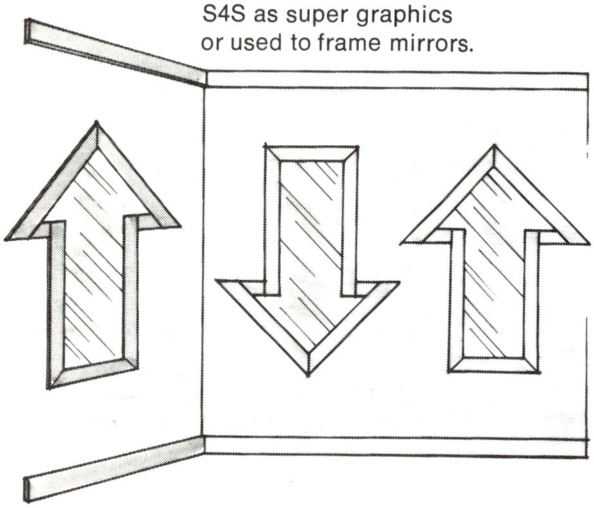

Wall medallion or super graphics
made from lattice.

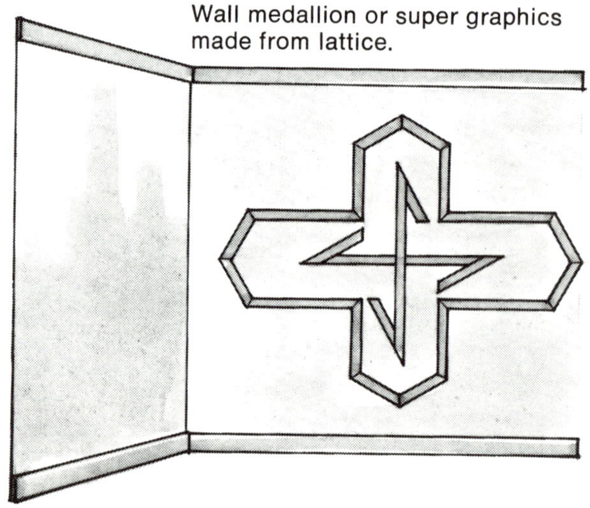

Rosette rows below a chair rail.
For rosette designs, see page 88.

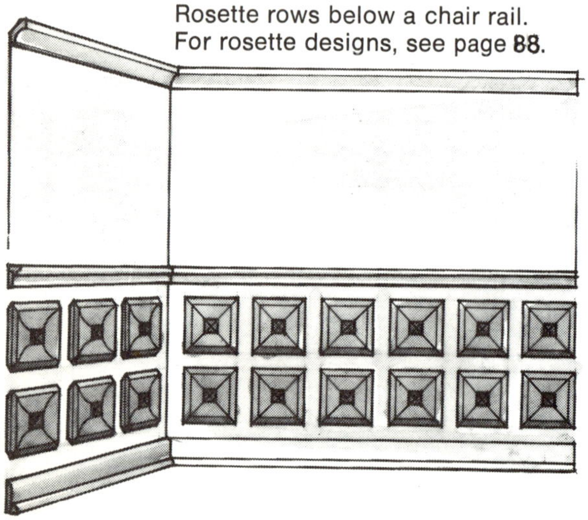

Wall is made up of S4S squares
framed with cove or quarter rounds.
For ceiling variations, see page 15.
For base variations, see page 119.

WASTE BASKETS

Why should something as ordinary as a waste basket look ordinary? It doesn't really have to. Give it some style with wood mouldings. What type of style is up to you. Here are four suggestions, but they are just starter ideas.

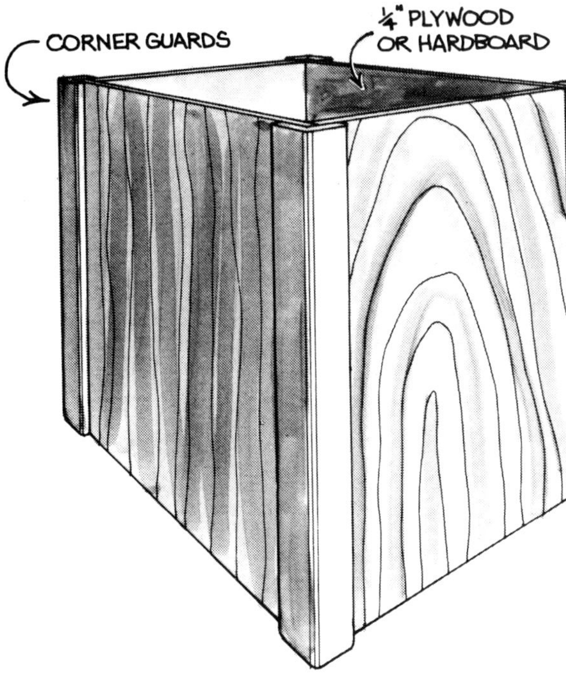

The first one is just a ¼″ plywood box with corner guards. The bottoms should be a heavier plywood, such as ½″.

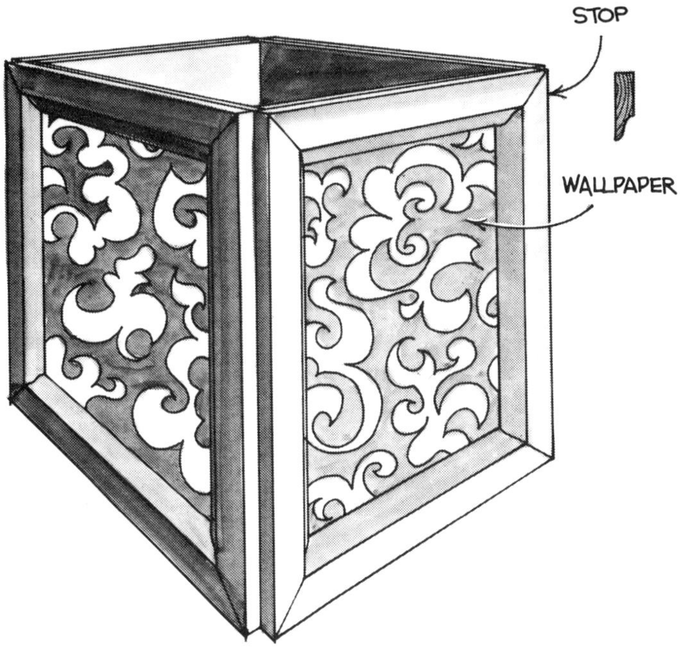

The second waste basket is again a plywood box, but this time it has mitered frames of stop moulding on each side. Casing, lattice, drip cap or base cap could also be used. Before attaching the frames apply wall paper or vinyl wall covering.

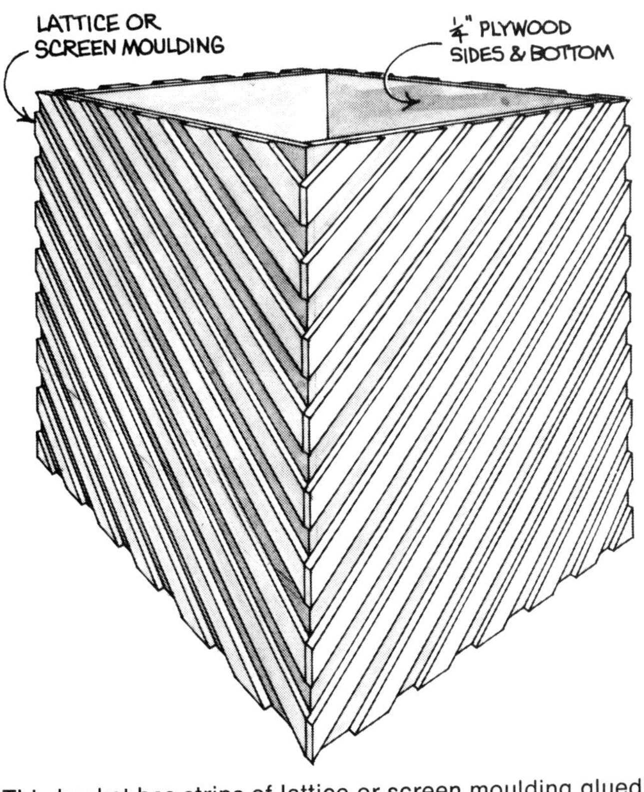

This basket has strips of lattice or screen moulding glued and tacked to the plywood box at 45° angles.

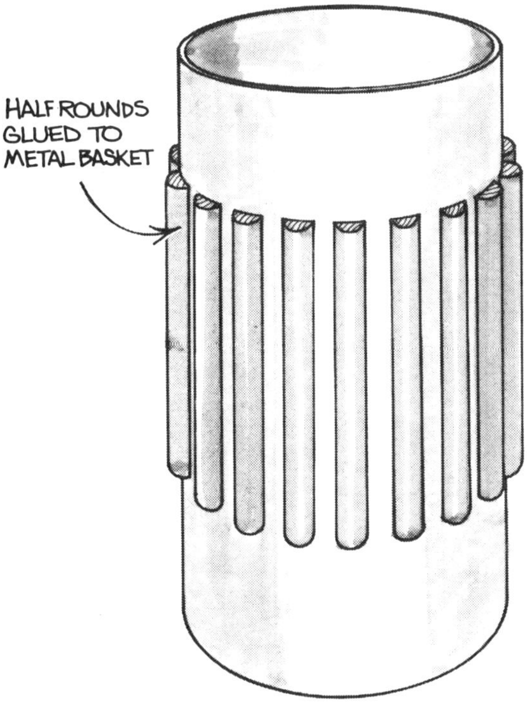

An existing metal basket, whether it's round, oval, oblong or square can be dressed up just by applying strips of half round. Use glue, then nail from the inside. Spray paint it a fresh new color.

WEAVERY/STITCHERY FRAME

To the weaving or stitchery artist, a ready-made factory-built operating frame can be a costly little tool, especially when you'd rather be buying yarns or thread with that hard-earned money. The do-it-yourself hobbyist can save a lot by building their own frame with wood mouldings. Once you've devised the size you need, its just a matter of getting the materials and putting it together, which can be done in about an hour's time. The four legs are cut from moulding squares. Nail or screw on the S4S side pieces and then attach the lattice side braces at 45° angles. The ends of the lattice are also mitered at 45°. For added stability, add inside corner braces of lattice. That's all there is. It now awaits your work of art.

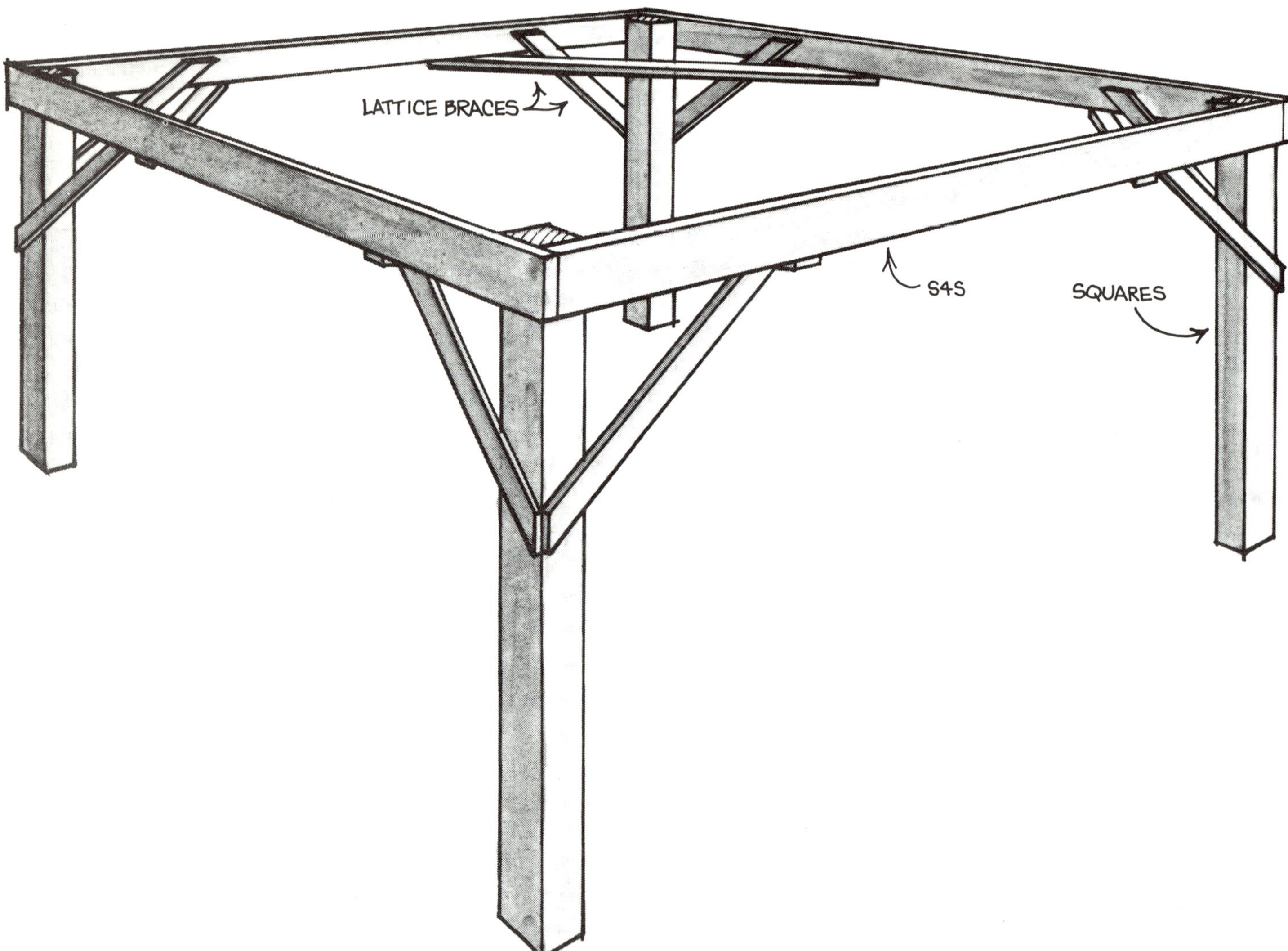

WIND CHIME

Create your own open-air concert with an orchestra of lattice, screen, rounds, half rounds, quarter rounds, dowels, squares, base caps, coves and stops, all clustered together to form a melodious wind chime. The tones will vary according to the weight, length and patterns of wood mouldings you choose. After you've made your selection, attach small eye hooks at the end and suspend them from a frame of squares with dowel cross pieces. The distance between them should be far enough apart to swing freely, but close enough to touch in a gentle breeze. Stain or oil as you desire, then hang it from your porch or a nearby tree and await the first musical selection.

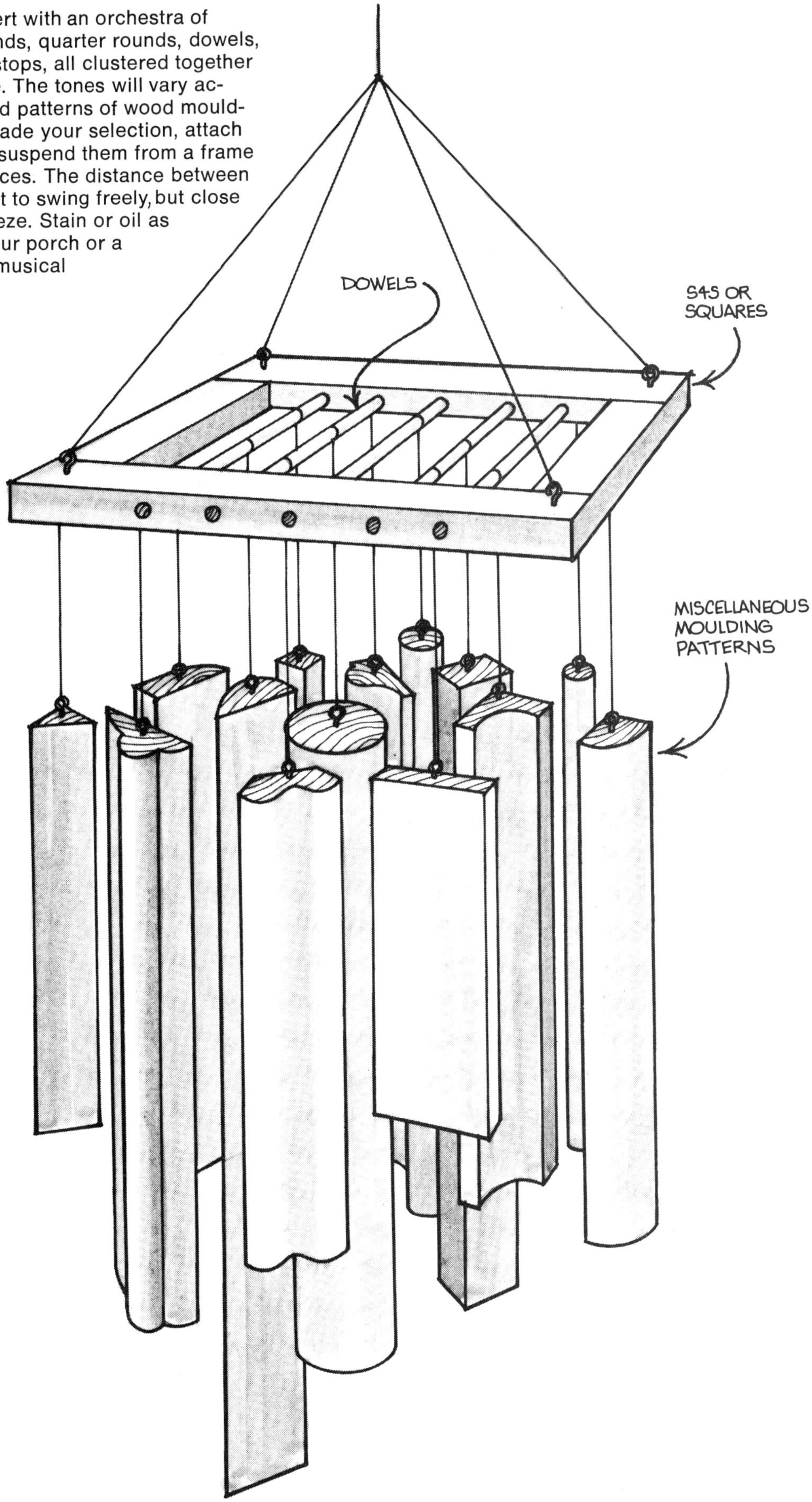

WINE RACKS/WINE GLASS HOLDER

Good wine deserves to be displayed in a wine rack that reflects the charm and quality of the wine it holds. The warm richness of wood moulding makes it a perfect material for building your own wine rack.

The first unit pictured below holds twelve bottles and is made of three shelves of casing, notched to hold four bottles each. The back notches are 3½″ half circles and the front notches are 1½″. A saber saw or jig saw will do the job easily. A coping saw will work too, if the others are unavailable. S4S or lattice is used to connect the shelves and dowels are used to fasten the front to the back. Nail and glue securely. A rich oil stain finish will complement the wine bottles beautifully.

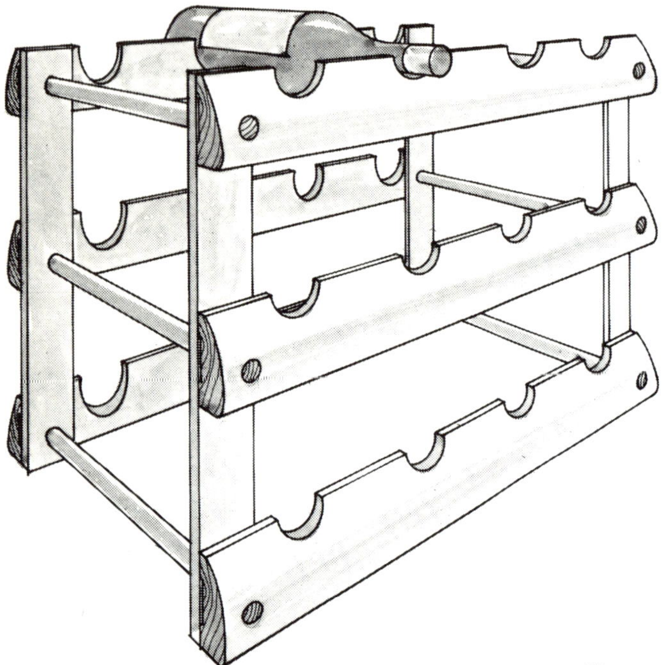

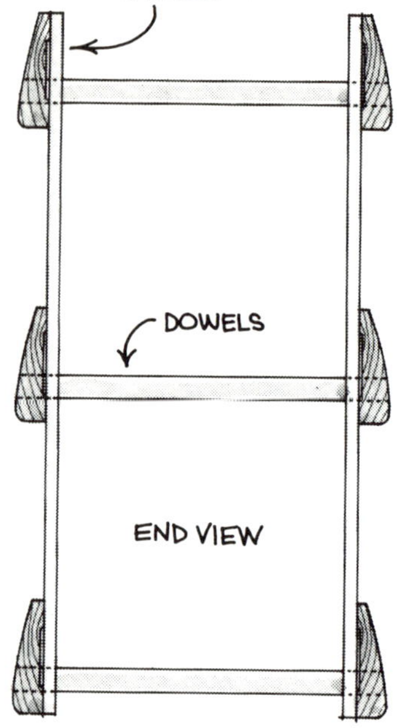

The rack on the right can be shortened or lengthened to any size you wish and may have as many shelves as you need. The wine bottles sit on dowels inserted into lattice. The shelves are connected at the corners by pieces of corner guard. The end dowels are fastened to both the lattice and the corner guards.

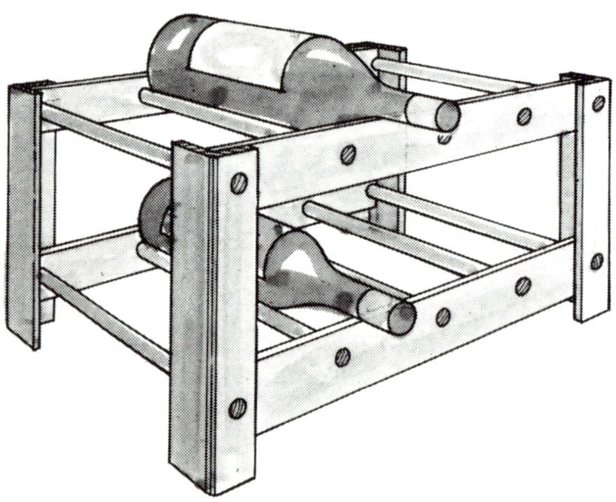

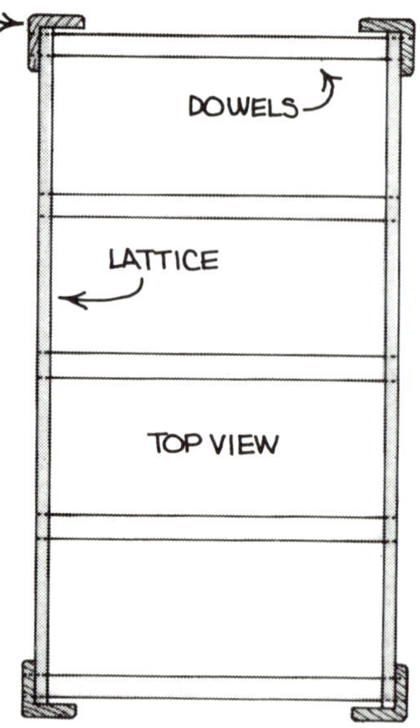

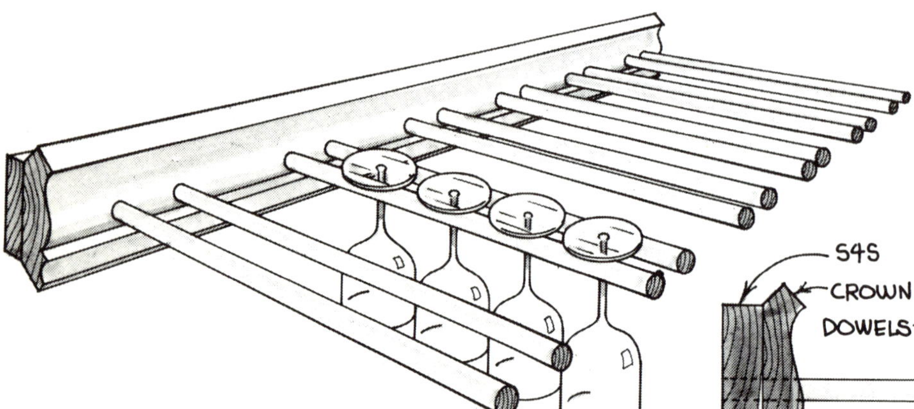

Wine glasses are nice to look at, so why not hang them out in the open. They won't take up valuable cupboard space either. Build the rack to fit your glasses and the space you have available. For this holder, crown moulding on a piece of S4S was used, with ¼″ dowels inserted at appropriate distances. Mount it in place with countersunk screws. Under a cabinet is an ideal spot.

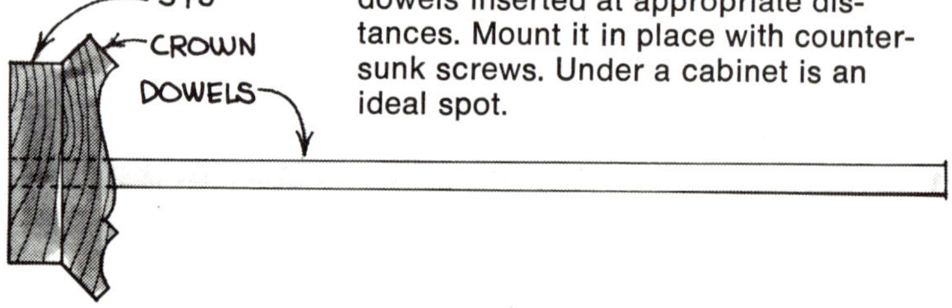

WOOD SPOON RACK & HOLDERS

Like most utensils used in the kitchen, wooden spoons have a habit of getting lost in a drawer or misplaced somewhere unless they're out in the open where they can be seen when needed. The answer, of course, is to build a container for them.

The wooden spoon wall rack shown here is one piece of S4S connected to another with appropriate sized holes for the spoons. To dress up the rack, flat astragal has been attached to the front, with shingle and quarter round used below. Ply cap and quarter round have been applied to the top. Your spoon supply and kitchen wall space will determine the length of the rack.

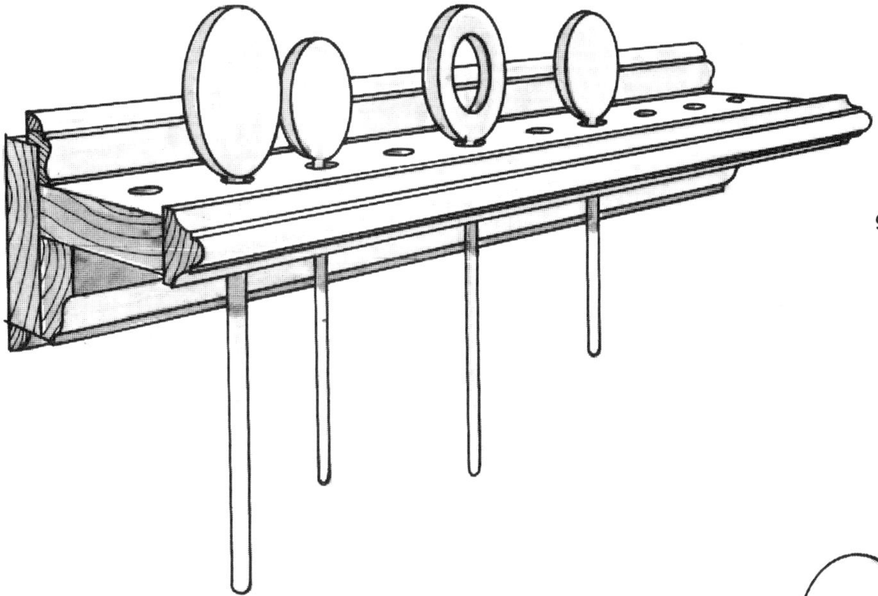

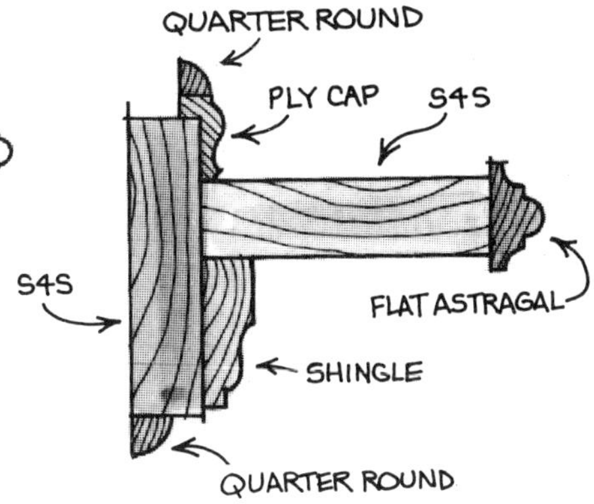

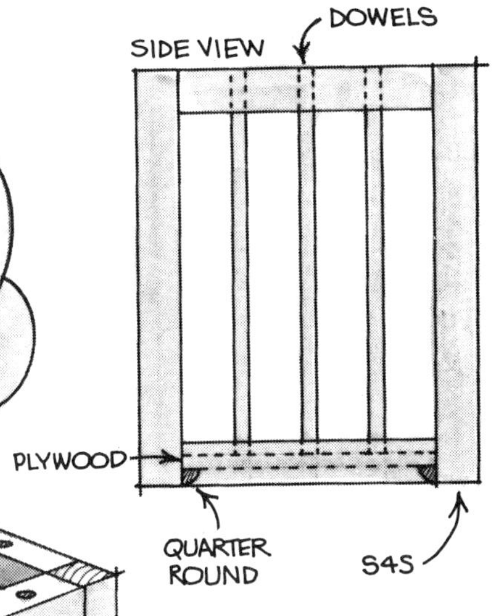

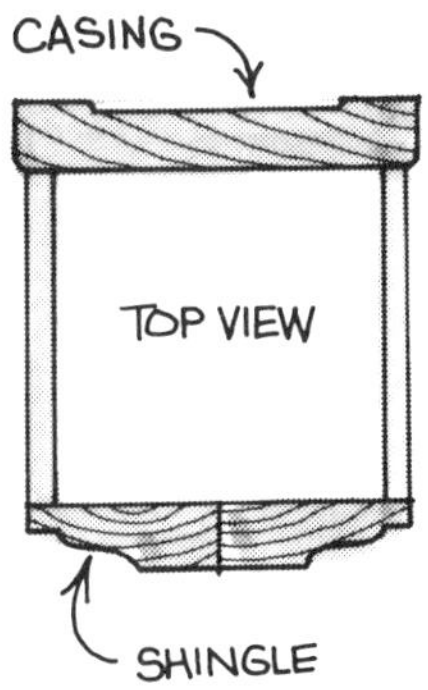

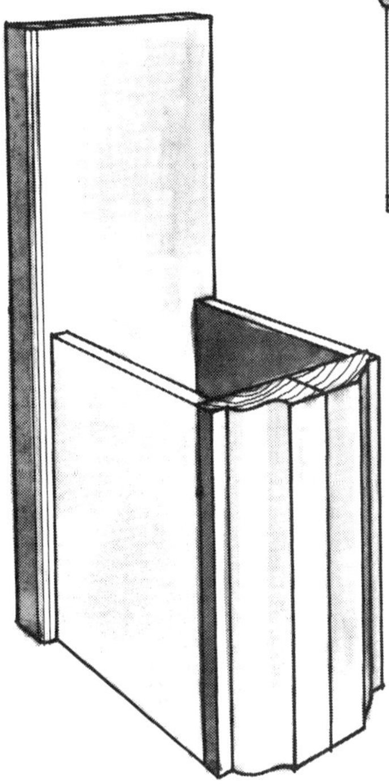

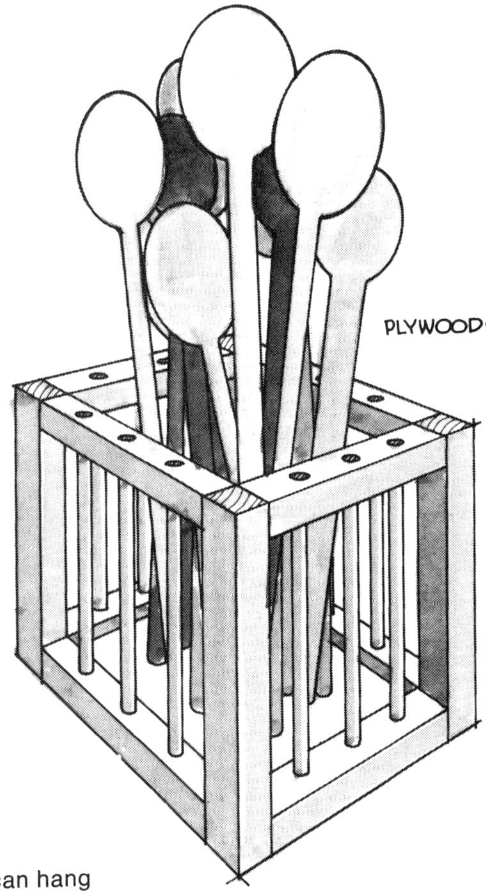

This one can either sit on the stove or counter, or it can hang on the wall. The back is a tall piece of round edge casing and the sides can be either lattice or S4S with the floor made of the same material. The front section is two pieces of shingle moulding or two stops joined together with glue. If you want to hang this one, drill a small hole in the back casing.

This container is a box-like structure made of square frames with dowels inserted in the sides. A floor of plywood is supported from underneath with quarter rounds. These racks can be stained or left natural to match the wooden spoons.

The traditional uses for wood mouldings have not changed much over the years. Mouldings have always been used to cover up imperfections at corners and seams of walls, ceiling and floors. Their most popular use, however, has been decorative, improving an existing piece of furniture, wall, ceiling or divider. But you can go beyond that and use wood mouldings for creating whole new projects. That is the purpose of this book. To show you new ways of using wood mouldings. To give you new ideas or to inspire you to create your own ideas. However, it must be emphasized that this is not a detailed book of construction plans. It is an IDEA book. In many cases, the project sizes and suggested patterns are optional, usually depending on your own preferences or moulding availability. The versatility of wood moulding and the ease with which it can be worked, makes it fun for everyone. So go ahead. Create a project or beautify a room. That satisfactory feeling of accomplishment will be the result.

ACKNOWLEDGEMENTS

Designed, illustrated and written by Ray Robbins.
With special help from Warren Jimerson, Jim Snodgrass, Neal Heflin, Bernard Tomasko, Gloria Gilbert, Joseph Erceg, Lorraine Robbins, Jim's Woodshop, Photo-Art Commercial Studios and Alderman Studios, Inc.

"To err is human" a word moulder once said. "To err is human" a wood moulder also said. Every effort has been made to produce this book with accuracy. However, when working with this many projects, the chance of error becomes a distinct possibility. Therefore, check carefully before beginning a project.